Mitigation and Adaptation of Urban Overheating

The Impact of Warmer Cities on Climate, Energy, Health, Environmental Quality, Economy, and Quality of Life

Mitigation and Adaptation of Urban Overheating

The Impact of Warmer Cities on Climate, Energy, Health, Environmental Quality, Economy, and Quality of Life

Edited by

Nasrin Aghamohammadi
School of Design and the Built Environment, Curtin University
Sustainability Policy Institute, Bentley, WA, Australia

Mattheos Santamouris
Faculty Arts Design and Architecture, School of Built Environment,
University of New South Wales, Sydney, NSW, Australia

ELSEVIER

Elsevier
Radarweg 29, PO Box 211, 1000 AE Amsterdam, Netherlands
125 London Wall, London EC2Y 5AS, United Kingdom
50 Hampshire Street, 5th Floor, Cambridge, MA 02139, United States

Notices
Knowledge and best practice in this field are constantly changing. As new research and experience broaden our understanding, changes in research methods, professional practices, or medical treatment may become necessary.

Practitioners and researchers must always rely on their own experience and knowledge in evaluating and using any information, methods, compounds, or experiments described herein. In using such information or methods they should be mindful of their own safety and the safety of others, including parties for whom they have a professional responsibility.

To the fullest extent of the law, neither the Publisher nor the authors, contributors, or editors, assume any liability for any injury and/or damage to persons or property as a matter of products liability, negligence or otherwise, or from any use or operation of any methods, products, instructions, or ideas contained in the material herein.

ISBN: 978-0-443-13502-6

For Information on all Elsevier publications
visit our website at https://www.elsevier.com/books-and-journals

Publisher: Candice Janco
Acquisitions Editor: Jessica Mack
Editorial Project Manager: Aleksandra Packowska
Production Project Manager: Rashmi Manoharan
Cover Designer: Vicky Pearson Esser

Typeset by MPS Limited, Chennai, India

Working together
to grow libraries in
developing countries

www.elsevier.com • www.bookaid.org

Contents

*Fabrizio Ascione, Nicola Bianco, Giacomo Manniti, Margherita
Mastellone, Francesco Tariello and Giuseppe Peter Vanoli*

Nasrin Aghamohammadi and Logaraj Ramakreshnan

List of contributors

Majed Abuseif School of Engineering and Built Environment, Griffith University, Gold Coast, QLD, Australia; Green Infrastructure Research Labs (GIRLS), Cities Research Institute, Griffith University, Gold Coast, QLD, Australia

Synnefa Afroditi School of Built Environment, Faculty of Arts, Design and Architecture, University of New South Wales, Sydney, NSW, Australia

Nasrin Aghamohammadi School of Design and the Built Environment, Curtin University Sustainability Policy Institute, Bentley, WA, Australia; Centre for Epidemiology and Evidence-Based Practice, Department of Social and Preventive Medicine, Faculty of Medicine, University of Malaya, Kuala Lumpur, Malaysia

Mavrogianni Anna Institute for Environmental Design and Engineering (IEDE), Bartlett School of Environment, Energy and Resources (BSEER), Bartlett Faculty of the Built Environment, University College London (UCL), London, United Kingdom

Fabrizio Ascione Department of Industrial Engineering, Piazzale Tecchio 80, Università degli Studi di Napoli Federico II DII, Napoli, Italy

Carlos Bartesaghi Koc Faculty of Science, Engineering and Technology, University of Adelaide, Adelaide, SA, Australia; Risk & Resilience, Governance and Legal, NSW Department of Planning and Environment, Sydney, NSW, Australia

Paolo Bertoldi European Commission, Joint Research Centre (JRC), Ispra, Italy

Nicola Bianco Department of Industrial Engineering, Piazzale Tecchio 80, Università degli Studi di Napoli Federico II DII, Napoli, Italy

Sofia Natalia Boemi Department of Mechanical Engineering, Process Equipment Design Laboratory (PEDL), Aristotle University of Thessaloniki, Thessaloniki, Greece; Cluster of Bioeconomy and Environment of Western Macedonia (CluBE), Kozani, Greece

Claudia Fabiani CIRIAF – Interuniversity Research Center on Pollution and Environment Mauro Felli – University of Perugia, Perugia, Italy; Department of Engineering, University of Perugia, Perugia, Italy

Lauren Ferguson Institute for Environmental Design and Engineering (IEDE), Bartlett School of Environment, Energy and Resources (BSEER), Bartlett Faculty of the Built Environment, University College London (UCL), London, United Kingdom

Ali Ghaffarianhoseini Department of Built Environment Engineering, School of Future Environments, Auckland University of Technology, Auckland, New Zealand

Amirhosein Ghaffarianhoseini Department of Built Environment Engineering, School of Future Environments, Auckland University of Technology, Auckland, New Zealand

M.E. González-Trevizo Facultad de Ingeniería, Arquitectura y Diseño, Universidad Autónoma de Baja California, Ensenada, Mexico

Elmira Jamei College of Engineering and Science, Victoria University, Melbourne, VIC, Australia; Institute of Sustainable Industries and Liveable Cities, Victoria University, Melbourne, VIC, Australia

Giacomo Manniti Department of Industrial Engineering, Piazzale Tecchio 80, Università degli Studi di Napoli Federico II DII, Napoli, Italy

Alberto Martilli CIEMAT, Madrid, Spain

K.E. Martínez-Torres Facultad de Ingeniería, Arquitectura y Diseño, Universidad Autónoma de Baja California, Ensenada, Mexico

Margherita Mastellone Department of Architecture, Via Forno Vecchio 36, Università degli Studi di Napoli Federico II DIARC, Napoli, Italy

Ilaria Pigliautile CIRIAF – Interuniversity Research Center on Pollution and Environment Mauro Felli – University of Perugia, Perugia, Italy; Department of Engineering, University of Perugia, Perugia, Italy

Anna Laura Pisello CIRIAF – Interuniversity Research Center on Pollution and Environment Mauro Felli – University of Perugia, Perugia, Italy; Department of Engineering, University of Perugia, Perugia, Italy

Logaraj Ramakreshnan Institute for Advanced Studies, University of Malaya, Kuala Lumpur, Malaysia

J.C. Rincón-Martínez Facultad de Ingeniería, Arquitectura y Diseño, Universidad Autónoma de Baja California, Ensenada, Mexico

Mattheos Santamouris Faculty Arts Design and Architecture, School of Built Environment, University of New South Wales, Sydney, NSW, Australia

Hideki Takebayashi Department of Architecture, Graduate School of Engineering, Kobe University, Kobe, Japan

Francesco Tariello Department of Agricultural, Environmental and Food Sciences, Via Francesco De Sanctis 1, Università degli Studi del Molise DiAAA, Campobasso, Italy

Aldo Treville European Commission, Joint Research Centre (JRC), Ispra, Italy

Giulia Ulpiani European Commission, Joint Research Centre (JRC), Ispra, Italy

Giuseppe Peter Vanoli Department of Medicine and Health Sciences, Via Francesco De Sanctis 1, Università degli Studi del Molise DiMeS, Campobasso, Italy

Konstantina Vasilakopoulou School of the Built Environment, Faculty of Arts, Design and Architecture, University of New South Wales, Sydney, NSW, Australia; UNSW Ageing Futures Institute, UNSW Sydney, NSW, Australia

Nadja Vetters European Commission, Joint Research Centre (JRC), Brussels, Belgium

Komali Yenneti School of Architecture and Built Environment, University of Wolverhampton, Wolverhampton, United Kingdom

Foreword

"Mitigation and Adaptation of Urban Overheating: The impact of warmer cities on climate, energy, health, environmental quality, economy, and quality of life." It sounds like a manual for life, the future, and everything. Perhaps it is, because we need one!

In my perspective from being involved in the climate movement since the first Earth Day in 1970, its about time we had a manual that looks at the problems associated with climate change and turns it into a manual for how we can create a better world with a better economy and a better quality of life.

It is a remarkable period of history where we find ourselves. Not only were the climate scientists, engineers, and planners right about climate change happening as we dramatically left the safe operating space of an extra 1 degree rise in global temperatures. But they were also right to show that the new renewable systems may be cheaper and better than how we lived and created economies for the past few hundred years of fossil fuel—based civilization.

The weather patterns in 2023 have shown a consistent pattern of increases in the intensity of "heat, high water and hurricanes" as the New York Times has labelled it. Global average temperatures broke world records. None of this is unexpected as the greenhouse effect has been understood since the 1890s, and it simply shows that if we keep using fossil fuels and clearing land more than we revegetate it, then more energy will flow into the atmosphere and from there into the land and oceans.

That is the scary bit and we must find out new ways of how to adapt to this.

But the most amazing thing about our period of history is that we now have the best source of energy available and it is cheaper than any previous source of energy: sunshine. Solar and wind are now cheap, easy to install in days, in comparison to the years for fossil fuels and decades for nuclear. The applications to cities, transport, agriculture, mining, and industry are flowing into commercial activity that outstrips most government planning.

This is the driving force that is recreating our economies faster than anyone predicted, other than a few who understood disruptive innovation and exponentially declining cost curves. Tony Seba has now predicted by 2030 we are likely to have solar 70% cheaper than fossil fuel power and new electric vehicles will be driving 90% of the land transport market.

But its not quite as simple as letting such technology solve it all for us.

For a start, the sunshine can only be tapped if we have smart systems that manage how we integrate it into batteries, appliances, and electric vehicles, and have the right governance that enables these changes to make the most of the fantastically cheaper and more reliable solar and wind systems. And these vary with the functions and geography of the applications.

But mostly we will need to see how we can integrate these new renewable and smart systems into our cities. They work best at different scales to how we made power for our urban economies before—with big power stations next to big coal fields and huge transmission lines that lost 40% of the power from their origins, before distributing it to those at their ends. The new economy favors favor that can be distributed locally, and that fortunately is what we need to improve our quality of life as well as all those other things in this book.

Cities are made up of different urban fabrics, and they are being rebuilt along with each technological era. The old walking cities that were preindustrial had dense centers where people could easily meet, talk, play, and exchange goods. These centers have been regenerated in recent decades to become critical parts of our urban economies. The United States recently found that the 0.25% of their cities which are walkable spaces created 20% of US GDP! This fabric will need lots of solar and vegetation to cool the urban heat island effect, but it must remain dense and have few EV cars and lots of good EV transit and bikes.

The old train and tram fabrics from the 19th and 20th centuries are the corridors of medium density, and they also are having a big revival and extension as new rail systems have signaled a second rail revolution and have become the basis of much educational and health-oriented economic activity as well as having their own walkable centers of place-based human activity around stations. This fabric will take more solar and more EV cars but mostly will need to be regenerated around new electric transit systems like Trackless Trams with net zero precincts around every station.

The past 70 years have also seen huge sprawling car−based urban areas created that consume three times the amount of fuel and power, so it needs a lot of work to reduce its need for energy as well as decarbonizing it. The extra space means more room for local solar, trees and gardens, and the more local the better in all parts of its economy.

Such ideas are being trialed in demonstration net zero precincts and corridors, and the research will be invaluable as cities around the world take up the grand solar and greening opportunities. This book helps us to begin to see what may emerge as the solutions we desperately need for the mitigation and adaptation to urban overheating.

Peter Newman AO
Sustainability Curtin University

Preface

As urban landscapes across the globe continue to evolve, with burgeoning skylines and sprawling cityscapes, a formidable challenge has emerged on the horizon—the phenomenon of urban overheating. This book, *Mitigation and Adaptation of Urban Overheating: The Impact of Warmer Cities on Climate, Energy, Health, Environmental Quality, Economy, and Quality of Life*, delves into the intricate web of issues posed by the warming urban environment and offers comprehensive insights into strategies for mitigation and adaptation.

For urban climatologists, architects, urban planners, environmental engineers, energy scientists, healthcare practitioners, policymakers, and urban governing authorities, this work represents a vital compass in navigating the complex landscape of urban overheating. It is our sincere hope that this book serves as a valuable resource, enriching your understanding and empowering your decision-making.

The pages herein provide a panoramic view of the multifaceted challenges posed by urban overheating. We delve into the cascading impacts on climate patterns, the burgeoning energy demands, the growing burden on public health, the ramifications for environmental quality, the intricate dance with economic dynamics, and, perhaps most importantly, the profound influence on the quality of urban life.

Through the collaboration of experts and thought leaders from diverse backgrounds, this book amalgamates knowledge, research findings, and practical solutions. It is a testament to the interconnectedness of the challenges we face and the need for holistic, interdisciplinary approaches to address them.

Our primary goal is to equip you, the reader, with a comprehensive understanding of urban overheating and arm you with the knowledge needed to formulate effective strategies for mitigation and adaptation. In the rapidly urbanizing world we inhabit, where cities are both crucibles of innovation and crisis, the urgency of addressing urban overheating cannot be overstated.

We extend our gratitude to the countless individuals and organizations whose dedication, expertise, and collaborative spirit have contributed to this endeavor. Without their invaluable contributions, this book would not have been possible.

As you embark on this journey through the intricacies of urban overheating, we invite you to consider the profound impact that cities have on our planet and our collective future. May the knowledge contained within these pages inspire informed action and usher in a future where cities not only thrive but also stand as bastions of sustainability, resilience, and quality of life.

Thank you for joining us on this crucial exploration of urban overheating and its far-reaching implications.

With warm regards,

Nasrin Aghamohammadi
Mattheos Santamouris

Acknowledgments

This monumental work, *Mitigation and Adaptation of Urban Overheating*, has been a collaborative endeavor that would not have been possible without the dedication, support, and expertise of numerous individuals and organizations. We extend our heartfelt gratitude to all those who contributed to this project, directly or indirectly, and helped bring it to fruition.

We express our gratitude to the dedicated urban climatologists, architects, urban planners, environmental engineers, energy scientists, healthcare practitioners, and policymakers whose tireless efforts in their respective fields continue to drive progress in understanding and addressing urban overheating.

Our sincere appreciation goes to the urban governing authorities and municipalities that have shown commitment to sustainability and resilience in the face of urban challenges. Their dedication to implementing policies and practices that mitigate the impacts of urban overheating is commendable.

We acknowledge the invaluable contributions of the academic and research communities, whose relentless pursuit of knowledge has laid the foundation for the insights presented in this book. Their groundbreaking research has illuminated the complexities of urban overheating and has been instrumental in shaping the discourse.

We also extend our thanks to the healthcare professionals whose insights into the health implications of urban overheating have highlighted the urgency of our collective efforts. Their expertise is vital in safeguarding public health in urban environments.

We appreciate the environmental engineers and energy scientists whose innovative solutions and technologies offer hope for a more sustainable urban future. Their contributions to reducing energy consumption and enhancing environmental quality are indispensable.

To the countless individuals who work tirelessly behind the scenes in various capacities, providing support, encouragement, and resources, we offer our sincere thanks. Their contributions may be less visible, but they are no less critical to the success of this endeavor.

Lastly, we dedicate this work to the cities and urban communities around the world. They are at the heart of the challenges and opportunities presented by urban overheating. May the knowledge contained within these pages inspire positive change and foster urban environments that are not only resilient but also conducive to the well-being and quality of life of their inhabitants.

We thank them for their unwavering dedication to addressing the complex issues of urban overheating and their commitment to creating a more sustainable and livable urban future.

With deepest appreciation,

Nasrin Aghamohammadi
Mattheos Santamouris

Chapter 1

Urban overheating and its impact on human beings

Mattheos Santamouris

Faculty Arts Design and Architecture, School of Built Environment, University of New South Wales, Sydney, NSW, Australia

1.1 Introduction

Because of the positive thermal balance of the urban environment, cities experience higher ambient temperatures than the surrounding suburban and rural spaces (Santamouris, 2001). The phenomenon is known as urban heat island, while the term urban overheating is also used to include the synergetic impact of both the global and local climate change, and it is linked with the increase of the frequency of heat waves and the prolongation of the hot spells (Paolini & Santamouris, 2022).

According to Oke et al. (1991), the development of urban heat island is influenced by the following factors:

1. The thermal characteristics and properties of the materials used in cities. The most common materials used for the buildings, envelope and urban fabric, present a very high absorption to solar radiation. As a result, their surface temperature is high, while the stored heat is released to the atmosphere as sensible heat, increasing the ambient temperature.
2. Decrease of the evaporation processes in cities. The replacement of natural soil and vegetation in cities by nonevaporating materials has decreased the release of the latent heat in cities and increased the ambient temperature.
3. High anthropogenic heat generated in cities. Transport, industry, energy systems, and other combustion processes generate additional heat overheating cities.
4. The radiative geometry of urban canyons. The emitted infrared radiation by buildings, pavements, and streets cannot escape because of the geometry of urban canyons. The infrared radiation reflected and reabsorbed

Mitigation and Adaptation of Urban Overheating. DOI: https://doi.org/10.1016/B978-0-443-13502-6.00001-4

inside the canyon increases the surface temperature of the materials and leads to the corresponding release of sensible heat.
5. The reduced turbulent transfer of heat from the reduced turbulent transfer of heat within streets.
6. The urban greenhouse effect. Given that the concentration of atmospheric pollutants in cities is high, part of the emitted infrared radiation from the ground surfaces is reflected back to the earth's surface.

The thermal balance of cities is changing as a function of several critical parameters like the specific land use, the urban density and size, the local topography, the levels of the urban green infrastructure, the optical and thermal properties of the used materials, and the landscape characteristics of the cities (Santamouris, 2015). The magnitude of the urban overheating highly depends on the specific synoptic conditions of a place. The development of an urban heat island is favored under anticyclonic conditions, while cyclonic conditions usually correspond to lower intensities of urban overheating (Giannopoulou et al., 2014).

Urban overheating is the most severe and documented phenomenon of climate change. According to Tuholske et al., 2021, there are more than 13,000 cities exhibiting overheating problems, and there are more than 1.7 billion people living under severe overheating conditions, while there are almost three times more overheating hours since 1980.

There are important synergies between urban overheating and regional scale heat waves. During the period of extreme climatic conditions like heat waves, high pressure weather systems are prevailing, delivering clear skies and warm air from the troposphere, affecting the magnitude of the released sensible and latent heat, as well as the advective, anthropogenic, and storage heat fluxes. As a result, the urban temperature is increasing and the urban−rural thermal contrast is affected (Pyrgou et al., 2020). Several studies have demonstrated that during the period of heat waves, the magnitude of the urban heat island is increasing considerably (Founta & Santamouris, 2017; Kassomenos et al., 2022; Ngarambe et al., 2020; Saeed Khan et al., 2020).

Higher ambient temperatures in cities have a serious impact on human life (Santamouris & Vasilakopoulou, 2021). Urban overheating increases the cooling energy consumption of buildings, peak electricity demand, making utilities to build additional power plants, and the concentration of harmful pollutants, in particular that of the ground level ozone, deteriorates the living conditions of low income and vulnerable population, affects human health, increases the levels of heat-related mortality and morbidity as well as the problems related to mental health, and decreases considerably the productivity of humans (Santamouris, 2020).

To counterbalance the impact of urban overheating, several mitigation technologies have been developed and implemented in large-scale urban renewal projects (Akbari et al., 2016). Mitigation technologies are based on the use of reflecting and photonic materials to be implemented in roofs, facades, and building structures (Santamouris & Yun, 2020), greenery on open urban spaces and buildings (Santamouris et al., 2018), and solar control and dissipation techniques based on the use of low temperature heat sinks (Agas et al., 1991). Monitoring of the existing large-scale projects has demonstrated that the available mitigation techniques are able to decrease the peak temperature of cities up to 2.5 C−3 C (Santamouris et al., 2017). In parallel, it has been documented that implementation of urban mitigation technologies decreases considerably the energy consumption of urban buildings (Garshasbi et al., 2023), while it seriously decreases heat-related mortality and morbidity (Santamouris & Fiorito, 2021, Santamouris & Osmond, 2020).

This chapter aims to present and discuss the main impact of urban overheating on human life, in particular, the impact of higher urban temperatures on the energy consumption of buildings, the peak electricity demand, the concentration of pollutants, the quality of life of low income and vulnerable population, and health and economy.

1.2 On the impact of urban overheating on the energy demand of buildings

Increase of the urban ambient temperature has a serious impact on the cooling energy consumption of buildings. Several studies have shown that during the summer period, urban zones characterized by an important magnitude of urban overheating suffer from a serious increase of the cooling demand of buildings (Hassid et al., 2000; Santamouris et al., 2001). Estimation of the cooling demand of a typical residential building in Athens, Greece, has shown that overheated urban areas present almost the double cooling demand than the coolest parts of the city (Santamouris et al., 2001). Recent studies in Seoul, Korea, found that changes in average UHI intensity of 0.5K correspond to an increase in monthly cooling energy consumption in the range of 0.17−1.84 kWh/m^2 (Mi et al., 2021).

A classification of all existing studies assessing the energy impact of urban overheating has allowed to evaluate the range of the potential impact of overheating on the energy demand of buildings under various climatic and urban conditions (Santamouris, 2014). Comparison of the energy consumption of reference buildings located in rural and/or urban zones, presented by 24 different studies, has shown that the average penalty of cooling demand induced by the urban overheating is close to 12% and varies between 0.1 and 20 kWh/m^2/y with an average close to 2.4 kWh/m^2/y. This corresponds to about 2.7 kWh/m^2/y per degree of temperature increase (Santamouris, 2014).

The temporal increase of the cooling demand of urban buildings has been evaluated by 18 studies using a long series of climatic data recorded by the same meteorological station. Numerous studies have evaluated the UHI-induced temporal increase of the urban cooling energy demand using long climatic data series from the same meteorological station. It is found that between 1970 and 2010, the cooling demand of buildings increased by 23% in average, or 11 kWh/m²/y, while the corresponding heating demand has decreased by 19%, and the sum of the cooling and heating load has increased by 11% (Santamouris, 2014).

In parallel, the cooling penalty induced by the urban overheating on the total urban building stock is evaluated for several world cities. It is reported that the average global energy penalty caused by the urban overheating is close to $0.73 \pm (0.64)$ kWh/m²/C, the average global energy penalty per person is close to $230 \pm (120)$ kWh/p, and the average global energy penalty per person and degree of temperature increase is $78 \pm (47)$ kWh/p/C (Santamouris, 2014).

Further analysis of the existing studies resulted in the following conclusions (Santamouris, 2014):

1. The energy penalty induced by urban overheating in cooling-dominated climates is much higher than the corresponding decrease of the heating load.
2. In heating-dominated climates, the decrease of the heating demand is much higher than the corresponding increase of the cooling penalty.
3. In climatic zones presenting an average summer ambient temperature below 23 C, urban overheating tends to decrease the total building energy consumption, while when the average ambient temperature is higher than 27 C, the global energy consumption is increasing considerably.
4. The urban overheating-induced cooling penalty per degree of the average overheating intensity is found to be correlated against the logarithm of the corresponding reference cooling demand.
5. The estimated cooling penalty induced by urban overheating is a strong function of the specific local characteristics of the overheating as well from several operational parameters like the selected set point temperature, the ventilation rate, and the quality of the building envelope.

Apart from the energy impact induced by urban overheating, global climate change is also affecting the energy consumption of buildings. Simulation data for the period 1990–2010 and the future (2030–2100) energy consumption of office buildings from 144 projects all around the world have concluded to the following (Santamouris, 2016):

The magnitude of the future cooling penalty induced by the global climate change depends strongly on the characteristics of the current climate and the considered climatic scenario. When the actual cooling demand is quite low, the foreseen penalty is also limited. On the contrary, the expected future cooling penalty is greater in climatic zones presenting a high current cooling demand. Quite a strong nonlinear relation between the reference cooling load, Q, and the relative increase of the cooling demand per degree of temperature rise is observed.

1.3   On the impact of urban overheating on the peak electricity demand

Higher urban temperatures reduce considerably the potential of passive cooling techniques like day and night-time ventilation and force consumers to use air conditioning more frequently (Santamouris et al., 2010). Increase in the peak electricity demand during the summer period induced by the extensive use of air conditioning makes utilities to build additional power plants to be operated for a limited period, increasing the cost of the produced electricity (Santamouris et al., 2015).

The additional power required to satisfy the increased demand for air conditioning depends on the thermal quality of the building stock, the set point temperature, and the operational characteristics of the local electricity network (Akbari et al., 1992; Colombo et al., 1999; Parkpoom & Harrison, 2008; Yabe, 2005). Analysis of the existing studies evaluating the additional peak electricity demand induced by the potential increase of the ambient temperature has shown that the penalty per degree of temperature increase varies between 0.45% and 12.3%, while the average urban penalty on the electricity demand is 3.7% or 215 MW per degree of temperature increase and the power penalty per person is close to $21.9 \pm (11.8)$ W/C/person (Santamouris et al., 2015). This corresponds to an additional electricity penalty of about 21 ($\pm$ 10.4) W per degree of temperature increase per person. Future projections of the peak electricity demand and the corresponding required investments show a very considerable increase (Fig. 1.1). It is reported that in India, the additional electricity consumption to satisfy the cooling demand by 2030 is close to 239 TWh/y, equivalent to an additional power installation of 143 GW (Phadke et al., 2012) or 300 new coal fired electricity power plants of 500 MW each. In parallel, Downing et al. (1995, 1996) estimated that increase of the average ambient temperature by 1K can result in an additional energy consumption for cooling purposes costing around to 75 billion dollars (Fig. 1.2).

Besides the increase of the peak electricity demand, higher ambient temperatures affect significantly the carrying capacity of the electricity transmission networks because of the power line sagging, and it is foreseen that by 2040−60, the average summertime electricity transmission capacity in the United States may decrease between 1.9% and 5.8%, relative to 1990−2020 (Bartos, 2016). In parallel, higher ambient temperatures increase the losses of transformers and substations (Mikellidou et al., 2018). It is estimated that because of the local and global climate change, almost 14%−23% additional investments on electricity capacity will be required in the United States between 2010 and 2055 (Linder & Inglis, 1989).

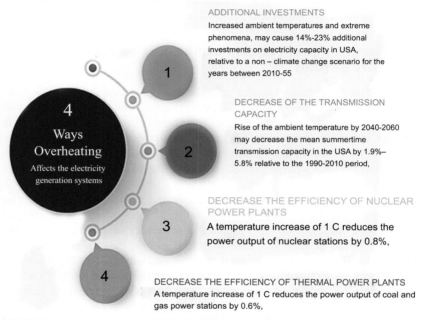

FIGURE 1.1 Fours ways that overheating affects the electricity generation system.

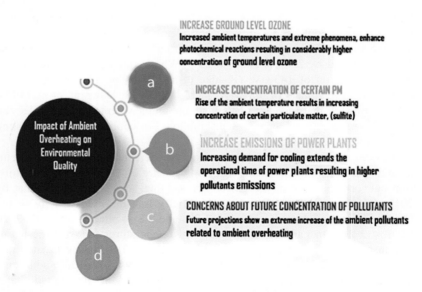

FIGURE 1.2 Impact of ambient overheating on urban environmental quality.

1.3.1 On the impact of urban overheating on the energy performance of power plants

Increase of the ambient temperature significantly affects the electricity generation performance of the thermal and nuclear power plants. It is reported that the power output of nuclear power plants decreases by 0.8% per degree of temperature increase (Davcock et al., 2004). It is foreseen that because of the increase of the ambient and water temperatures, the electricity generation capacity of the nuclear power plants may decrease worldwide up to 6 GW (Rubbelke & Vogele, 2011). It is characteristic that during the severe 2022 heat wave in France and the corresponding rise of the rivers' water temperature, the day-ahead baseload power prices were almost 10 times higher than the prices from 2017 to 2021 (Bloomberg Green Newsletter, 2022).

In parallel, the generation capacity of thermal power plants based on coal and gas and operating under Brayton and Rankine cycles depends on ambient temperature, humidity water availability, and pressure (Arrieta & Lora, 2005). Operation under high ambient temperature affects the heat rate and the delivered power, decreases the electricity generation efficiency, and affects the reliability of supply (Paeth et al., 2007; Schaeffer et al., 2012) (Table 1.1).

1.4 On the impact of urban overheating on the concentration of urban pollutants

Increased urban temperatures accelerate photochemical reactions and the corresponding generation of ground level ozone and affect the air flow and the turbulent exchange in cities, increasing the concentration of pollutants. In parallel, they slow down the flow of sea breeze in coastal areas and block the pollutants (Meier, 2017). Table 1.2 summarizes the various implications of urban and ambient overheating on the environmental quality and the concentration of pollutants.

Because of the longer period of heat waves and the extensive need for air conditioning, utilities must operate power plants for an extended period to satisfy the demand. Increased operation of the power plants rises the emission of several pollutants. It is reported that for each degree of temperature rise in the Eastern United States, the emissions of SO_2, CO_2, and NOx increase by 3.35%/°C ± 0.50%/°C, 3.32%/°C ± 0.36%/ °C, and 3.60%/°C ± 0.49%/°C, respectively (Abel, 2017; Meier, 2017). Forecasts of the electricity consumption in the Eastern United States, considering a temperature increase between 1 C and 5 C, show that the electricity demand may increase on average close to 7% and the noncoincident peak electricity demand by 32%, resulting in an increase of the NOx and SO_2 emissions by 16% and 18%, respectively.

Overheating in cities can increase the concentration of sulfate-based particulate matter components because of the faster SO_2 oxidation (Dawson et al., 2007; Jacob & Winner, 2009). The impact of high ambient

TABLE 1.1 Implications of ambient and urban overheating on demand and supply side components of electricity.

Demand system component	Ambient and urban overheating effect	Implications
Cooling load of buildings, peak electricity demand, load duration curves, non-temperature-sensitive demand	Higher ambient temperature in summer, higher ambient temperature in summer, important change of air conditioning profile, increased cooling water temperatures	Increase of the cooling demand of buildings, increase of the peak electricity demand, higher demand curve peaks, and much greater load variability may increase chances of breaching the market price cap generation curtailments and potential interruption of power to avoid blackouts
Implications of ambient and urban overheating on supply-side components of electricity		
Supply system component, thermal electricity generation, plants and components, transmission network, substations and transformers, fuel stock, power plants	Overheating effect, increased ambient temperatures, increased water temperatures, higher ambient temperatures and longer spells of dry weather, high ambient temperatures, increased ambient temperature, and extreme events increase the peak electricity demand	Implications: Decreased efficiency of electricity-generating equipment like gas turbines, coal power plants, and so forth. Power disruptions and increased cost of adaptation designs. Reduced equipment lifetime. Reduced power carrying capacity of transmission lines may cause disruptions because of the power line sagging. Increased losses within substations and transformers. Coal stocks may spontaneously combust or self-ignite. Utilities must build additional power plants to cover the peaks. Increased cost of electricity production during the peak hours.

Source: Adapted from ([Santamouris, M. (2020). Recent progress on urban overheating and neat Island research. Integrated Assessment of the Energy, Environmental, Vulnerability and Health Impact Synergies with the Global Climate Change, Energy and Buildings, 207, 109482; Chandramowli, S. N., & Felder, F.A. (2014). Impact of climate change on electricity sys- tems and markets —A review, of models and forecasts, Sustainable Energy Technologies and Assessments 5 (2014) 62—74]).

TABLE 1.2 Quantified implications of ambient and urban overheating on ambient air quality.

Air quality component	Overheating effect	Implication
Ozone concentration	Global overheating—heat waves	(a) Increase of the ozone concentration between 9.6%–20% during heat (a) waves,
Ozone concentration	Combined impact of urban heat island and global climate change	(b) Increase of the ambient temperature by 1% increases the number of days exceeding the threshold of ozone concentration by 10%. (c) Urban Heat island increases the number of days exceeding the threshold ozone concentration by 18%, (d) Urban overheating increases the ozone concentration between 10–30 and 1–13 ppb during the night and day time, (e) About 67%–84% of the annual variability of the ozone concentration is due to the change of temperature and other meteorological parameters, (f) Predicted future concentration of the ozone ranges between 6 to 12 ppb, (g) Predicted percentage future increase of the concentration between 20%–60% in 2050, 80% in 2080, and 400% by 2100.
Emissions of pollutants by power plants	Combined impact of urban heat island and global climate change	(h) Increased emission of pollutants by power plants per degree of temperature increase: 3.32%lQC ± 0.36%10 C increase in CO_2 emissions, 3.35%10 C ± 0.50%10 C increase in SO_2 emissions, and a 3.60%/ 0 C ± 0.49%1 0c increase in NOX emissions
Future emissions of pollutants by power plants	Combined Impact of Urban Heat Island and Global Climate Change	(i) Estimated increase of emissions by power plants: Plus 16% NOX emissions and plus 18% SO2 emissions by 2050.

Source: Adapted from (Santamouris, M. (2020). Recent progress on urban overheating and ηeat Island research. Integrated Assessment of the Energy, Environmental, Vulnerability and Health Impact Synergies with the Global Climate Change, Energy and Buildings, 207, 109482).

temperatures on nitrates and organic volatile components is not found to be significant (Aw & Kleeman, 2003).

High temperatures speed up photochemical reactions in the atmosphere because of the chemical interactions with nitrogen oxides and hydrocarbons, resulting in the formation of ground level ozone. (Coates et al., 2016). Ozone is a toxic pollutant and an oxidant affecting seriously the human cardiovascular and respiratory systems by irritating the lungs and is strongly related to elevated heat-related health problems like heat-related morbidity and mortality (Reid et al., 2012), while its formation depends on the magnitude of solar radiation, ambient temperature, and the concentration of NOx and VOCs and the ratio of VOCs and NOx (Stathopoulou et al., 2008). Ozone is classified as a causal intermediate in the heat-related mortality association (Wang et al., 2017).

Numerous studies have documented the strong association between the increased ambient temperature and the concentration of ground level ozone (Lai & Cheng, 2009; Pyrgou et al., 2018). The impact is found to be stronger in polluted urban zones, where the concentration of ozone may exceed 60 ppb (Sillman & Samson, 1995).

Despite the global decrease of the atmospheric pollutants in cities of the developed world because of the intensive measures to fight urban pollution, ground level ozone seems to increase constantly during the recent years. Data from 74 cities in China during the years 2013−15 show a significant increase of the concentration (Wang et al., 2013). It is characteristic that the fraction of noncompliant cities to ozone concentration threshold increased from 23% to 38% during the same period, while the average concentration increased from 69 to 75 ppbv (Wang et al., 2013).

The foreseen increase of the urban ambient temperature is expected to further increase the concentration of the ground level ozone and the frequency of severe ozone episodes (Cheng et al., 2007; Patz et al., 2005). It is reported that the frequency of severe ozone episodes in four Canadian cities may increase up to 50% by 2050 and 80% by 2080 (Cheng et al., 2007), while in Tuscon, Arizona, USA, the severe ozone episode may increase up to 400% by 2100, causing a very serious health impact (Wise, 2009). In parallel, it is predicted that in the North Eastern part of the United States, the potential increase of the ground level ozone may vary between 10% and 30% by 2020 and double by 2050 (Lin et al., 2007).

1.5   On the impact of urban overheating on health

It is well documented that exposure of human beings to high ambient temperatures is a very serious health hazard, as it may deregulate their thermoregulation system, resulting in serious health problems or even in deaths (Patz et al., 2005). The spectrum of the direct and indirect impact of higher ambient temperatures are well investigated and documented, as shown in

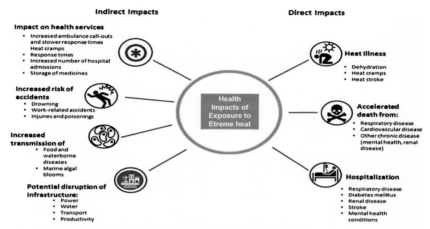

FIGURE 1.3 Impact of overheating on health. *WHO: Heat and health (who.int).*

Fig. 1.3 (WHO, 2023). Elderly population, people with the preexisting health problems, those using medication, and those lacking in economic assets and access to public support systems seem to be the most vulnerable groups (Paravantis et al., 2017).

Numerous studies have documented the impact of high ambient temperature on morbidity and mortality related to cardiovascular problems, respiratory diseases, heat stroke, mental illnesses, diabetes, dehydration nephric symptoms, and so forth (Li et al., 2019; Lin et al., 2009). It is reported that heat-related hospital admissions increase during heat waves between 1% and 11%, on average, per degree of temperature increase, while during the rest of the period, it ranges between 0.05% and 4.6% (Santamouris, 2020). Given the magnitude of urban overheating, heat-related hospital admissions are much higher in cities than in rural areas, and low income neighborhoods present up to 50% higher morbidity risk than the average (Bassil, 2009; Hondula & Barnett, 2014). During the 1995 heat wave in Chicago, USA, the relative risk of heat related hospital admissions was almost 3.9 times higher than in the suburban areas (Rydman et al., 1999).

Socioeconomic and biophysical parameters that determine the levels of urban vulnerability and deprivation strongly affect the magnitude of heat-related mortality (Smargiassi et al., 2009). Recent metaanalysis of numerous articles analyzing the impact of local overheating on heat-related mortality concluded that population living in the warmer urban neighborhoods presents almost 6% higher risk of mortality than those living in the cooler neighborhoods (Schinasi et al., 2018). The same study concluded that those living in less vegetated urban areas present 5% higher mortality risk than those living in the greener urban districts (Schinasi et al., 2018). Other studies have concluded that urban overheating increases the background heat-related

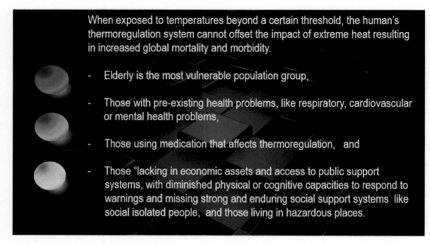

When exposed to temperatures beyond a certain threshold, the human's thermoregulation system cannot offset the impact of extreme heat resulting in increased global mortality and morbidity.

- Elderly is the most vulnerable population group,

- Those with pre-existing health problems, like respiratory, cardiovascular or mental health problems,

- Those using medication that affects thermoregulation, and

- Those "lacking in economic assets and access to public support systems, with diminished physical or cognitive capacities to respond to warnings and missing strong and enduring social support systems like social isolated people, and those living in hazardous places.

FIGURE 1.4   Impact of overheating on human health.

mortality between 1 and 27 deaths per million of population on average (Santamouris, 2020) (Fig. 1.4).

1.6   On the impact of urban overheating on vulnerable and low income population

Urban overheating has a serious impact on vulnerable and low income population, as it affects the magnitude of the urban socioeconomic and biophysical vulnerability (Kolokotsa & Santamouris, 2015). The impact is well documented in numerous studies performed mainly in developed countries, while information regarding the population in developing countries is missing, as shown in Table 1.3.

It has been shown that low income and vulnerable population is living in urban districts characterized by disproportionately high UHI intensity and excess heat stress. As a result, indoor and outdoor discomfort levels are higher than the average, while there is a higher risk of heat-related morbidity and mortality (Klein et al., 2014; Lapola et al., 2019; Santamouris & Kolokotsa, 2015; Taylor, 2015). It is characteristic that in 108 US cities, urban neighborhoods redlined in the 1930s are found to present considerably higher surface temperature profiles than the rest of the urban zones (Hoffman et al., 2020; Nazarian, 2022), while data from six Brazilian metropolitan cities show that the heat stress vulnerability index presents high association with existing socioeconomic differences (Lapola et al., 2019). Sociodemographic data collected from 175 large US cities associated with high resolution surface temperature data have shown that the average person of color lives in neighborhoods with a higher intensity of the surface UHI intensity than non-Hispanic whites (Hsu et al., 2021).

TABLE 1.3 Impact of overheating on vulnerable and low income population.

Component	Overheating ambient effect	Implication
Cooling energy cost	Combined impact of global and regional climate change	(a) The cost of air conditioning of low income households may increase up to 100% relative to the average conditions
Indoor temperature	Combined impact of global and regional climate change	(b) Indoor peak temperatures in low income houses during heat waves may exceed 40 C
Cooling needs	Combined impact of global and regional climate change	(c) Low income households cover a very small part (even 3%) of their cooling needs
Urban heat island and vulnerability	Combined impact of global and regional climate change	(d) Low income population is living in deprived neighborhoods characterized by excess heat stress and high urban heat island intensity(e) Districts of high vulnerability levels are usually associated with a higher risk of heat-related mortality

Source: Adapted from (Santamouris, M. (2020). Recent progress on urban overheating and heat Island research. Integrated Assessment of the Energy, Environmental, Vulnerability and Health Impact Synergies with the Global Climate Change, Energy and Buildings, 207, 109482).

Low income population lives in buildings of low energy and thermal quality (Santamouris, 2014). As a result, indoor temperatures during extreme climatic events may exceed considerably the set threshold for comfort and health (Sakka et al., 2012). Indoor temperatures exceeding 35 C are reported by numerous monitoring experiments (Haddad et al., 2022; Mavrogianni et al., 2010; Summerfield et al., 2007). In parallel, the required electricity load to satisfy the cooling needs is excessive and almost double the average cooling demand in the same city (Synnefa et al., 2017).

1.7   On the impact of urban overheating on human productivity

Exposure of humans to excessive ambient temperatures is causing heat exhaustion, reduced human performance, and limited working capacity and is associated with a serious increase of occupational injuries and declining labor productivity (Binazzi, 2019; Kjellstrom et al., 2009). When temperature exceeds a threshold sweating, the main heat rejection human mechanism is not sufficient, resulting in body temperatures that may rise close to 39 C, causing serious health and wellbeing problems (Bridger, 2008). For social and economic reasons, human beings must work under conditions of very high heat stress, although productivity and work intensity need to drop under such conditions (Wästerlund, 1998). Results of metaanalysis of the existing research articles show that productivity in either indoor or outdoor works under severe heat stress conditions may decrease by 30%, while the workers' productivity decreases by 2.6% for each degree above 24 C wet bulb globe temperature (DARA, 2012). Numerous physiological studies have concluded that there is a highly nonlinear association between the maximum ambient temperature and the supplied labor time, characterized by a very rapid decrease of the supplied labor above 30 C (Zivin & Neidell, 2014). Numerous studies have concluded that workers working a single shift under severe heat stress conditions are almost four times more likely to suffer from occupational heat strain than those working under thermoregulated conditions (Flouris, 2018), while losses of labor time between 40% and 60% are reported under ambient temperatures between 35 C and 40 C (Zivin & Neidell, 2014). Analysis of the heat-related losses in the United States during the last 40 years shows an increase close to 9% (Parsons et al., 2022), while in several countries of Southeast Asia, about 15%−20% of the annual labor time may already be lost regarding heat-exposed jobs (Kjellstrom & Meng, 2016).

High ambient temperatures may cause serious occupational injuries (Xiang et al., 2014). It is reported that industrial workers present odds ratios of acute injuries close to 2.28 and 3.52 for temperature ranges between 32°C and 38°C and above 38°C compared to ambient temperatures between 10°C and 16°C (Fogleman et al., 2005).

1.8 Conclusions

Overheating of cities is one of the most important environmental problems causing a very considerable impact on various aspects of human life. During the recent years, significant progress has been achieved regarding the characteristics, the impact, and the mitigation of the phenomenon. Important mitigation technologies that can highly counterbalance the impact of urban overheating have been developed (Santamouris, 2014). The existing mitigation technologies are based on the use and implementation of cool and super cool materials (Garshasbi & Santamouris, 2022; Khan et al., 2023), greenery, evaporation, and so forth (Wang et al., 2022).

There are several challenges associated with the development of proper scientific knowledge to fight urban overheating. Challenges are related to technological, economic, and political issues that obstruct the development and implementation of the proper policies to fight urban overheating. Given the very rapid increase of the temperature of our cities and the serious impact of the global climate change, it is more than urgent to accelerate and deepen research on the topic.

References

Abel, D., et al. (2017). Response of power plant emissions to ambient temperature in the Eastern United States. *Environmental Science & Technology*, *51*(10), 5838−5846. Available from https://doi.org/10.1021/acs.est.6b06201.

Agas, G., Matsaggos, T., Santamouris, M., & Argyriou, A. (1991). On the use of the atmospheric heat sinks for heat dissipation. *Journal of Energy and Buildings*, *17*, 321−329.

Akbari, H., Davis, S., Dorsano, S., Huang, J., & Winnett, S. (1992). *Cooling our communities: A guidebook on tree planting and light-colored surfacing. Environmental Protection Agency.* USA: EPA.

Akbari, H., Cartalis, C., Kolokotsa, D., Muscio, A., Pisello, A. L., Rossi, F., Santamouris, M., Synnefa, A., Wong, N. H., & Zinzi, M. (2016). Local climate change and urban heat island mitigation techniques − The state of the art. *Journal of Civil Engineering and Management*, *22*(1), 1−16.

Arrieta, R. P., & Lora, E. E. S. (2005). Influence of ambient temperature on combined-cycle power-plant performance. *Applied Energy*, *80*, 261−272.

Aw, J., & Kleeman, M. J. (2003). Evaluating the first-order effect of intraannual temperature variability on urban air pollution. *Journal of Geophysical Research: Atmospheres*, *108*(12). Available from https://doi.org/10.1029/2002jd002688.

Bartos, M., et al. (2016). Impacts of rising air temperatures on electric transmission ampacity and peak electricity load in the United States. *Environmental Research Letters*, *11*, 114008.

Bassil, K. L., et al. (2009). Temporal and spatial variation of heat-related illness using 911 medical dispatch data. *Environmental Research*, *109*(5), 600−606. Available from https://doi.org/10.1016/j.envres.2009.03.011.

Binazzi., et al. (2019). Evaluation of the impact of heat stress on the occurrence of occupational injuries: Meta-analysis of observational studies. *American Journal of Industrial Medicine*, *62*(3), 233−243. Available from https://doi.org/10.1002/ajim.22946.

Bloomberg Green Newsletter. (2022).

Bridger, R.S. (2008). Introduction to ergonomics, international edition.

Chandramowli, S. N., & Felder, F. A. (2014). Impact of climate change on electricity systems and markets − A review, of models and forecasts. *Sustainable Energy Technologies and Assessments*, *5*, 62−74.

Cheng, M., Campbell, Q., Li, G., Li, H., Auld, N., Day, D., Pengelfly, S., & Gin-grich, D. Y. (2007). A synoptic climatological approach to assess climatic impact on air quality in south-central Canada. Part II: Future estimates. *Water, Air, and Soil Pollution*, *182*, 117−130.

Coates, J., Mar, K. A., Ojha, N., & Butler, T. M. (2016). The influence of temperature on ozone production under varying NOx conditions - A modelling study. *Atmospheric Chemistry and Physics*, *16*, 11601−11615.

Colombo, A. F., Etkin, D., & Karney, B. W. (1999). Climate variability and the frequency of extreme temperature events for nine sites across Canada: Implications for power usage. *Journal of Climate*, *12*(8), 2490−2502. Available from https://doi.org/10.1175/1520-0442(1999), PART 2.

DARA. (2012). Climate vulnerability monitor 2012: A guide to the cold calculus of a hot planet (2nd ed.). DARA International. Madrid, Spain. Available from https://daraint.org/wp-content/uploads/2012/10/CVM2-Low.pdf.

Davcock, C., DesJardins, R., & Fennell, S. (2004). in: Generation of cost forecasting using online thermodynamic models, proceedings of electric power, Baltimore, MD.

Downing, T. E., Greener, R. A., & Eyre, N. (1995). The economic impacts of climate change, assessment of fossil fuel cycles for the externe project. *Oxford and Lonsdale: Environmental Change Unit and Eyre Energy Environment*, 1−48.

Downing, T. E., Eyre, N., Greener, R., & Blackwell, D. (1996). *Projected costs of climate change for two reference scenarios and fossil fuel cycles*. Oxford: Environmental Change Unit.

Dawson, J. P., Adams, P. J., & Pandis, S. N. (2007). Sensitivity of PM2.5 to climate in the Eastern US: A modeling case study. *Atmospheric Chemistry and Physics*, *7*(16), 4295−4309. Available from https://doi.org/10.5194/acp-7-4295-2007.

Flouris, A. D., et al. (2018). Workers' health and productivity under occupational heat strain: A systematic review and meta-analysis. *The Lancet Planetary Health*, *2*(12), e521−e531. Available from https://doi.org/10.1016/S2542-5196(18)30237-7.

Fogleman, M., Fakhrzadeh, L., & Bernard, T. E. (2005). The relationship between outdoor thermal conditions and acute injury in an aluminum smelter. *International Journal of Industrial Ergonomics*, *35*(1), 47−55. Available from https://doi.org/10.1016/j.ergon.2004.08.003.

Founta, D., & Santamouris, M. (2017, December 1). Synergies between urban heat island and heat waves in Athens (Greece), during an extremely hot summer. Scientific Reports − Nature, 7(1), Article number 10973.

Garshasbi, S., & Santamouris, M. (2022, May 15). Adjusting fluorescent properties of quantum dots: Moving towards best optical heat-rejecting materials. Solar Energy, 238, 272−279.

Garshasbi, S., Feng, J., Paolini, R., Duverge, J.J., Bartesaghi-Koc, C., Arasteh, S., Khan, A., & Santamouris, M. (2023, January 1). On the energy impact of cool roofs in Australia. Energy and Buildings, 278, 112577.

Giannopoulou, K., Livada, I., Santamouris, M., Saliari, M., Assimakopoulos, M., & Caouris, Y. (2014, February). The influence of air temperature and humidity on human thermal comfort over the greater athens area. Sustainable Cities and Society, 10.

Haddad, S., Paolini, R., Synnefa, A., De Torres, L., Prasad, D., & Santamouris, M. (2022). Integrated assessment of the extreme climatic conditions, thermal performance, vulnerability, and well-being in low-income housing in the subtropical climate of Australia. *Energy and Building*, *272*, 112349. Available from https://doi.org/10.1016/j.enbuild.2022.112349, vol.

Hassid, S., Santamouris, M., Papanikolaou, N., Linardi, A., Klitsikas, N., Georgakis, C., & Assimakopoulos, D. N. (2000). The effect of the athens heat island on air conditioning load. *Journal of Energy and Buildings*, *32*(2), 131−141.

Hoffman, J. S., Shandas, V., & Pendleton, N. (2020). The effects of historical housing policies on resident exposure to intra-urban heat: A study of 108 U.S. urban areas. *Climate*, *8*, 12.

Hondula, M. D., & Barnett, A. G. (2014). Heat-related morbidity in Brisbane, Australia: Spatial variation and area-level predictors. *Environmental Health Perspectives*, *122*(8), 831−836. Available from https://doi.org/10.1289/ehp.1307496.

Hsu, A, Sheriff, G., Chakraborty, T., & Manya, D. (2021). Disproportionate exposure to urban heat island intensity across major US cities. *Nature Communications*, *12*, 2721.

Jacob, D. J., & Winner, D. A. (2009). Effect of climate change on air quality. *Atmospheric Environment (Oxford, England: 1994)*, *43*(1), 51−63. Available from https://doi.org/10.1016/j.atmosenv.2008.09.051.

Kassomenos, P., Kissas, G., Petrou, I., Begou, P., Khan, H. S., & Santamouris, M. (2022). The influence of daily weather types on the development and intensity of the Urban Heat Island in two Mediterranean coastal metropolises. *Science of the Total Environment*, *819*.

Khan, A., Khorat, S., Doan, Q.-V., Khatun, R., Das, D., Hamdi, R., Carlosena, L., Santamouris, M., Georgescu, M., & Niyogi, D. (2023). Exploring the meteorological impacts of surface and rooftop heat mitigation strategies over a tropical city. *Journal of*

Geophysical Research — Atmospheres, *128*. Available from https://doi.org/10.1029/2022JD038099, e2022JD038099.

Klein, J., Rosenthal, P. L., Kinney., & Metzger, K. B. (2014). Intra-urban vulnerability to heat-related mortality in New York City, 1997−2006. *Health Place*, *30*, 45−60. Available from https://doi.org/10.1016/j.healthplace.2014.07.014, vol.

Kjellstrom, T., Holmer, I., & Lemke, B. (2009). Workplace heat stress, health and productivity-an increasing challenge for low and middle-income countries during climate change. *Global Health Action*, *2*(1), 1−6. Available from https://doi.org/10.3402/gha.v2i0.2047.

Kjellstrom, T., & Meng, M. (2016). Impact of climate conditions on occupational health and related economic losses, Asia Pacific Journal of Public Health. *Supplement Issue: Global Environmental Change and Human Health*, 28S−37S.

Kolokotsa, D., & Santamouris, M. (2015). Review of the indoor environmental quality and energy consumption studies for low income households in Europe. *The Science of the Total Environment*, *536*, 316−330. Available from https://doi.org/10.1016/j.scitotenv.2015.07.073, vol.

Lai, L.-W., & Cheng, W.-L. (2009). Air quality influences by urban heat island coupled with synoptic weather patterns. *The Science of the Total Environment*, *407*, 2724−2732.

Lapola, D. R., Braga, G. M., Di Giulio, R. R., Torres., & Vasconcellos, M. P. (2019). Heat stress vulnerability and risk at the (super) local scale in six Brazilian capitals. *Climatic Change*, *154*(3−4), 477−492. Available from https://doi.org/10.1007/s10584-019-02459-w.

Li, M., Shaw, B. A., Zhang, W., Vásquez, E., & Lin, S. (2019). Impact of extremely hot days on emergency department visits for cardiovascular disease among older adults in New York State. *International Journal of Environmental Research and Public Health*, *16*(12). Available from https://doi.org/10.3390/ijerph16122119.

Lin, S., Luo, M., Walker, R. J., Liu, X., Hwang, S. A., & Chinery, R. (2009). Extreme high temperatures and hospital admissions for respiratory and cardiovascular diseases. *Epidemiology (Cambridge, Mass.)*, *20*(5), 738−746. Available from https://doi.org/10.1097/EDE.0b013e3181ad5522.

Lin, C.-Y.C., Mickley, L.J., Hayhoe, K., Maurer, E.P., & Hogrefe, C. (2007, February 20−21). Rapid calculation of future trends in ozone exceedances over the Northeast United States: Re- sults from three models and two scenarios. Presented at the consequences of global change for air quality festival, EPA, Research Triangle Park, NC.

Linder, K., & Inglis, M. (1989). *The potential impact of climate change on electric utilities, Regional and National Estimates*. Washington DC: US Environmental Protection Agency.

Mavrogianni, A., Davies, M., Wilkinson, P., & Pathan, A. (2010). London housing and climate change: Impact on comfort and health - Preliminary results of a summer overheating study. *Open House International*, *35*(2), 49−59.

Meier, P., et al. (2017). Impact of warmer weather on electricity sector emissions due to building energy use. *Environmental Research Letters*, *12*(6). Available from https://doi.org/10.1088/1748-9326/aa6f64.

Mi, A.S., Ngarambe, J., Santamouris, M., & Yun, G.Y. (2021, May 21). Empirical evidence on the impact of urban overheating on -building cooling and heating energy consumption i-SCIENCE-D-21−00934, 24, 102495.

Mikellidou, C. V., Shakou, L. M., Boustras, G., & Dimopoulos, C. (2018). Energy critical infrastructures at risk from climate change: A state of the art review. *Safety Science*, *110*, 110−120.

Nazarian, N., et al. (2022). Integrated assessment of urban overheating impacts on human life. *Earth's Future*, *10*(8). Available from https://doi.org/10.1029/2022ef002682.

Ngarambe, J., Nganyiyimana, J., Kim, I., Santamouris, M., & Yun, G. Y. (2020). Synergies between urban heat island and heat waves in Seoul: The role of wind speed and land use characteristics. *PLoS*, *1*. Available from https://doi.org/10.1371/journal.pone.0243571, December 7.

Oke, T. R., Johnson, G. T., Steyn, D. G., & Watson, I. D. (1991). Simulation of surface urban heat islands under 'ideal' conditions at night - Part 2: Diagnosis and causation. *Boundary Layer Meteorology*, *56*, 339–358.

Paeth, A., Scholten, P., & Friederichs, A. (2007). Hense, Uncertainties in climate change prediction: El-Niño southern oscillation and monsoons. *Global Planetary Change*, *60*, 265–288.

Phadke, A., Abhyankar, N., & Shah, N. (2012). Avoiding 100 new power plants by increasing efficiency of room air conditioners in India: Opportunities and challenges. Berkeley, USA: Lawrence Berkeley National Laboratory.

Paolini, R., & Santamouris, M. (2022). *Urban climate change and heat islands: Characterization.* Elsevier, ISBN: 9780128189771.

Paravantis, I., Santamouris, M., Cartalis, C., Efthymiou, C., & Kontoulis, N. (2017). Mortality associated with high ambient temperatures, heatwaves, and the urban heat island in Athens, Greece. *Sustainability*, *9*, 606. Available from https://doi.org/10.3390/su9040606.

Parkpoom, S., & Harrison, G. P. (2008). Analyzing the impact of climate change on future electricity demand in Thailand. *IEEE Transactions on Power Systems*, *23*(3), 1441–1448. Available from https://doi.org/10.1109/TPWRS.2008.922254.

Parsons, L. A., Masuda, Y. J., Kroeger, T., Shindell, D., Wolff, N. H., & Spector, J. T. (2022). Global labor loss due to humid heat exposure underestimated for outdoor workers. *Environmental Research Letters*, *17*, 014050.

Patz, J. A., Campbell-Lendrum, D., Holloway, T., & Foley, J. A. (2005). Impact of regional climate change on human health. *Nature*, *438*, 310–317.

Pyrgou, A., Hadjinicolaou, P., & Santamouris, M. (2018). Enhanced near-surface ozone under heatwave conditions in a Mediterranean island. *Scientific Reports*, *8*(1), 9191.

Pyrgou, A., Hadjinicolaou, P., & Santamouris, M. (2020). Urban-rural moisture contrast: Regulator of the urban heat island and heatwaves' synergy over a Mediterranean city. *Environmental Research*, *182*.

Reid, C. E., Snowden, J. M., Kontgis, C., & Tager, I. B. (2012). The role of ambient ozone in epidemiologic studies of heat-related mortality. *Environmental Health Perspectives*, *120*(12), 1627–1630. Available from https://doi.org/10.1289/ehp.1205251.

Rubbelke, D., & Vogele, S. (2011). Impacts of climate change on European critical infrastructures: The case of power sector. *Environmental Science & Policy*, 53–63.

Rydman, R. J., Rumoro, D. P., Silva, J. C., Hogan, T. M., & Kampe, L. M. (1999). The rate and risk of heat-related illness in hospital emergency departments during the 1995 Chicago heat disaster. *Journal of Medical Systems*, *23*(1), 41–56.

Sakka, A., Santamouris, M., Livada, I., Nicol, F., & Wilson, M. (2012). On the thermal performance of low income housing during heat waves. *Energy and Buildings*, *49*, 69–77, Volume.

Santamouris, M. (2001). *Energy and climate in the urban built environment.* James & James (Science Publisher)/Earthscan. Available from http://doi.org/10.4324/9781315073774.

Santamouris, M., Papanikolaou, N., Livada, I., Koronakis, I., Georgakis, C., Argiriou, A., & Assimakopoulos, D. N. (2001). On the impact of urban climate to the energy consumption of buildings. *Solar Energy, 70*(3), 201−216.

Santamouris, M., Sfakianaki, A., & Pavlou, K. (2010). On the efficiency of night ventilation techniques applied to residential buildings. *Energy and Building, 42*, 1309−1313.

Santamouris, M. (2014). On the energy impact of urban heat island and global warming on buildings. *Energy and Buildings, 82*, 100−113, October 2014.

Santamouris, M. (2014). Cooling the cities − A review of reflective and green roof mitigation technologies to fight heat island and improve comfort in urban environments. *Solar Energy, 103*, 682−703, 2014B.

Santamouris., et al. (2014). Freezing the poor - Indoor environmental quality in low and very low income households during the winter period in Athens. *Energy and Building, 70*, 61−70. Available from https://doi.org/10.1016/j.enbuild.2013.11.074, vol.

Santamouris, M. (2015). Analyzing the heat island magnitude and characteristics in one hundred Asian and Australian cities and regions. *The Science of the Total Environment, 512−13*, 582−598.

Santamouris, M., Cartalis, C., Synnefa, A., & Kolokotsa, D. (2015). On the impact of urban heat island and global warming on the power demand and electricity consumption of buildings−A review. *Energy and Buildings, 98*, 119−124. Available from https://doi.org/10.1016/j.enbuild.2014.09.052, 1 July.

Santamouris, M., & Kolokotsa, D. (2015). On the impact of urban overheating and extreme climatic conditions on housing, energy, comfort and environmental quality of vulnerable population in Europe. *Energy and Building, 98*, 125−133. Available from https://doi.org/10.1016/j.enbuild.2014.08.050, vol.

Santamouris, M. (2016). Cooling of buildings. Past, present and future. *Energy and Buildings, 128*, 617−638, 2016.

Santamouris, M., Ding, L., Fiorito, F., Oldfield, P., Osmond, P., Paolini, R., Prasad, D., & Synnefa, A. (2017). Passive and active cooling for the outdoor built environment − Analysis and assessment of the cooling potential of mitigation technologies using performance data from 220 large scale projects. *Solar Energy, 154*, 14−33, 15 September.

Santamouris, M., Ban−Weiss, G., Cartalis, C., Crank, C., Kolokotsa, D., Morakinyo, T. E., Muscio, A., Ng, E., Osmond, P., Paolini, R., Pisello, A. L., Rossi, F., Sailor, D., Synnefa, A., Taha, H., Takebayashi, H., Zinzi, M., Zhang, J., & Tan, Z. (2018). Progress in urban greenery mitigation science − Assessment methodologies advanced technologies and impact on cities. *Journal Civil Engineering and Management, 24*(8), 638−671, 2018.

Santamouris, M. (2020). Recent progress on urban overheating and ηeat island research. Integrated assessment of the energy, environmental, vulnerability and health impact synergies with the global climate change. *Energy and Buildings, 207*, 109482, 15 January.

Santamouris, M., & Osmond, P. (2020). Increasing green infrastructure in cities-impact on ambient temperature, air quality and heat related mortality and morbidity. *Buildings, 10*, 233. Available from https://doi.org/10.3390/buildings10120233.

Santamouris, M., & Yun, G. Y. (2020). Recent development and research priorities on cool and super cool materials to mitigate urban heat island. *Renewable Energy*, 792−807.

Santamouris, M., & Fiorito, F. (2021). On the impact of modified urban albedo on ambient temperature and heat related mortality. *Solar Energy, 216*, 493−507, March.

Santamouris, M., & Vasilakopoulou, K. (2021). Present and future energy consumption of buildings and opportunities towards decarbonisation. In-Prime. Available from https://doi.org/10.1016/j.prime.2021.100002.

Saeed Khan, H., Paolini, R., Santamouris, M., & Caccetta, P. (2020). Exploring the synergies between urban overheating and heat waves (HWs), in Western Sydney. *Energies*, *13*, 470. Available from https://doi.org/10.3390/en13020470.

Schaeffer, R., Szklo, A. S., de Lucena, A. F. P., Borba, B. S. M. C., Nogueira, L. P. P., Fleming, F. P., Troccoli, A., Harrison, M., & Boulahya, M. S. (2012). Energy sector vulnera- bility to climate change: A review. *Energy*, *38*, 1e12.

Schinasi, L. H., Benmarhnia, T., & De Roos, A. J. (2018). Modification of the association between high ambient temperature and health by urban microclimate indicators: A systematic review and meta-analysis. *Environmental Research*, *161*, 168−180. Available from https://doi.org/10.1016/j.envres.2017.11.004, November 2017.

Sillman, S., & Samson, P. J. (1995). The impact of temperature on oxidant formation in urban, polluted rural and remote environments. *Journal of Geophysical Research*, *100*, 11497−11508.

Smargiassi, A., Goldberg, M. S., Plante, C., Fournier, M., Baudouin, Y., & Kosatsky, T. (2009). Variation of daily warm season mortality as a function of micro-urban heat islands. *Journal of Epidemiology and Community Health*, *63*(8), 659−664. Available from https://doi.org/10.1136/jech.2008.078147.

Stathopoulou, E., Mihalakakou, G., Santamouris, M., & Bagiorgas, H. S. (June 2008). Impact of temperature on tropospheric ozone concentration levels in urban environ- ments. *Journal of Earth System Science*, *117*(3), 227−236.

Summerfield, R. J., Lowe, H. R., Bruhns, J. A., Caeiro, J. P., Steadman., & Oreszczyn, T. (2007). Milton keynes energy park revisited: Changes in internal temperatures and energy usage. *Energy and Building*, *39*(7), 783−791. Available from https://doi.org/10.1016/j.enbuild.2007.02.012.

Synnefa, A., Vasilakopoulou, K., Kyriakodis, G.-E., Lontorfos, V., De Masi, R. F., Mastrapostoli, E., Karlessi, T., & Santamouris, M. (2017). Minimizing the energy consumption of low income multiple housing using a holistic approach, Original Research Article. *Energy and Buildings*, *154*, 55−71, Volume.

Taylor, J., et al. (2015). Mapping the effects of urban heat island, housing, and age on excess heat-related mortality in London. *Urban Climate*, *14*, 517−528. Available from https://doi.org/10.1016/j.uclim.2015.08.001, vol.

Tuholske, C., Caylor, C., Funk, C., Verdin, A., Sweeney, S., Grace, K., et al. (2021). Global urban population exposure to extreme heat. *PNAS*, *118*(41), e2024792118.

Wang, Y., Shen, L., Wu, S., Mickley, L., He, J., & Hao, J. (2013). Sensitivity of surface ozone over China to 2000−2050 global changes of climate and emissions. *Atmospheric Environment (Oxford, England: 1994)*, *75*, 374−382.

Wang, T., Xue, L., Brimblecombe, P., Lam, Y. F., Li, L., & Zhang, L. (2017). Ozone pollution in China: A review of concentrations, meteorological influences, chemical precursors, and effects. *The Science of the Total Environment*, *575*, 1582−1596.

Wang, J., Meng, Q., Tan, K., & Santamouris, M. (2022). Evaporative cooling performance estimation of pervious pavement based on evaporation resistance. *Building and Environment*, *217*, 109083, 1 June.

Wästerlund, D. S. (1998). A review of heat stress research with application to forestry. *Applied Ergonomics*, *29*(3), 179−183. Available from https://doi.org/10.1016/S0003-6870(97)00063-X.

WHO. (2023). Heat and health (who.int).

Wise, E. K. (2009). Climate-based sensitivity of air quality to climate change scenarios for the southwestern United States. *International Journal of Climatology*, *29*(1), 87−97, January.

Xiang, J., Bi, P., Pisaniello, D., & Hansen, A. (2014). Health impacts of workplace heat exposure: An epidemiological review. *Industrial Health, 52*(2), 91–101. Available from https://
doi.org/10.2486/indhealth.2012-0145.

Yabe, K. (2005). Evaluation of energy saving effect for the long-term maximum power forecast
(title only in original language).

Zivin, G., & Neidell, M. (2014). Temperature and the allocation of time: Implications for climate
change. *Journal of Labor Economics, 32*(1), 1–26. Available from https://doi.org/10.1086/
671766.

Chapter 2

Urban heat mitigation and adaptation: the state of the art

Giulia Ulpiani[1], Komali Yenneti[2], Ilaria Pigliautile[3,4], Anna Laura Pisello[3,4], Alberto Martilli[5], Carlos Bartesaghi Koc[6,7], Claudia Fabiani[3,4], Aldo Treville[1], Nadja Vetters[8] and Paolo Bertoldi[1]

[1]*European Commission, Joint Research Centre (JRC), Ispra, Italy,* [2]*School of Architecture and Built Environment, University of Wolverhampton, Wolverhampton, United Kingdom,* [3]*CIRIAF — Interuniversity Research Center on Pollution and Environment Mauro Felli — University of Perugia, Perugia, Italy,* [4]*Department of Engineering, University of Perugia, Perugia, Italy,* [5]*CIEMAT, Madrid, Spain,* [6]*Faculty of Science, Engineering and Technology, University of Adelaide, Adelaide, SA, Australia,* [7]*Risk & Resilience, Governance and Legal, NSW Department of Planning and Environment, Sydney, NSW, Australia,* [8]*European Commission, Joint Research Centre (JRC), Brussels, Belgium*

Adapting to and mitigating heat has become a necessity, rather than a commodity in many cities worldwide. Here, we analyze the state of the art along four axes that are pivotal in the design of more resilient and heat-safe urban futures: (1) knowledge of the arena of possible technologies, techniques, and pathways to heat adaptation and mitigation and ability to apply combinatorial approaches that look into co-benefits and adverse secondary effects; (2) assessment of risks and vulnerabilities, as a way to embed the human dimension in the design of responsive actions; (3) high-quality monitoring data for evidence-based action; and (4) urban heat modeling and mitigation/adaptation scenario-making to inform sturdy policies and coping strategies that target the local causes of heat. We zoom in on the role that greenery and materials play in heat mitigation and adaptation by highlighting most recent advancements and perspectives. Lastly, we focus on how cities are advancing in heat mitigation and adaptation within the frameworks of two major global and European initiatives devoted to cities and climate action (the Covenant of Mayors for Climate & Energy and the 100 Climate-Neutral and Smart Cities Mission) to better identify assistance needs and inspire future services and research strands.

Mitigation and Adaptation of Urban Overheating. DOI: https://doi.org/10.1016/B978-0-443-13502-6.00002-6

2.1 Urban heat: technologies and pathways toward mitigation and adaptation

Against the backdrop of rapid urbanization, aging populations, and escalating extreme heat events, the dual pressure of greenhouse gas-driven climatic warming and locally induced urban overheating is threatening citizens worldwide, with severe consequences especially for vulnerable groups and communities (Leal Filho et al., 2018; Voelkel et al., 2018; Wilhelmi & Hayden, 2010). Nevertheless, by knowing the drivers for exacerbated heat, it becomes possible to adjust, mitigate, even reverse such threats and protect those most at risk. Precondition is a well-informed knowledge of the local territory and of the arena of mitigation and adaptation options available.

Heat mitigation translates into overheating attenuation by reducing the number and magnitude of global and local sources of heat, by modulating the causes of heat entrapment (e.g., urban materials and anthropogenic heat), or by leveraging heat sinks and dispersion mechanisms. Mitigation techniques and technologies aim at guaranteeing suitable levels of ventilation, irradiation, permeability, and transpiration across the urban fabric. On the other hand, adapting means alleviating the vulnerability to heat (e.g., by limiting exposure or increasing the coping capacity). In its most natural connotation, heat adaptation comes as a change in behavior, habits, responsiveness, and awareness to adjust to actual or expected climate conditions and their effects. However, nowadays, a variety of instruments for adaptation exist: (1) "hard" forms, such as heat-safe building envelopes or restoration techniques for urban wetlands, (2) "soft" forms, such as insurance schemes or subsidies, or (3) a mix of both such as heatwave early warning systems (Martinez-Juarez et al., 2019). Adaptation solutions typically come with a high degree of flexibility and wide social and legal acceptance, as they tend to be capital-nonintensive, amenable to small-scale/individual interventions, and malleable to local circumstances. However, mitigation technologies have the potential to eradicate locked-in vulnerabilities and emissions in a long-lasting and fully-fledged fashion, thus standing as the backbone of deep-rooted climate resilience in the built environment.

Co-acting on heat adaptation and mitigation and mastering the way they interact (cobenefits, synergies, trade-offs, and spill over effects) is critical to create sustainable and salutogenic urban futures (Sharifi, 2021). There are many solutions at the watershed between adaptation and mitigation, including (2) modifying the urban albedo, (2) increasing the urban green infrastructure (UGI), (3) enhancing heat transpiration and evaporation, and (4) using low temperature heat sinks. These are displayed in Fig. 2.1 together with the main drivers for urban overheating.

One of the most efficient solutions to urban overheating is renaturalization. By bringing back green and blue elements into the city, the ratio of sensible heat exchange to latent heat exchange close to the surface

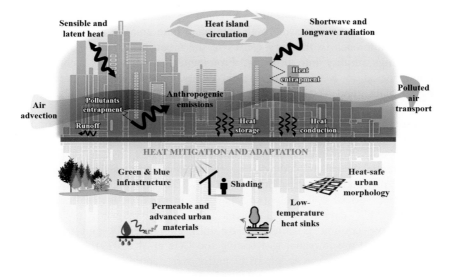

FIGURE 2.1   Urban overheating: typical causes (on the top) and solutions (on the bottom).

(Bowen ratio) can be rebalanced for the reduction induced by urban impervious materials and by the aerodynamic roughness of the urban canopy. Well-watered and transpiring surfaces produce urban airsheds that are armored against overheating and extreme events (e.g., heatwaves, droughts, fires, flooding, and pollution episodes), all the while promoting biodiversity in the built environment (Gao & Santamouris, 2019; Ulpiani et al., 2020).

Vegetation cools down the air through a combination of evapotranspiration, shade provision, and increased albedo, reducing peak surface temperature by 2°C−17°C depending on configurational and territorial determinants (Wong et al., 2021). It can be introduced in the cityscape in a variety of forms to create public green spaces of all sizes, from scattered trees to parks, from meadows to urban forests, while buildings can have their envelope turned into finely engineered green variants (green roofs, roof gardens, and green facades). Water-based solutions are complementary, as they cool down the air through evaporative cooling and enhanced convective heat transfer (Yang, Liu et al., 2020), reaching air cooling of few degrees to over 10°C in their very proximity (Chatzidimitriou et al., 2013; Gao & Santamouris, 2019; Meng et al., 2022; Ulpiani, 2019). Beyond natural water bodies (such as rivers and lakes) that are a prerogative of few privileged cities, evaporative cooling can be emphasized via (1) constructed wetlands like canals, ponds, drains, reservoirs, artificial lakes, and stormwater treatment sites or (2) capillary dissemination of point-like artificial water features such as fountains, sprinklers, and misting systems (Alikhani et al., 2021). Furthermore, an effective means of increasing both green and blue areas is urban farming, a reversed approach to the very causes of the urban heat island (UHI) effect that brings rurality into urbanity. Prominent urban farming technologies, implemented in a variety of urban centers, are aquaponics, aeroponics, hydroponics, and vertical farming (Carolan, 2020; Ladan et al., 2022). Another form of urban agriculture that also serves as urban blight counter-measure is the transformation of vacant lots into community gardens, multi-functional green areas that catalyze a variety of socioenvironmental cobenefits (Smith et al., 2021). Green and blue infrastructure develops its full potential when coupled with water-sensitive urban design that aims at permeability and reduced runoff.

Indeed, surface conditions (i.e., materials predominantly) arbitrate which and to what extent natural forces are stored, retained, or rejected. Evaporative cooling is enhanced by infiltration-friendly and evaporable materials that are remedial to flooding too (Xie et al., 2019), reaching sur-face cooling in the order of 10°C under wet conditions (Vujovic et al., 2021), yet rapidly diminishing after irrigation (Santamouris, 2013). Beyond biologically active surfaces (e.g., groundcover vegetation, wood chips, reten-tion basins, rain gardens, and bioswales), professionally crafted pervious pavements allow water to flow through their porous structures into the under-lying soil partly contributing to groundwater recharge and partly to second-ary evaporative cooling (Jerzy et al., 2020; Wang, Meng et al., 2022). The three most common technological variants are pervious interlocking pave-ments, pervious concrete pavements, and pervious asphalt pavements (Wang, Meng et al., 2022). However, the water budget is not the only critical equi-librium to maintain at the surface level in order to control urban overheating. The interaction between urban materials and solar radiation is commonly the prime cause of extreme urban heat. In this regard, a decisive technological contribution comes from recent advancements in thermooptics and materials science aimed at increasing the urban albedo and/or the heat reemission. Cool and supercool materials are revolutionizing the shortwave and long-wave heat transfer at city scale (Santamouris & Yun, 2020), producing mean surface temperature reductions between 1.4°C and 4.7°C and energy savings between 15% and 35.7% in different climatic zones (Rawat & Singh, 2021). Passive daytime radiative coolers (PDRCs) are among the envisioned game changers, as they transfer the heat directly to the massive heat sink repre-sented by the outer space, thus offering great potential for urban heat mitiga-tion (Bartesaghi-Koc et al., 2021; Carlosena et al., 2020; Feng et al., 2021; Khan et al., 2022), outdoor comfort, and cooling energy conservation (Anand et al., 2021; Carlosena et al., 2021; Feng et al., 2022; Mokhtari et al., 2022), especially when precautions are taken against power instability and overcooling (Ulpiani, Ranzi, Shah et al., 2020, 2021). Synergistic combi-nations with evaporative cooling or with solar heating have also proved to be extremely efficient (Li, Sun et al., 2020; Sun et al., 2022). Fluorescent and phosphorescent materials can also reduce the heat gains owing to their unique ability to re-emit a portion of the absorbed photons as visible light rather than longwave radiation through the so-called photoluminescence effect, thus complementing the cooling potentiality of high albedo materials through a two-pronged solar rejection mechanism (Garshasbi & Santamouris, 2019; Kousis et al., 2020; Xue et al., 2020). Cool facades have also a critical role to play in avoiding excessive solar gains (Mourou et al., 2022). However, in deep urban canyons and packed urban settings where the use of cool materials may aggravate the local heat entrapment due to multiple inter-reflection and absorption processes, the best practice is the implementation of retroreflective facades, as they reflect solar radiation back along the inci-dent direction, thus acting like cool surfaces while still allowing the heat to escape (Manni et al., 2020; Wang, Liu et al., 2021).

Beating the urban heat may also require breaking natural forces, for instance, by blocking direct solar radiation from reaching the surface. Trees are certainly an effective means of providing solar shading and preserve out-door comfort especially in the presence of deciduous species; however, high-performance and esthetically pleasant technological solutions are also being implemented where time to mature, intrusive root systems, and initial nurtur-ing process are limiting factors to green alternatives. Fixed structures are most commonly examined in literature; however, strong consensus arises on the advantages of using variable shading levels—according to seasonal, diur-nal, or even hebdomadal variations—through dynamic or movable structures that can vary their design parameters (e.g., rotational and folding angles) (Chi et al., 2021). Shape-wise, fractal forms (e.g., Sierpinski Tetrahedron fractal canopies) are most promising, as they mimic trees and provide sun-shade without being heated themselves (Ella et al., 2018). Another natural force that might be broken in order to alleviate urban overheating is wind. While increasing the ventilation rate is commonly conducive to reduced heat and pollutants, wind may also be the very reason for urban overheating. This is the typical case of cities swapped by desert or Foehn-like winds, where the advection of extremely hot air is the prime cause of heat stress (Haddad, Ulpiani et al., 2020; Hirsch et al., 2021; Yun et al., 2020). Again, urban veg-etation comes in handy, with windbreak tree belts being a major ally. Further, technological solutions developed for noise shielding can be repur-posed for heat adaptation and mitigation. These include vertical green wall systems, earth berms, and acoustic screens (Van Renterghem et al., 2015).

In counteracting overheating, cities can also take advantage of low-temperature heat sinks (e.g., water bodies and soil) that, by virtue of their high thermal capacity and lower temperature, can accommodate excess heat (Zoras & Dimoudi, 2016). Earth-to-air heat exchangers (EAHE, i.e., buried pipes at certain soil depths) and heat storage modified pavements (e.g., high-conduction, phase change materials (PCMs) doped, and energy harvesting pavements) can be used to lower the air or surface temperature by transferring the heat to lower soil layers (Peretti et al., 2013; Qin, 2015; Yinfei et al., 2015). EAHE have been found to cool down the inlet air by 10°C, which could result in ambient temperature drops of over 2°C (Gaitani et al., 2017). Modified pavements can mitigate the surface temperature by 2.5°C−9°C depending on the specific technology (Efthymiou et al., 2016; Jiang et al., 2018; Yinfei et al., 2015).

Despite the potential of individual solutions, two general considerations apply: (1) ample scientific evidence demonstrates that combinatorial approaches, tailor-made to the local specificities, tend to achieve greater impacts and synergies as compared to single-measure or unfitted strategies (Qi et al., 2020; Yenneti et al., 2020) and (2) the local performance strongly depends on the urban design and land use planning. Efficient strategies thrive on the enabling environment generated by "smart growth" urban planning, based on principles of (1) mixed land use, (2) compact design, and (3) capillary heat-safe pockets. Appropriate levels of density, promotion of public transportation, and urban greenery are measures that are more likely to provide synergistic benefits if combined with other adaptation and/or mitigation measures (Sharifi, 2021). However, notwithstanding the tremendous cobenefits across all sectors vital to a healthy urban metabolism, determining the right level of compactness, density, and land use mix still stands as a very challenging endeavor (Yan et al., 2021).

Another general warning is that extreme urban heat is not a self-contained issue. Rather, it is particularly life-threatening because (1) it can sustain and empower its mechanisms in vicious feedback loops (e.g., excess heat call for more air conditioning which pours more exhaust heat in the outdoors) and (2) it can trigger/exacerbate/negatively interact with other environmental hazards (e.g., drought, heatwaves, wildfires, and pollution) (Dong et al., 2018; He, Wang et al., 2021; Ulpiani et al., 2020; Ulpiani, 2021) and socioeconomic dynamics (human and economic losses on the healthcare system, reduced industrial output value, and loss of productivity) (Adélaïde et al., 2022; He et al., 2022; Kuznetsov & Tomitsch, 2018; Whiteoak & Saigar, 2019). Even urban heat mitigators can have adverse secondary effects when improperly planned. For instance, tree planting can worsen pollution levels if its arrangement blocks the advection of cleaner air and if the selected species feature high emission rates of biogenic volatile organic compounds (Santiago et al., 2017, 2019; Ulpiani, 2021). In the same vein, cool and supercool materials must be distributed with attention to (1) the alteration of UV reflectance, which may lead to increased ozone production (Ulpiani, 2021), and (2) the changes induced in the lower atmosphere such as inversion phenomena that may exacerbate air stagnation (Khan et al., 2021). As such, urban overheating presents an entanglement of challenges that do not necessarily converge onto a single solution (Kuznetsov & Tomitsch, 2018).

On account of this complexity, the study of heat mitigation and adaptation is rapidly evolving into a highly transdisciplinary research field, touching on many emerging areas in materials sciences, physiology and chemistry, epidemiology, and computer science. Despite the extensive topical coverage, literature tends to (1) disproportionately focus on materials and vegetation, while overlooking water features and urban form, (2) underrepresent the interaction between micro-, meso-, and macroscale phenomena that contribute to urban overheating, (3) provide an incomplete understanding of heat-related impacts leaving social and economic externalities unaddressed, and (4) overlook the challenges in implementation, urban financing, and social acceptance (He et al., 2023). A deeper understanding could better serve the purpose of turning science into policy and could assist in solving conflicts and disputes among the variety of stakeholders in urban planning. Further, the assessment of urban overheating has so far focused on aspects related to the built environment (indoor/outdoor temperatures, energy use, thermal comfort, and peak electricity demand) with increasing interest in the health implications (morbidity and mortality) (He et al., 2023). However, urban heat is impactful on a broad range of urban metabolic flows and activities related to, for example, tourism, transportation, food, construction, and agriculture and eventually on the achievement of zero-emission targets (He et al., 2023; Kolbe, 2019; Roxon et al., 2020; Vetters et al., 2021). Indeed, counteracting urban heat is effective at both climate adaptation (i.e., building resilience against climate change) and climate mitigation (i.e., reducing greenhouse gas emissions). On the other hand, heat mitigation and adaptation come with the risk to trigger social injustice and inequities by favoring wealthier communities and causing forms of "green" gentrification (Anguelovski et al., 2022; Bouzarovski et al., 2018). Policymaking and planning play a delicate role in handling competing riskscapes so as to guarantee that the benefits of climate change counteraction eliminate, not merely reconfigure, climate vulnerabilities (He, Zhao et al., 2021; Shokry et al., 2022).

Overall, while we expand our understanding of urban heat dynamics and its ramifications, there are still many open questions for investigation.

2.2 Multidimensional approaches to heat vulnerability

High temperatures and frequent extreme heat waves can affect a large urban population around the world and pose significant risks to urban resilience and governance (Fischer et al., 2007; Luber & McGeehin, 2008; Yenneti et al., 2020). High urban temperatures are already affecting human health by exposing people to heat-related illnesses such as heat exhaustion and heat stroke (Santamouris et al., 2020). Heat-related mortality, morbidity, and other indirect health impacts have increased significantly over the past two decades (Williams et al., 2012; Yu et al., 2010). More than half of the world's population is now vulnerable to extreme temperatures and heat-related health impacts (Li, Yuan et al., 2020). In the recent years, significant efforts have also been taken to define the potential direct and indirect effects of heat on human health (Dovie et al., 2017; Franchini & Mannucci, 2015). Extreme heat can have a major impact on the global economy and productivity, with research estimating the cost at around $16 trillion since the 1990s (Quaglia, 2023). These costs are relatively high for cities and may be disproportionately distributed. Managing extreme heat requires a deeper understanding of vulnerability and how it manifests in spaces and places, as it affects populations unevenly (Cheng, 2016; Kim & Ryu, 2015). Heat vulnerability refers to the degree of heat risk to human life and may depend not only on physiological and behavioral characteristics, but also on social, economic, and health opportunities. Human vulnerability to extreme heat varies globally, regionally, and within populations (Sillmann & Roeckner, 2008), being a location-specific, time-varying, and complex multiscalar concept (Demuzere et al., 2014; Hilhorst & Bankoff, 2013). It spans across multiple disciplines, such as physiology, psychology, climate change, urban planning, and building physics. This section aims to provide some understanding of the concept of heat vulnerability and the multidimensional approaches and techniques to assess heat vulnerability. The concepts and approaches discussed here can be applied to both cities and regions around the world and fall within two broad perspectives: (1) end point approach and (2) starting point approach. Assessing the adverse effects of heat stress on human life has been the focus of the end point approach, whereas the starting point approach focuses on a combination of environmental and social aspects.

2.2.1   End Point Approach

This approach focuses on the physiological and psychological dimensions of human vulnerability to heat stress and is grounded on the premise that human exposure to extreme climatic conditions is based on their individual physiological and behavioral characteristics (Nag et al., 2013; Parsons, 2014), which can vary even over the same temporal and spatial scales (McMichael et al., 2006). In general, human vulnerability to adverse climatic conditions is assessed in the form of thermal stress and strain. Under changing environmental conditions, the adaptive capacity of people is based on the thermoregulatory processes of the human body (Chen & Ng, 2012; Vanos et al., 2010) and is influenced by intrinsic and extrinsic factors such as external climate variability, environmental stimuli, microclimate conditions, thermal history, and human behavior patterns (Aljawabra & Nikolopoulou, 2018; Miller & Bates, 2007). Vulnerability assessments based on this argument can be dynamic and subjective, dynamic in the sense that adaptation to ambient thermal conditions is gradual and thermal sensation is largely influenced by previous experience and subjective in the sense that the assessment of thermal comfort conditions is not always consistent with objective climatic or biometeorological conditions.

The current knowledge of human vulnerability to thermal stress began in the late 18th century, with the efforts of engineers, building physicists, and physiologists in developing various thermal comfort indices (Binarti et al., 2020; Epstein & Moran, 2006; Jendritzky & Tinz, 2009) and standards that appropriately reflect a person's characteristics in a given environmental situation (Eludoyin & Oluwatumise, 2021). Over the last century, a large number of human thermal indices (e.g., heat stress index, thermal strain index, and universal thermal climate index) have been proposed to assess human exposure and vulnerability to thermal stress (see Carlucci & Pagliano, 2012; de Freitas & Grigorieva, 2015; Yasmeen & Liu, 2019 for comprehensive lists of thermal indices). The most influential parameters used in these indices are air temperature, humidity, solar radiation, wind speed (Davtalab & Heidari, 2021; Rodríguez Algeciras et al., 2016), clothing, body metabolism and other behavioral parameters (Fang et al., 2018; Havenith & Fiala, 2011) that together mold the magnitude of respiratory, evaporative, conductive, convective, and radiative thermal processes.

2.2.2   Starting point approach

In recent years, the approach to heat vulnerability has shifted from an impact-led approach (i.e., projecting impacts of heat through thermal modeling) to a social vulnerability approach (i.e., addressing various social and economic criteria) (Bao et al., 2015; Reid et al., 2009). The latter takes into account human exposure, risk, sensitivity, and adaptive capacity to determine the degree of vulnerability to heat (Adger, 2006; Cutter et al., 2003), as displayed in Fig. 2.2. Through this lens, heat vulnerability is socially

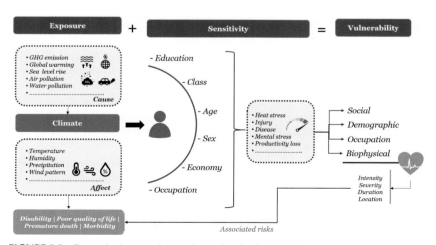

FIGURE 2.2   Dynamics between heat and associated vulnerability.

constructed and manifested with stratification and inequality across populations and places (Yenneti et al., 2016). The Intergovernmental Panel on Climate Change reports and recent burgeoning empirical literature on vulnerability to climate change and natural hazards (Füssel & Klein, 2006; Kantamaneni, 2019; Thornton et al., 2014) have also been significantly influenced by social vulnerability.

Drawing on this approach, the recent literature on heat vulnerability (Bao et al., 2015; Conlon et al., 2020; Karanja & Kiage, 2021; Li et al., 2022; Reid et al., 2009) recognizes the need for social construction of heat vulnerability. Heat vulnerability is generally related to the likelihood that a population will be adversely affected by extreme heat events in space and time (Kershaw & Millward, 2012; Leal Filho et al., 2018) and is underpinned by economic and social structures (Flanagan et al., 2011). Exposure to heat increases stress in a population and is influenced by factors such as local climate (Cai et al., 2019), built environment, and socioeconomic disparities (Johnson et al., 2012; Reid et al., 2009). Sensitivity is the degree to which a community is affected by heat and is related to intrinsic characteristics of the population (Johnson et al., 2012; Wolf & McGregor, 2013). The adaptive capacity of populations depends on the resources available to reduce the impacts of extreme heat events (Tapia et al., 2017; Yenneti et al., 2016).

Over the years, a range of methods and tools have evolved to analyze vulnerability (Harlan et al., 2006; Johnson et al., 2012; Reid et al., 2009; Wolf & McGregor, 2013). Indicator- or index-based methods applying principal component analysis (Conlon et al., 2020; Johnson et al., 2012) and expert judgement (Shih & Mabon, 2021) are more commonly used in heat vulnerability assessment studies (Bao et al., 2015; Romero-Lankao et al., 2012). Other heat vulnerability studies have used equal weightage (Liu, Yue et al., 2020; Vincent & Cull, 2010) and the Pareto ranking method (Qi et al., 2022) to avoid discrepancies. Some of the indicators used for heat vulnerability assessment are related to local climate, local nature, and built environment (e.g., greenspaces, water bodies, and sealed surfaces), sociodemography (e.g., population density and composition, mortality and morbidity, and education), economy, and governance (e.g., employment, technologies, and physical and social infrastructure) (Cheng et al., 2021; Nayak et al., 2018; Zhang et al., 2018). This approach has been applied in a variety of city contexts (e.g., New York, Santiago, Toronto, and London) (Inostroza et al., 2016; Nayak et al., 2018; Rinner et al., 2010; Wolf & McGregor, 2013).

An overall analysis of ongoing research on human vulnerability to heat suggests that the assessment of human susceptibility and vulnerability to extreme temperatures and heat is influenced by a range of factors, depending on the methodology. The resultant effect of environmental, physiological, behavioral, and psychological characteristics is generally ascertained in the end point approach, whereas social dynamics, economy, demography, and climate have a profound influence on the starting point approach.

Overall, these lines of research emphasize that people's susceptibility to extreme climatic conditions in outdoor settings is influenced by both intrinsic and extrinsic factors. It is important to note, however, that the starting point approach has recently received serious attention due to its increased societal relevance, applicability to cities, and potential for developing place-based solutions.

2.3   Urban heat monitoring: excursus on different techniques

The layout, structure, and purpose of urban areas, along with the natural and human-related patterns that occur throughout the day and year, play a crucial role in determining how much energy a city uses. These factors combine to create a distinctive thermal environment in urban areas (Wang, Xie et al., 2022; Yang, Yang et al., 2021; Yuan et al., 2022). Such inter- and intraurban complexity and heterogeneity calls for hyperlocal investigations (Pigliautile & Pisello, 2020). Furthermore, the UHI interacts synergistically with extreme weather events such as heatwaves (He, Wang et al., 2021), which are expected to be more frequent and intense in the future due to ongoing climate change (Li & Bou-Zeid, 2013). To better understand how the thermal environment in urban areas is affected by various features and to improve our ability to adapt to climate change, it is crucial to monitor it properly. This entails collecting relevant data at different resolutions in both space and time, in order to account for temporal and spatial variability. By doing so, we can gain valuable insights into the relationship between urban features and the thermal environment, ultimately improving our ability to adapt to a changing climate.

To identify the most appropriate techniques for urban heat monitoring, it is crucial to understand what and where to measure (Kim & Brown, 2021). Oke et al. identified three types of UHI based on the scale of analysis (Oke et al., 2017). The first type is the boundary layer heat island, which focuses on the entire urban boundary layer that typically extends between 250 m and 2500 m in height. This UHI typology covers a horizontal range from macro- to mesoscales, spanning tens to hundreds of kilometers, and representing the regional impact of urban areas on thermal environments. The canopy layer heat island, on the other hand, focuses on the energy balance within the urban canopy defined by building roofs. This typology is used to assess the impact of building density on local thermal environments in neighborhoods, but it does not represent outdoor thermal comfort conditions. However, data collected at the street level can be used to refine analysis down to the microscale (hundreds of meters). The urban surface heat island focuses on the surface temperature of urban features that are mostly impervious, compromising the evaporative cooling potential typical of natural surfaces.

By selecting the appropriate scale of observation and urban heat monitoring infrastructure/technique, critical areas that pose health risks to citizens can be identified. This information can guide decision-making and tailored adaptation actions to enhance the livability of cities. This section provides an overview of urban heat monitoring techniques that have been developed and applied over the last century. The article also highlights upcoming opportunities due to advances in sensing, information, and communication technologies. The two main techniques analyzed are ground-based observations and satellite imaging.

2.3.1   Ground-based observations

Ground-based observations have traditionally been used to detect the UHI phenomenon by collecting air temperature data in urban and rural areas. This method allows for analysis of long-term trends at a fine temporal resolution, typically with subhourly data collection rates (Bassett et al., 2016). Data collected through this technique provide insights into the UHI intensity variations over the years (due to urbanization and climate change), seasons (in response to weather boundary conditions and metabolic activity of vegetation), and 24-hour cycles (highlighting differences between urban and rural thermal environments during day and night). Each monitoring spot is representative of its surroundings as the thermal source area accounts for both the three-dimensional structure of the built environment and the prevailing weather conditions, that is, retrieved data are representative of the upwind area (Núñez Peiró et al., 2019). In this view, Stewart and Oke (Stewart & Oke, 2012) defined the local climate zones (LCZs) to contextualize urban heat measurements and provide standards for the intercomparability of the observations. Each LCZ is given by a unique combination of surface structure, cover, and human activity, becoming a reference for the local thermal climate investigation through fixed weather station networks (Yang, Peng et al., 2020). The monitoring stations are generally located above rooftops. Despite the contribution in assessing the temporal evolution of the UHI, this ground-based monitoring technique lacks precision for describing the urban climate at the microscale. A dense network of weather stations, also below the building canopy, would be needed to adequately map and comprehensively describe the heterogeneous urban thermal environment at the pedestrian level, as advocated and demonstrated by Ulpiani et al. (2022), Warren et al. (2016). The sensor network should be designed based also on variables that may influence the need for monitoring, such as presence of vulnerable population groups and gaps in current regulatory monitoring networks (Ulpiani et al., 2022). However, setting up and maintaining a high-density weather station network is challenging and expensive so that ground-based observations through fixed weather stations typically consist of limited data points (Stewart, 2011). Moreover, the continuous urbanization process implies ongoing changes in the urban LCZ definition that would require a certain level of flexibility in the network design that is incompatible with the structure of fixed weather stations, unless predicted urban developments are duly incorporated in the network design (Ulpiani et al., 2022). In this view, crowdsourcing represents a great opportunity to address spatial granularity without huge investments in the monitoring infrastructure. Moreover, crowd-sourced weather stations are generally closer to anthropogenic sources and urban fabrics providing a deeper look into the city (Potgieter et al., 2021).

Since the 1970s, mobile monitoring units have been used to enhance the quality of ground-based urban heat monitoring techniques down to the microscale (Conrads & Van Der Hage, 1971). This approach involves setting up dedicated environmental monitoring units on various carriers (such as automobiles, trains, buses, pedestrians, bicycles, and trolleys) along with GPS units to track collected variable profiles in specific urban areas. The carrier selection determines the monitoring speed and spatial density of the collected data points, as well as the extent of the investigated area. The carrier's characteristics also impose specific constraints on the instruments' exposure in order to prevent data distortion, calling for e.g. shading and proper ventilation of the air temperature probes and for appropriate setups that avoid proximity to vehicle exhaust or engine heat in case of motorized carriers. Mobile monitoring units capture intraurban variabilities with hyper-local spatial granularity (Kousis et al., 2021). The fine spatial resolution of this technique allows to shift the focus of the analysis from the characterization of the urban canopy layer to the assessment of citizen well-being and environmental health risk (Pioppi et al., 2020). This is particularly evident in the case of wearable systems that capture the real heat exposure of pedestrians in the urban settlement (Cureau et al., 2022). However, comparing air temperature data from different urban areas crossed by the sensing unit is not straightforward due to the elapsed time. Therefore, preprocessing the retrieved air temperature data is essential. The data correction should consider the air temperature variation occurring during the monitoring, which can be obtained from a reference fixed weather station in the same area (Pigliautile et al., 2021). In the absence of fixed weather stations, the monitoring route should be circular or include several intersections, so that consecutive observations at the same points provide time-dependent temperature variation. It is important to note that the proposed correction assumes the applicability of linear cooling throughout the area, which may not be appropriate, especially for prolonged monitoring campaigns. Despite the great potential of ground-based mobile monitoring techniques in capturing the hyperlocal variability of the urban thermal environment, these are still not systematically implemented as part of the urban infrastructure and practices. A recent review by Kousis et al. (2022) highlights that the technique is currently only adopted in academic research and there is a lack of common protocols and standardized procedures necessary for the implementation of mobile heat monitoring in the real world. Standards should be developed for the positioning of the monitoring unit on the selected agent, the assessment of the agent's impact on retrieved data, the data collection procedure considering reliable and feasible collection rates with respect to agent speed, and data processing for elapsed time correction. These standards would spur the deployment of mobile monitoring networks on, for example, public transportation. Equipped fleets could become an intervehicle network for microclimate monitoring, increasing the granularity of the whole urban heat mapping system, as proposed by Rodriguez et al. (2020).

Although mobile monitoring allows for a finer spatial resolution compared to fixed weather station networks, the comprehensive spatial mapping of the thermal environment is limited by the fact that data collection is restricted to points and lines within the three-dimensional urban environment. To address this limitation, various spatial interpolation methods have been used to analyze heat measurements from both fixed weather stations and mobile traverses. This research is essential because it allows for a more detailed understanding of the thermal environment in urban areas, which has significant implications for urban planning, design, and public health. By utilizing advanced spatial interpolation methods, researchers can create more accurate and comprehensive maps of UHIs, which can help identify areas that require intervention and adaptation measures to mitigate the adverse effects of high temperatures. Ultimately, this research can help create more sustainable and resilient cities that promote the health and well-being of their residents (Colaninno & Morello, 2022). In analyzing heat measurements from fixed weather stations and mobile traverses, researchers may employ data-driven approaches such as artificial neural networks or methods that account for weighting. Weighting methods are typically divided into two groups: (1) deterministic interpolation methods, such as inverse distance weighting or spline techniques and (2) statistical interpolation methods that encompass different types of kriging (Touati et al., 2020).

2.3.2 Remote sensing

Satellite imaging is a powerful complement to ground-based monitoring of urban heat, as it captures the thermal radiation emitted by surfaces from a unique vantage point. This technique provides a wider spatial coverage compared to ground-based observations, allowing for the detection of urban heat patterns across larger areas, such as entire cities or regions. Furthermore, satellite data can be utilized to analyze changes in urban land use and land cover over time and to compare heat patterns in different urban areas, ultimately extending the analysis to transboundary administrative regions. A notable example of such potential is the global urban footprint tool presented by Esch et al. (2017), which employs satellite imagery to map urban areas and analyze their evolution over time. The ability to collect and analyze data at this scale is essential for understanding the complex interplay between urbanization and the thermal environment and can inform policies and strategies to enhance the resilience and livability of cities. The continued development and application of satellite-based techniques in urban heat monitoring is thus critical for building sustainable and adaptive cities. The spatial and temporal resolution associated with the information retrieved by satellite imaging depends on the remote sensing sources. In their systematic review, Deilami et al. (2018) pointed out that the Landsat satellite constellation is the main source for research aimed at addressing the existing link between spatiotemporal factors and UHI effect. Among the main reasons for adopting Landsat images, the authors recognized the following: (1) Landsat images are freely available, (2) they offer a world-wide coverage with a reasonable spatial resolution of 30 by 30 m, (3) they present long-term temporal coverage and a revisit time of 16 days, and (4) thermal and thematic spectral bands are simultaneously delivered since Landsat 4. These four aspects summarize the most relevant characteristics of a remote images source to be suitable for urban heat monitoring. Indeed, free and open access policies for satellite data greatly benefitted operational applications and research advances worldwide (Zhu et al., 2019).

The Landsat program includes a series of Earth-observing satellites operated by the United States Geological Survey (USGS) since 1972 (the last launch was in September 2021, Landsat 9). In 2008, USGS made Landsat data available via internet for free. Other Earth-observation projects are currently proving satellite images for free that could support the assessment of urban heat, including Terra and Copernicus programs. Terra (EOS AM-1) is a multinational scientific research satellite lead by NASA (US National Aeronautics and Space Administration) that was launched in 1999 for a better understanding of the Earth's climate. Among the onboard instruments, Moderate Resolution Imaging Spectroradiometer (MODIS) and Advanced Spaceborne Thermal Emission and Reflection Radiometer (ASTER) provide useful information for the investigation of land use and land cover effects on urban heat. MODIS has a viewing swath width of 2330 km and a short revisit time of $1-2$ days, but it has a lower spatial resolution compared to other satellite sensors, that is, from 1000 down to 250 m. On the contrary, ASTER provides high-resolution images, from 90 down to 15 m^2 per pixel, to create detailed maps of land surface temperature (LST), emissivity, reflectance, and elevation, but it does not collect data continuously providing an average of 8 minutes of data per orbit. Whenever available, the stereoscopic images from ASTER are useful for monitoring small-scale surface temperature patterns as Nath et al. did across different districts in Manhattan pointing out the role of building height in surface UHI patterns (Nath et al., 2021). In 2014, the European Commission established the Copernicus program as a global and continuous European Earth observation system that includes both satellite and ground-based observations. Within this program, the Sentinel missions have been developed, with a forthcoming mission (LSTM—Land

Surface Temperature Monitoring) designed to address the UHI challenge set to launch in 2028. The current available data from Sentinel, including those from Sentinel-3A and Sea and Land Surface Temperature Radiometer (SLSTR), provide observations in the wavelength range of 0.55 to 12.0 μm, with a global coverage revisit time between 0.5 and 1.9 days and a spatial resolution of 1000 down to 500 m.

Zheng et al. have developed a procedure for extracting LST from SLSTR images and obtained an estimation error ranging between 1.5K and 2.5K (Zheng et al., 2019). The availability of such advanced satellite-based observations can be crucial in monitoring the urban thermal environment and mitigating the adverse impacts of UHIs. This highlights the importance of continued development and investment in Earth observation systems such as Copernicus to support informed decision-making and actions toward building sustainable and resilient cities (Zheng et al., 2019). However, given the different spatial resolution and accuracy of available remote sensing technologies, the usage of different data sources could generate slightly different outcomes. Hidalgo García compared the accuracy in applying split window algorithms for Landsat-8 and Sentinel-3 images, respectively, and achieved slightly better results with the Sentinel data (R^2 equal to 0.94 compared to the 0.93 of the Landsat-based model), highlighting greatest discrepancies between the two models during dry periods (Hidalgo García, 2021). The adoption of remote satellite images retrieved from different sources could be combined to improve the accuracy in urban heat assessment. Amazirh et al. used Sentinel-1 radar data to improve the disaggregation of MODIS LST data (Amazirh et al., 2019). Abunnasr and Mhawej recently published a proposal for improving remote LST measurement by combining Landsat-8 and Sentinel-2 data in a Google Earth Engine-based system, where machine learning and regressions methods were included to downscale surface temperature resolution down to 10 m (Abunnasr & Mhawej, 2022).

Overall and regardless of the source, satellite imaging has limitations. First, the quality of retrieved images is highly dependent on weather conditions and uncloudy images are necessary for a detailed analysis. This limits remote images availability, especially in specific regions and for specific times of the year. Moreover, LST measurements are obtained through emissivity calculations, which in turn depend on atmospheric effects, sensor noise, and concentration of aerosols, all factors contributing to potential uncertainties and inaccuracies. Image processing and analysis are therefore necessary to extract useful information, but this could be a time-consuming process that requires specialized software and expertise to avoid the introduction of errors or biases. Finally, satellite observations limit the investigation to a two-dimensional space. In this regard, remote/aerial sensing techniques, running on instruments different from satellites to provide thermal images of the three-dimensional space, are being increasingly used. These include drone- or aircraft-assisted monitoring (Martin et al., 2022; Yang, Zhu et al., 2021).

Overall, both ground-based and infrared remote observations have their unique strengths and limitations when it comes to monitoring urban heat. Combining these techniques can provide a more comprehensive understanding of the UHI phenomenon and its impacts on urban environments. For instance, Alvi et al. proposed a method for mapping air temperature data based on correlations between few air temperature measurements and remotely observed LST, along with emissivity and land cover information (Alvi et al., 2022). Further, Haddad et al. explored the combination of aerial and ground-based mapping of surface and atmospheric UHI, with the ground-based mapping consisting of a short-term mobile and a long-term fixed-station investigation; this enabled a holistic assessment of the cobenefits of local climate mitigation in a hot humid region of Australia (Haddad, Paolini et al., 2020).

2.3.3 Future avenues

Effective urban planning strategies to mitigate UHI require detailed urban heat mapping, which can be achieved through cost-efficient methods. With the increasing availability of open data and advances in data processing techniques, it is now possible to generate interpretation, prediction, and interpolation models based on limited and easy-to-access information. The use of multi-source data fusion approaches, which rely on databases of ground-based measurements from environmental and meteorological agencies, satellite imagery repositories, and socio-economic data, is promising for the future of urban heat monitoring and mapping. These solutions can identify critical areas for citizens' health risks and guide decision-making for tailored adaptation actions to enhance the livability of cities (Shen et al., 2020). Moreover, advances in sensing and communication technologies are opening new perspectives on urban heat-related data availability at a hyperlocal scale thanks to citizen science and crowdsourcing (Potgieter et al., 2021; Venter et al., 2021). While data quality could be a concern, procedures for data quality checks have been developed to reduce the risk of statistically implausible air temperature due to misplacement of sensors, including the solar exposition and related radiative errors, inconsistent metadata, and device malfunctions (Meier et al., 2017; Napoly et al., 2018). The crowdsourced data quality check should include the following steps: outliers detection, spatial correlation checks within the same crowdsourced database, and cross-calibration with reference meteorological stations (Potgieter et al., 2021). However, raising citizens' awareness on proper urban heat monitoring processes and need for action would also enhance the quality of the measurements derived from this novel approach. In this view, detailed metadata on station placement are critical for proper interpretation and data correction (Ulpiani et al., 2022).

Finally, there is a growing interest in employing multidomain investigation approaches in environmental studies. Such an approach focuses on observing the interactions among various physical environmental domains, including thermal, visual, acoustic, and air quality domains, and their impact on human responses. Although there are only a few studies that have applied this approach in outdoor settings, they are promising and could help determine the potential use of nonthermal factors as proxies for urban thermal conditions. For instance, in Ulpiani et al. (2022), the authors combine thermal and air quality monitoring at the subcanopy level through a citizen-science approach to raise the awareness of vulnerable populations around the risks of combined environmental hazards, notably on the interlacement of urban and pollution heat islands (Ulpiani, 2021). These multidimensional and multidisciplinary approaches would guarantee comprehensive datasets to cotarget different urban stressors and provide the evidence needed to create future-proof built environments.

2.4 Urban heat modeling: capitalizing on lessons learnt

Historically, the main tools used to simulate the urban climate have been mesoscale atmospheric models. Such models run over a domain of 10−100 km, with a horizontal resolution in the order of 1 km. The advantage of these models is that whole cities as well as the rural surroundings can be simulated, which is essential to obtain a faithful climate representation. Moreover, these models can be run for periods ranging from weeks to years. The disadvantage, however, is that at this spatial resolution, the phenomena in the urban canopy cannot be explicitly resolved, and their effect must be parametrized. With this aim, at beginning of the 21st century, several so-called urban canopy parameterizations have been developed (Kusaka et al., 2001; Martilli et al., 2002; Masson, 2000), taking advantage of the knowledge on the behavior of the urban canopy layer built on the experimental work of the last decades of the previous century (Oke, 1988; Rotach, 1993) and the increase of computational power. All these schemes have in common an idealization of the urban morphology, represented by a series of parallel urban canyons of constant width, separated by buildings also of constant width and height. The Town Energy Balance (Masson, 2000) and the Single Layer Urban Canopy Model (Kusaka et al., 2001) are single-layer models, hence they simply account for the fluxes at the lower boundary, while the Building Effect Parametrization (BEP) model (Martilli et al., 2002) is multilayer, so the effect of the buildings is taken into account in several atmospheric layers, allowing variable building height and very tall buildings to be accounted for. These approaches enabled the resolution of separate energy budgets for roofs, roads, and walls, considering shadowing and radiation trapping effects in the canyon.

Recent lines of development have evolved essentially in two directions: (1) obtaining realistic urban morphological input data and (2) introducing in the parametrizations the possibility to simulate mitigation/adaptation strategies. These strands are discussed in Section 2.4.1, while future lines of development are outlined in Section 2.4.2.

2.4.1   Current lines of development

2.4.1.1   Urban morphology

When the first urban canopy parametrizations were developed, information about the urban morphology, from which street and building widths and building heights could be derived, were scarce, and these parameters were determined based on simple expert judgement. The first attempt to provide quantitative information about urban morphology was the National Urban Database and Access Portal Tool (Ching et al., 2009), where LIDAR data were used to estimate a range of morphological parameters for different cities in the United States. Since then, this technique, which is becoming increasingly affordable, has been used in several cities worldwide. However, the fraction of cities for which this information exists is still small and mainly concentrated in the "Global North," and often not freely accessible. To fill this gap, the community project World Urban Database and Access Portal Tools (WUDAPT, Ching et al., 2018) was created. Within this project, a methodology to produce LCZ (Stewart & Oke, 2012) maps based on freely available satellite data and google street images has been developed. Here, the user defines so-called training zones, for example, areas belonging to a specific LCZ, which are then combined with a random forest classification method with satellite images to create the LCZ maps (Demuzere et al., 2021). In a participatory process, LCZ maps of different cities have been uploaded to the WUDAPT webpage by many users worldwide, and recently a global map of LCZ at 10 m resolution has been made public (Demuzere et al., 2022). This approach takes advantage of the fact that for each one of the 10 urban LCZ, a range of morphological values that can be used by the urban canopy parametrization were defined. This approach was firstly used by Brousse et al. (2016) and then improved by Zonato et al. (2021). Now, the global LCZ map has been introduced in the standard database provided with the mesoscale model WRF (Kawamoto, 2017) and can then be used to simulate any city in the world with the urban canopy schemes implemented in that model. One of the advantages of the LCZ approach is that users can adapt the morphological parameters of each urban LCZ to their city. It is this flexibility that makes the universal LCZ useful to represent different cities. Recently, new high resolution datasets are becoming available, like those provided by Bing, Microsoft (https://github.com/microsoft/GlobalMLBuildingFootprints), representing building footprint for the majority of urban areas in the world, which can be combined with other information like population density to derive the morphological parameters (Kamath et al., 2022). Being able to represent the interurban variability of morphology is crucial not only to simulate faithfully the urban climate and in particular the urban overheating, but also to provide relevant information for urban planning by selecting the type of morphologies that adapt better to the climate where the city is located.

2.4.1.2 Mitigation/adaptation strategies

While the motivations for the development of the first urban canopy parametrizations range from the study of urban climatology (Masson, 2000), UHI (Kusaka et al., 2001), or air quality (Martilli et al., 2002), these parameterizations have been progressively used to investigate urban overheating and to evaluate different mitigation/adaptation strategies. For this reason, these schemes evolved by including different features relevant for this type of applications. Given that one of the most relevant consequences of the urban overheating is the need to consume energy to keep the interior of the buildings in a thermally comfortable range (e.g., by using air conditioning), one of the first developments that have been introduced included simple building energy models (BEMs, Bueno et al., 2011; Kikegawa et al., 2003; Krpo et al., 2010; Salamanca et al., 2010). Such models, based on the resolution of an energy budget for the indoor air temperature, made it possible not only to estimate the amount of energy needed to extract the excess of heat from the interior of the buildings, but also to estimate the amount of heat added to the atmosphere, opening the way to investigate the interactions between urban meteorology/climatology and building energy consumption.

In terms of strategies to reduce the impact of urban overheating, changes in the properties of urban surfaces (building and roads) are among the most common. While some are relatively simple to be considered, like cool roofs and roads, which require only a change in the optical properties of such surfaces (albedo), others, like green roofs or roof top solar panels, are more complex to account for. For this reason, specific submodules of green roofs and roof top solar panels have been implemented in urban canopy parametrizations of mesoscale models (Yang et al., 2015; Zonato et al., 2021). Here, rooftop solar panels have been included among the strategies to reduce the impact of urban overheating, not because of their direct impact on the outdoor thermal environment, but because they can provide clean energy to run air conditioning and keep the indoor in the comfortable range. In this regard, an important result is the one presented by Zonato et al. (2021), which used WRF-BEP-BEM (Martilli et al., 2002; Salamanca et al., 2010) to analyse the relationship between urban structure and efficiency of different rooftop adaptation/mitigation strategies. By means of a series of idealized simulations, this work shows that the efficiency of different rooftop strategies (cool roofs, green roof, and rooftop solar panels) strongly depends on the urban structure, namely, the mean building height and the building density. The denser the buildings, the higher the roof surface area that can be modified and therefore the overall impact. Similarly, the taller the buildings, the smaller the impact at street level. This highlights the importance of a detailed description of the urban canopy parameters and therefore justifies all the work described in Section 2.4.2.

Another very common adaptation/mitigation strategy to reduce the impact of urban overheating is the inclusion of street trees. Accounting for them in an urban canopy parametrization is significantly more complex than considering green roofs, for example, mainly due to the need to represent the interaction of the trees with radiation within the urban canyon. Schemes of different complexities have been proposed, both for the single layer approach (Lee et al., 2016; Meili et al., 2020) and for the multilayer scheme (Krayenhoff et al., 2020). The latter is probably the most advanced and is based on a ray tracing technique that allows to estimate view factors between vegetation and building and road surfaces. Such an approach has been implemented in an offline version of BEP, called BEP-tree (Krayenhoff et al., 2020), as well as in the mesoscale model COSMO (Mussetti, 2019). Ongoing works aim to complete the introduction of this approach in WRF and improve the coupling with the hydrological part to account for variability in soil moisture content.

2.4.2 Future lines of development

2.4.2.1 Modeling across scales

As explained in Section 2.4.1, the idealization of the urban morphology was a key step in the development of urban canopy parameterizations, because it made it possible, for the first time, to represent urban canopy features in mesoscale models. It was a needed step considering the characteristics of the computational resources and data availability at the beginning of the 21st century. Today, such idealization is an obstacle to the future development of the field. Thanks to the increase of computational power available, much more sophisticated approaches can be devised, which can better resolve the subgrid spatial heterogeneity present in urban areas and therefore resolve also spatial scales that current urban canopy schemes can only parameterize. In particular, three approaches can be imagined, of increasing complexity, which are described as follows:

- Mosaic approach: It is one of the most common approaches used, from mesoscale to global models, to account for the land use subgrid heterogeneity. It consists of estimating different surface fluxes of heat, moisture, and momentum based on the different land use characteristics present within the grid cell. Those fluxes are then weighted averaged, considering the percentage of the different land uses present in the grid cell, to obtain a representative flux that is then used by the atmospheric model. A similar approach can be used for urban areas, leveraging, for example, the high resolution 100 m LCZ global map (Demuzere et al., 2022; see Section 2.4.1), which features much higher resolution than that of typical mesoscale models (1 km). However, the main issue to be solved with this approach is that the fluxes are all estimated using the grid values for air temperature and wind speed, losing in this way the information on the subgrid spatial variability of atmospheric fields close to the ground. It is in fact this subgrid variability that is often the most important information for many applications to estimate the air temperature value that is representative of a certain urban morphology within the grid cell. A possible solution to this problem is the introduction of a column model to vertically solve the turbulence transport for each urban land use present in the grid cell and in this way estimate air temperature and wind at pedestrian level for each land use (e.g., each LCZ). This approach may be useful in particular for continental scale simulations, which can be run only at coarse resolution (10–20 km).

- Fast, highly parameterized microscale models: After the 9/11 terrorist attack in New York City, significant efforts have been taken to develop very fast microscale models, which are able to resolve explicitly the flows in urban areas, in order to simulate, in real time, the dispersion of toxic gases suddenly emitted by an accident. This type of models does not resolve explicitly the Navier–Stokes equations—it would be too computationally expensive. Instead, it uses a series of simple, empirical rules (often based on more sophisticated simulations) to represent building wakes and street canyon vortexes. By combining these rules with detailed urban morphology and by imposing mass conservation, these models can produce reasonably reliable flow fields at 1 m resolution in urban areas. One example of this type of models is QUIC-URB (Pardyjak & Brown, 2003). With the current computational power, this type of models could be implemented in a mesoscale model below the blending height (up to 2–3 times the mean building height) to resolve the flow in the urban canopy at 1 m resolution, which would represent a significant improvement compared to the current approaches that can only estimate spatially averaged wind, even if the thermal effects on the flow would be neglected.
- Full microscale model implemented in a mesoscale model: In the last decade, in parallel with the work done with mesoscale models, several microscale models have been developed. Such models run at a spatial resolution in the order of 1 m and can explicitly resolve the building flows, but cannot run over domains much larger than few square kilometers. Recently, versions that account also for the heat exchanges between the urban surfaces and the atmosphere have been created (PALM-4U; Maronga et al., 2019). In most of the cases, the

computational time is still too high to fully couple these models with a mesoscale model, but new avenues may become available with more advanced computational techniques, like graphical processing units, as shown by the first tests with Fast-Eddy (Sauer & Muñoz-Esparza, 2020).

2.4.2.2 Coupling with other models of the urban system

The need to have a comprehensive understanding of how the whole urban system will behave in the context of a changing climate will motivate studies where the atmospheric model will be either used in a modeling chain with other models of the urban system, or directly include new components of the system. The following can be foreseen:

- Building models: The work done by introducing simple BEMs in the urban canopy parametrization will continue, both in terms of improving the characterization of the building materials for cities, and in terms of introducing new features about building energy use. This activity will benefit from close interactions with the building energy field, which advanced significantly in the recent years.
- Biometeorology: Heat stress depends on many different meteorological parameters, as well as human conditions (Guzman-Echavarria et al., 2022). Atmospheric numerical models will evolve to produce better estimates of the meteorological variables needed (e.g., mean radiant temperature, among the others) at the spatial and temporal resolution required by biometeorological models.
- Coupling with agent-based models: New models, which simulate the movement of the people in the city, are being developed. This information may be combined with crowdsourcing from many different sources (cell phones, car's computers, etc.) to generate models of people behavior. Such models can be used to investigate the interactions between citizens and the urban atmosphere. In particular, this approach can be used to improve the estimation of the building energy use, and eventually the anthropogenic heat, but also the exposure to heat stress and pollution that affect people. Ultimately, this can be used to understand how a changing climate affects people's behavior and vice versa.
- Air quality: Any evaluation of the efficiency of heat mitigation/adaptation strategies must consider also the interactions with air quality, and therefore coupled studies will become more and more common in the future. The studies conducted so far indicate that heat mitigation strategies that are based on a reduction of the sensible heat emitted to the atmosphere (cool roofs, green roofs, cool pavements, etc.) induce a reduction of planetary boundary layer height and therefore a potential increase of the concentration of primary pollutants. On the other hand, secondary pollutants like ozone that form with temperature-dependent chemical reactions may be reduced by the implementation of these strategies, as already indicated

by the pioneer studies of Taha (1997). Given that the amount, distribution, and type of traffic emissions in urban areas will change drastically in the next decades, due to the introduction of electric cars, and the change in travel habits (mainly due to the increase in popularity of remote working), it will be very important to use modeling to evaluate different urban configurations to adapt cities to future climate and human conditions.

2.5 Urban climatology of greenery

In response to increasing global urbanization, climate change, and global warming, cities and governments are gradually implementing UGI and nature-based solutions to prepare urban communities for current and future environmental challenges (adaptation) and reduce the impacts of climate change and urban overheating (mitigation) (Teixeira et al., 2022; Zölch et al., 2016). UGI—also referred to as "urban vegetation" or "urban greenery"—is capable of delivering multiple ecosystem services and functions to urban areas including economic and social benefits (Bartesaghi-Koc et al., 2022; Wong et al., 2021). The knowledge of the impacts of UGI on the thermal environment is ample; however, research outputs are diverse, fragmented, heterogeneous, and nonstandardized, hindering the comparability of results (Santamouris & Osmond, 2020). Moreover, there is a persistent gap between scientific research and the development of evidence-based guidelines to plan, implement, and manage UGI for an integrated mitigation and adaptation of urban heat (Graça et al., 2022). This is mostly because of the uncertainty and ambiguity of future climatic changes and the context specificity of such effects (Hami et al., 2019; Sharifi, 2021; Teixeira et al., 2022).

In the latest decades, intensive research has been conducted to measure, evaluate, and decode the different mechanisms of greenery-related cooling such as shade provision, evapotranspiration, increased albedo, and wind modification (Wong et al., 2021). Based on the analysis of recent research and review papers (Aflaki et al., 2017; Bartesaghi Koc et al., 2018; Charoenkit & Yiemwattana, 2016; Fu et al., 2022; Graça et al., 2022; Hami et al., 2019; Santamouris & Osmond, 2020; Shao & Kim, 2022; Sharifi, 2021; Teixeira et al., 2022; Wong et al., 2021; Yu et al., 2020), five key knowledge areas are identified, in which novel findings and significant progress have been achieved. As shown in Fig. 2.3, these areas relate to the spatiotemporal scales and extent of analysis and correspond to the following:

1. The evaluation of the heat-related adaptation and mitigation capacity of different types of "on-ground UGI" (i.e., greenspaces, tree canopy) and "on-building UGI" (green roofs and vertical systems), including the challenges for a holistic assessment of urban greenery in combination with other technologies.

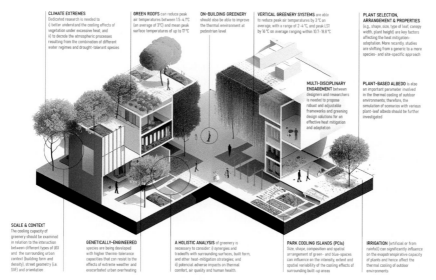

FIGURE 2.3 A summary of key knowledge areas and significant progress regarding the urban climatology of greenery.

2. The assessment of plant-related effects in relation to different—and sometimes extreme—climatic conditions, scales of approximation, and urban contexts, including the role of irrigation and the development and adoption of drought-tolerant species.
3. The influence of size, shape, composition, and spatial configuration of green and blue spaces on the intensity, extent, and spatial variability of the cooling effect on surrounding built-up areas.
4. The importance of the selection, combination, and arrangement of plants, the modification of their physiological operation, and the development of genetically engineered species.
5. The barriers and enablers for a more effective implementation of design guidelines for urban heat adaptation and mitigation.

In the following sections, the main advancements and achievements from recent studies are discussed within each of the above mentioned categories, with a critical analysis of the limitations and potential avenues for future research.

2.5.1 Heat-related adaptation and mitigation capacity of green infrastructure

In literature, the terms "adaptation" and "mitigation" are frequently used interchangeably when referring to the climate-related impacts of UGI; and in many cases, a simultaneous integration of adaptation–mitigation factors, measures, synergies, and cobenefits from urban greenery is observed (Sharifi, 2021). Generally, these impacts can be grouped into three main categories: (1) atmospheric cooling, (2) LST or skin cooling; and (3) outdoor thermal comfort improvement; not always well-differentiated (Feng et al., 2023; Graça et al., 2022).

Regardless of the specific focus of the researched phenomena, studies examining "on-ground" UGI have commonly evaluated the key mechanisms of greenery-related cooling by estimating the impacts of increased tree canopy (area or volume) and various green coverage ratios. Typically, biophysical parameters (e.g., plant size, plant type, crown size and density, type of leaf, and proportion/fraction of vegetation) and dimensionless indices including leaf area index, leaf area density, and normalized difference vegetation index have been used as indicators of ecological function and key factors influencing the provision of shade, the energy exchanges, and the evapotranspiration intensity (Bartesaghi Koc et al., 2018; Wong et al., 2021). In recent years, plant leaf albedo is being incorporated in the list of key parameters involved in the thermal cooling of outdoor environments, as different albedo values determine the amount of reflected radiation and thus impinge on the thermal budget (Daemei et al., 2018; Shahidan et al., 2012; Taleghani et al., 2014). Therefore, the simulation of scenarios with various plant leaf albedos should be further investigated to identify the most effective species for heat mitigation (Hami et al., 2019).

Although most studies reveal temperature reductions due to the presence of different types of UGI, the magnitude and intensity of these cooling effects vary significantly (Wong et al., 2021). For instance, some metastudies reporting on atmospheric cooling suggest a decrease of 0.94°C (Bowler et al., 2010), while others suggest 1.5°C–3.5°C (Saaroni et al., 2018) or 0.3°C–1.8°C (Santamouris & Osmond, 2020). These discrepancies can be attributed to differences in the methodologies adopted, urban contexts, climates, and types, proportion, and distribution of greenery. On the other hand, when referring to LST reductions, the reported cooling potential from greenspaces is considerably larger due to a combination of conductive, convective, and radiative heat transfer. It ranges between 1.9°C and 6.7°C (Wong et al., 2021), shifting to 2.0°C–8.0°C in case of well-irrigated vegetation (Bartesaghi-Koc et al., 2020).

"On-building" UGI has been mostly studied in compact city contexts with limited space for larger greenspaces or tree canopy cover and include intensive/extensive green roofs and vertical greenery systems (VGS) such as living walls and green facades (Pérez et al., 2014; Wong et al., 2010). Multiple studies have confirmed the cooling effects of rooftop and VGS systems, reporting peak air temperature reductions of 1.5°C–4.1°C (average of 3°C) and mean peak surface temperature reduction of 17°C in rooftops. In comparison, VGS are able to reduce the peak air temperatures by 3°C on average, ranging between 2°C and 4°C, and the peak LST by 16°C on average, ranging within 10.7°C−18.8°C (Wong et al., 2021). Nonetheless, as the effects are mostly evident in the immediate surroundings, the impact on pedestrians may be negligible if these systems are installed several meters above the ground (Santamouris et al., 2018). Furthermore, the magnitude of the cooling that greenery can deliver largely depends on prevailing boundary conditions, seasonal phenomena, morphological characteristics, and placement of vegetation and supporting systems (Morakinyo et al., 2019; Wong et al., 2021).

It is clear that UGI plays an important role on urban heat mitigation and adaptation as well as in carbon sinking; however, these impacts should not be assessed in isolation, rather in a holistic and site-specific way. For instance, where urban overheating is triggered or amplified by stalling air conditions, tree planting should be planned to avoid wind breaking. Similarly, vegetated species should be selected based on (1) their thermal response not to increase the stomatal resistance and induce heat-enhancing mechanisms under hot conditions, (2) their emission rate of biogenic volatile organic compounds (Ulpiani, 2021) and allergens (Ferrante et al., 2020) not to burden air quality, and (3) the impact of associated activities of landscape maintenance (such as irrigation, fertilization, lawn mowing, tree, and shrub pruning) in terms of carbon release (Jo & McPherson, 1995). Further, the implementation of additional greenery may result in outdoor discomfort in tropical contexts due to higher levels of relative humidity (Aflaki et al., 2017). Likewise, the impacts of UGI should be further examined in combination with other heat-mitigation strategies and technologies (Santamouris et al., 2018). A recent study showed that the atmospheric cooling within a single street canyon can be improved by 0.2°C when greenery is implemented in combination with reflective pavements and by 2.9°C in conjunction with cool pavements, artificial shading, water spray systems, and cool roofs (Bartesaghi-Koc et al., 2021; Haddad, Paolini et al., 2020). Since overlapping multiple solutions can cause complex, synergistic, and nonadditive effects on the thermal environment, it is critical to investigate the specific contributions of each mitigator as well as which factors influence the interaction, including differences in plant species (foliage, height, and plant physiology), placement of natural and artificial features/surfaces in close proximity, impacts from existing built form, and aging of materials (Bartesaghi-Koc et al., 2021; Fu et al., 2022; Howe et al., 2017; Kotthaus & Grimmond, 2014; Kyriakodis & Santamouris, 2018).

2.5.2   Cooling effects across different scales, climates, and contexts

The role and capacity of UGI on heat mitigation and adaptation is scale-, climate-, and context-dependent (Kong et al., 2014; Morakinyo et al., 2019; Oke et al., 2017). Most studies on optimized greenery configuration and on-building UGI for urban heat mitigation and thermal comfort improvement are conducted at the micro-scale level (i.e., site, street, urban canyon, and building), while the effects of size, composition, and configuration of greenspaces are typically studied at local- and meso-scale levels (i.e., from city block/district level to regional level) (Fu et al., 2022).

The cooling potential of UGI is influenced by diurnal synoptic weather conditions (Bartesaghi-Koc et al., 2021; Tsoka et al., 2018) and seasonal fluctuations (Oliveira et al., 2011). Peak temperature reductions associated with greenery are mostly occurring at daytime—compared to nighttime—as they are modulated by temperature gradients, direct solar exposure (including sunny vs cloudy conditions), rainfall, and amount of longwave radiation emitted back to the atmosphere and/or trapped within the street canyon and tree canopy (Morakinyo et al., 2019; Wong et al., 2021). Moreover, the cooling potential of vegetation is boosted under the presence of unrestricted water, while poor irrigation conditions and scarce precipitation hinder evapotranspiration (Bartesaghi-Koc et al., 2020; Gao et al., 2020). Similarly, seasonal cycles can cause significant variability in the cooling capacity as the shade potential depends on the proportion of deciduous vs evergreen types of foliage (Wong et al., 2021). This seasonal variability is more evident in rooftops and VGS, which have generally shown higher cooling potential in summer compared to winter periods (Bevilacqua et al., 2016; He et al., 2020).

Global climatic change is leading to the increment in the intensity, duration, and frequency of extreme weather conditions, heatwaves, droughts, and exacerbated urban overheating, which in turn are putting existing and newly implemented UGI under a significant stress, limiting the heat mitigation and adaptation capacity (Skoulika et al., 2014; Yao et al., 2013). Different plant species have characteristic responses to extreme conditions, yet there is no clarity on this in literature. Dedicated research is needed to (1) better understand the cooling effects of vegetation under excessive heat, with particular attention to stomatal conductance, photosynthetic rates and transpiration mechanisms (Rogers et al., 2017; Urban et al., 2017) and (2) decode the atmospheric processes resulting from the combination of different water regimes and drought-tolerant species (Gao et al., 2020; Vahmani & Ban-Weiss, 2016). Progress in this area includes the development and application of new stomatal conductance predictive models (Kala et al., 2016; Lin, Medlyn et al., 2015), species-sensitiveness to drought and heat (Bigras, 2000), analysis of the role of rooting systems under water stress (Drake et al., 2018), and optimization of irrigation systems (Broadbent et al., 2018).

The heat mitigation—adaptation capacity of plants should also be examined in relation to the interaction between different types of UGI and the surrounding urban context (building form and density), street geometry, and orientation (Morakinyo & Lam, 2016; Morakinyo et al., 2017, 2020; Tan et al., 2016). Some studies have demonstrated that areas fully covered by greenery are cooler compared to those incorporating a mixture of natural and artificial materials owing to their reduced heat storage and the presence of higher evapotranspiration during the day, which results in lower diurnal temperatures (Bartesaghi-Koc et al., 2020; Fenner et al., 2017; Geletič et al., 2018). On the contrary, the cooling potential of UGI is affected by reflected and emitted heat from surrounding surfaces and the presence of anthropogenic heat, particularly in compact urban arrangements and overexposed street canyons (Bartesaghi-Koc et al., 2021; Morakinyo & Lam, 2016; Morakinyo et al., 2017).

Furthermore, in-canyon tree canopy can affect overall ventilation and trap the upwelling long-wave heat emitted or reflected from different materials, a condition that might modify local heat fluxes and inhibit convective cooling in areas with decreased sky view factors (SVFs) (Tan et al., 2016), resulting in a warming effect mostly during the night (Arghavani et al., 2019; Balczó & Ruck, 2009; Imran et al., 2019; Morakinyo et al., 2017). Also, researchers have examined the thermal influence of generic trees in different settings with varied SVF and building compactness and found that the heat mitigation potential from trees reduces as urban density increases, as the shading effectiveness from trees is less effective at decreasing urban heat when these are self-shaded or overshadowed by nearby trees and buildings (Morakinyo et al., 2017; Tan et al., 2016; Thom et al., 2016). The incorporation of the effect of SVF in future studies can help identify suitable plant species for particular locations and define priority rankings for planting (Morakinyo et al., 2020).

2.5.3 Influence of size, shape, composition, and spatial arrangement of green and blue spaces

Evidence suggests that atmospheric temperatures of greenspaces can be between 1°C and 4°C cooler than surrounding urbanized zones, with the greatest zone of influence extending downwind (Bowler et al., 2010; Ellis et al., 2017; Skoulika et al., 2014). Cooling impacts outside greenspaces are commonly referred as to park cooling islands (PCIs) (Kong et al., 2014; Spronken-Smith & Oke, 1998). The proportion (ratio), selection, distribution, and optimization of tree coverage within the greenspace are important factors influencing PCI formation (Fu et al., 2022; Zhao et al., 2020).

Generally, larger, regularly shaped, polygonal (circular or rectangular), and clustered greenspaces tend to provide more pronounced cooling effects on the surroundings (Feyisa et al., 2014; Wong et al., 2021; Yu et al., 2017; Zhao et al., 2020), although in some cases, small well-distributed parks (Oliveira et al., 2011; Saaroni et al., 2018) and complex or irregular-shaped green patches (Asgarian et al., 2015; Du et al., 2017) can offer comparable temperature reductions. Moreover, it has been found that elongated (and narrow) greenspaces offer lower cooling effects than polygonal-shaped parks, mainly because linear greenspaces offer less opportunities for plant diversity and tend to be affected by heat transported from adjacent built-up areas (Chen et al., 2014). The threshold value of efficiency for green and blue space, that is, the size at which the cooling intensity per unit area reaches a plateau or starts dropping (Yu et al., 2020), ranges between 0.5 and 0.7 ha in temperate climates (Yu et al., 2018) and reaches 0.7−0.95 ha for tropical cities (Fan et al., 2019).

However, how far green and blue space cooling effects permeate the urban fabric varies considerably, as cooling distances are governed by size, characteristics (plant selection), arrangement (intervals and spacing) of greenspaces, surrounding urban geometry and built form, presence of anthropogenic heat, prevailing winds, synoptic weather, and seasonal cycles (Cao et al., 2010; Lin, Yu et al., 2015). For instance, cooling effects in the range of 2.6°C−4.8°C can stretch over 35 to 840 m away from the greenspace in some parks in Beijing (Lin, Yu et al., 2015), while a similar cooling intensity is registered at a distance of up to 330 m from greenspaces in London (Vaz Monteiro et al., 2016).

Optimized and targeted tree/vegetation configurations can increase the cooling potential of parks since particular arrangements of vegetation can maximize shading (Milošević et al., 2017), funnel air advection from parks into surrounding street canyons (Bernard et al., 2018; Takebayashi, 2017), or minimize wind disturbances (Tan et al., 2016). Furthermore, well-irrigated greenspaces with dense tree coverage typically develop strong daytime PCIs; however, warmer conditions at night can be recorded due to the heat trapped under the canopy and reduced advection. Conversely, parks sparsely covered by trees are warmer during the day and cooler at night due to long-wave radiation losses (Bartesaghi-Koc et al., 2019; Taha et al., 2018).

2.5.4 Plant selection, combination, placement, physiology, and engineering

Plant selection (e.g., shape, size, type of leaf, canopy width, and plant height) and placement are factors that play an important role in the heat mitigation−adaptation capacity of greenery by defining the amount of shading and evapotranspiration affecting air, surface, and mean radiant temperatures (Hami et al., 2019; Wong et al., 2021). For instance, the selection of plant species with higher stomatal conductance (or transpiration rate) can accelerate outdoor temperature reductions and diminish the heat transfer to buildings (Nagase & Dunnett, 2010). Furthermore, tall trees with greater crown diameters can improve thermal comfort in wide open streets (Sun et al., 2017); on the contrary, tall trees with lower crown diameters are more suitable for narrower street canyons (Morakinyo & Lam, 2016). Despite this, the study of the influence of the geometry, density, and shape of different tree crowns in different real-scale (or experimental) settings requires more attention and development in future (Hami et al., 2019). Regarding the influence of the type of foliage on microclimatic amelioration, it has been demonstrated that small-leaved species are relatively more effective at cooling the surrounding air (by regulating crown temperatures) compared to large-leaved species (Leuzinger et al., 2010). On average, the cooling potential of ever-green plants is greater than that of deciduous plants; however, the combinatory effects from plant species with different types of leaf and shapes is an aspect that has not been fully examined in literature (El-Bardisy et al., 2016). As a result, more recently, studies are shifting from a generic to a more species- and site-specific approach (i.e., "right tree at right place") (Morakinyo et al., 2020).

Optimized greenery configuration, particularly tree layout, is another important factor to consider when assessing the localized impacts of vegetation on the thermal environment (Fu et al., 2022). For instance, Tan et al. (2016) showed that strategic distribution of trees can maximize wind channeling and flow paths, which help in cooling high density and compact urban areas more efficiently. Other studies have examined how different planting schemes (i.e., clustered, rectangular, and double rows) can provide different thermal comfort outcomes (Abdi et al., 2020; Atwa et al., 2020; Bartesaghi-Koc et al., 2020; Thom et al., 2016).

In the context of accelerated climate change, researchers are developing genetically engineered species with higher thermotolerance capacity that can resist to the effects of extreme weather and exacerbated urban overheating (Santamouris et al., 2018). Several studies have observed the impact of climatic stressors on different tree genotypes and found that growth performance is significantly affected by heat stress; however, preconditioning can help minimize long-term damages (Bigras, 2000; Savva et al., 2007). Currently, heat-tolerant species are being developed by either traditional breeding or genetically engineered processes. Traditional breeding methods are slower and more challenging in comparison to genetically engineered species where wood properties, root formation, and stress tolerance are considerably improved (Bita & Gerats, 2013; Harfouche et al., 2011; WAHID et al., 2007). Other advancements in heat-tolerance include genomic, transcriptomic, and proteomic manipulation (Harfouche et al., 2011), the use of abiotic-stress-associated genes (Vinocur & Altman, 2005), and the identification of specific chromosome segments related to heat tolerance via quantitative trait locus mapping methods (Bita & Gerats, 2013). Nonetheless, more research is required to better understand the long-term effects of genetically modified plants, including the monitoring of stress−recovery cycles, rules and regulations that can help mitigate the risk of transgene spread (i.e., not-functional flowers), and other regulatory concerns for future commercialization (Santamouris et al., 2018).

2.5.5 Barriers and enablers toward effective design

Overall, there is a perceived lack of multidisciplinary engagement between designers and academic research in dealing with UGI, which results in limited collaborative research focusing on real-scale design interventions or on their transferability to other contexts (Grose, 2014). Graça et al. identified five key challenges hindering the implementation of evidence-based guidelines into design practice (Graça et al., 2022): (1) ambiguity in key definitions and operative levels of the design practice; (2) lack of connection between key mechanisms of specific types or design elements and their respective heat mitigation—adaptation outcomes; (3) lack of clarity in identifying trade-offs and synergies between morphology and composition of UGI impacting on the climate outcomes; (4) need for novel indicators and frameworks to monitor the performance of UGI types across temporal and spatial scales; and (5) necessity of codesigning research questions, methods, and outputs to better address the needs of professional practice. Initiatives, such as the "adaptative planting design and management framework" developed by Teixeira et al. (2022) or UGI multiscale approaches (Pauleit et al., 2021) are intended to address these gaps by providing strategies that enable learning from context-specific experiences to propose more robust, flexible, and adjustable greening solutions for heat mitigation and adaptation.

2.6 Urban heat: a materials perspective

When conducting research on urban overheating, the measurement of surface energy or heat balance is a crucial metric for determining the equilibrium between incoming and outgoing energy across the boundary. Several factors can influence the surface energy balance, including the properties of building materials, building orientation and shape, and the surrounding microclimate. The surface heat balance equation, expressed in its general form, is as follows:

$$R_n = \text{LE} + G + H \tag{2.1}$$

where R_n is the net radiation, that is the sum of the net shortwave and longwave components; H represents the sensible heat flux; and LE is the latent heat flux that arises from soil evapotranspiration and/or plant transpiration. Finally, G represents the heat storage. Each term in Eq. (2.1) corresponds to a specific heat transfer phenomenon and can be modified by adjusting the material characteristics and properties. By understanding the surface heat balance equation, we can gain insights into the complex interactions between the Earth's atmosphere, land, and oceans, and how they contribute to the overall energy budget of our planet. By continuously studying and refining our understanding of the surface heat balance equation, we can develop more precise and efficient strategies for managing our resources and mitigating the impacts of climate change, particularly in urban areas. Cities are crucial in the mosaic of urban materials, where human decisions have a significant impact. Improving our knowledge of the surface heat balance equation can contribute to achieving an urban vision that prioritizes sustainable practices and fosters healthier and more livable cities.

2.6.1 Materials affecting the net radiation

Radiation plays a crucial role in maintaining the surface energy balance. The sun is the primary source of energy that the surface receives, and building materials can reflect, transmit, or absorb this energy. The absorbed energy is then released as heat. Therefore, a deep understanding of the radiation balance is vital for combating urban overheating, maximizing energy efficiency, and reducing energy wastage. The net radiation term in Eq. (2.1) is the result of four contributions, as shown in Eq. (2.2).

$$R_n = R_{sw}^{\downarrow} + R_{sw}^{\uparrow} + R_{lw}^{\downarrow} + R_{lw}^{\uparrow} \tag{2.2}$$

The terms with a downward oriented arrow, $R_{sw}^{\downarrow}$ and $R_{lw}^{\downarrow}$, represent the incoming shortwave and longwave radiation, respectively, whereas the terms with the upward oriented arrow, $R_{sw}^{\uparrow}$ and $R_{lw}^{\uparrow}$, indicate the amount of shortwave and longwave radiation that is reflected or emitted by the surface. These last two terms are heavily influenced by two crucial material properties: solar reflectance and thermal emittance (Fabiani et al., 2018)[1]. Material scientists have developed several applications by specifically tailoring these properties to reduce surface overheating. These material-based technologies can significantly reduce the UHI effect and promote the development of sustainable and environmentally friendly urban surfaces. Three classes of materials can be identified as: (1) high reflectance and/or high emittance materials; (2) dynamic reflectance materials; and (3) dynamic emittance materials.

Type 1 materials are commonly referred to as cool or high reflectance materials and typically exhibit solar reflectance of 0.7 or higher and/or thermal emittance of 0.85 or higher (Pisello et al., 2017). This category of materials includes classic cool, colored, and directional applications as well as supercool materials. Classic cool materials have high reflectance throughout the shortwave and near-infrared range, and they primarily rely on the low light absorption and strong nonselective scattering in the visible range of white pigments. On the other hand, colored cool materials only exhibit high reflectance outside the visible range to blend better with the built environment. They mainly use mixtures of metal hydroxides, nitrates, acetates, and oxides such as titanium dioxide, zinc oxide, and bismuth vanadate (Capone et al., 2023). Directionally reflective solutions, such as retroreflective layers, utilize small glass beads, prisms, and recently orthogonal surfaces with aluminized Mylar film. These materials have very high specular reflectance across the solar spectrum and explore the potential of geometrically selective

---

1. Solar reflectance measures the amount of solar radiation that is reflected by a material's surface. A higher value indicates a greater ability to reflect sunlight, which is in the shortwave radiation range of 300−2500 nm. Thermal emittance, on the other hand, measures a material's ability to radiate heat. A material with high thermal emittance radiates more heat back into the environment, reducing the amount of energy stored into the surface.

reflection (Levinson et al., 2020). Finally, supercool materials, also known as PDRC, are being developed to enable subambient cooling under direct sunlight without consuming additional energy (Santamouris & Feng, 2018). These are investigated in more detail in Section 2.6.1.1.

Regardless of the specific material type, the principle of high reflectance and/or high emittance embeds a potential drawback that limits large-scale applicability, particularly in colder climates: the continuous rejection of solar radiation, which may cause overcooling and/or a heating penalty (i.e., increased heating loads and costs and reduced indoor/outdoor comfort), especially on annual basis. To address this issue, some degree of control or self-adjustment to the actual need for cooling should be incorporated. This has spurred the interest in the development of type 2 and 3 materials and their integration with type 1. Details are provided in the next subsections.

2.6.1.1 Passive daytime radiative coolers

PDRCs exhibit close-to-unity solar reflectance and close-to-unity emittance in specific wavelength ranges where the atmosphere is mostly transparent (notably, in the wavelength range $8-13$ μm called atmospheric window). This allows transferring the heat directly to the massive heat sink represented by the outer space resulting in subambient cooling even at peak hours. PDRCs can be categorized into several groups, including multilayered inorganic materials, metamaterials, polymer-based materials, biomimetic materials, switchable radiative cooling materials, and colored radiative cooling materials.

Multilayered inorganic films, pioneered by Raman et al. (2014), demonstrated a cooling power as high as 40 W/m^2 and subambient cooling of up to $8°C-9°C$ (Chae, Kim et al., 2020), but require advanced numerical simulations and complex fabrication procedures to optimize the thickness of each layer and the cooling potential (Zhai et al., 2017). As a result, they are typically expensive and fragile. The second category, metamaterials, possesses outstanding properties to manipulate electromagnetic waves in terms of absorption and reflection. Many nanophotonic structures, such as metal−dielectric conical and nonconical pillar and column arrays, metal-loaded dielectric resonator metasurfaces, metal−dielectric−metal resonators, multilayered pyramidal nanostructures, and metafabric, have been investigated (Liu, Li et al., 2020), showing subambient temperature drops as high as $9°C-12°C$ (Hossain et al., 2015). On the other hand, polymer-based materials have shown great potential due to their broad intrinsic selectivity in the infrared range, which matches the atmospheric window (Liu, Li et al., 2020). This makes them ideal candidates for PDRCs, in particular polytetrafluoroethylene, polydimethylsiloxane (PDMS), poly-vinylidene fluoride (PVDF), and polymethyl methacrylate (PMMA), due to their cost-effectiveness and scalability (Mandal et al., 2018). To date, three types of polymer-based radiative coolers have been explored: multilayered composites, particle hybrid coatings, and polymeric materials. Multilayered applications involve the integration of a thin metallic film, typically made of Ag or Al, to improve reflectivity (Zhai et al., 2017) or use SiO_2 or similar microparticles embedded in polymer-based layers to create strong phonon-polariton resonances in the infrared region (Chen et al., 2022). Particle hybrid coatings integrate micro- and nanosized particles made from various materials, including TiO_2, Al_2O_3, SiO_2, silica aerogel particles, and perovskite nanocrystals, to scatter light and increase solar reflection, while also enhancing the infrared emissivity of the hybrid material through intrinsic phonon—polariton resonance (Jeon et al., 2020; Qi et al., 2017). For these applications, daytime subambient temperature drops in the range $4°C-13°C$ have been reported in literature (Feng et al., 2020; Jing et al., 2021; Wang et al., 2020). While both approaches have shown promising results, they can be costly and complicated to prepare. Some studies have focused on engineering the polymer itself to achieve the desired effect, without the need for additional layers or particles (Bizheva et al., 1998). One innovative approach involves creating hierarchical nano/microscale pores in the polymer to enhance the total scattering efficiency and increase solar reflectivity. This is achieved by carefully optimizing the production process of the polymer, such as by using an innovative spray-phase separation strategy for PVDF-HFP (Song et al., 2022) or by producing hierarchically porous arrays of PMMA via the templating method (Wang, Wu et al., 2021). These simple structures make it possible to achieve subambient temperature drops of about $6°C$ and cooling powers of up to 100 W/m^2 (Mandal et al., 2018).

The fourth category of materials is biomimetic, drawing inspiration from nature's ingenious designs. Many creatures in the natural world utilize photonic structures for camouflage and thermal regulation. Organic matter often exhibits high thermal emittance, and with the right photonic structures, it can produce structural white or silver color, minimizing solar gains. Researchers worldwide have taken inspiration from nature's thermoregulation to develop fascinating bioinspired materials for PDRC. For instance, the silkworm cocoon of *Bombyx mori* has been mimicked by randomly stacking silk fibers using a melt-blown production process of polypropylene (MB-PP) and coating it with PDMS to improve its thermal emissivity in the atmospheric window (Yang & Zhang, 2021). The unique, optical properties of the SMB-PP resulted in a subambient temperature drop of $4°C$ in the daytime and $5°C$ in the night. Another example is the fabrication of a thin film with excellent radiative cooling and structural coloring by reconstructing the nanostructure of the dorsal wing of the *Archaeopreponademophon* (Lee et al., 2022). The biomimetic film exhibited impressive performance and produced a maximum temperature reduction of $8.4°C$.

Finally, switchable, asymmetric, and colored radiative cooling materials are currently the most innovative research frontiers in the field of PDRCs.

Various techniques can be applied for the energy modulation of radiative coolers, for instance, through thermoresponsive dynamic switch of the infrared emissivity, which comes in handy against overcooling (Ulpiani, Ranzi, Shah et al., 2020). One exciting application involves the development of advanced materials that can collect solar heat in cold weather and provide radiative cooling in hot weather. This has been achieved using vanadium-dioxide (VO_2)-based PCMs and thermochromic hydrogel (Wang, Jiang et al., 2021). Another line of research seeks to leverage asymmetric transmittance metamaterials and structures to solve the cooling instability that occurs in humid, polluted, and/or packed urban environments, where the magnitude of the radiation emitted by atmospheric particles and surrounding objects in the atmospheric window may impair the cooling performance of PDRCs (Ulpiani et al., 2021). Another promising research avenue involves the introduction in PDRCs of multilayered structured materials (Chen et al., 2020), core-shell nanoparticles (Min et al., 2022), nanocrystals (Chae, Son et al., 2020), or photoluminescent materials (Son et al., 2021), among others. These solutions aim to address the issue of ultrawhite appearance in broadband reflectance materials by selectively absorbing heat from the sun in narrow spectral bands in the visible range, while maintaining high reflection in the remaining part of the spectrum. This way, the inevitable cooling loss can be reduced and average temperature drops of $4°C-5°C$ can be achieved (Sheng et al., 2019). Overall, these cutting-edge developments in radiative cooling materials hold great potential for a wide range of applications, from building insulation to wearable technology.

2.6.1.2 Materials with tunable and adaptive spectral reflectance

Researchers have explored and implemented several dynamic reflectance or thermochromic applications (Fabiani, Castaldo et al., 2020). When heated or cooled, thermochromic materials absorb or reflect different wavelengths of light, resulting in a change of color. There are two broad categories of thermochromic materials: dye-based and non-dye-based. The former category exhibits thermochromism due to the interplay between its components, while the latter category derives its thermochromic properties from molecular rearrangements or nanoscale effects that modify the optical characteristics of the macrostructure (Garshasbi & Santamouris, 2019).

One of the earliest applications are dye-based systems, which rely on two distinct mechanisms to produce thermochromic dyes. The first mechanism—the interplay between the pH-indicator dye and the polymer matrix—occurs in a dye−polymer composite, while the second is a proton transfer reaction that occurs in leuco dye thermochromic systems. Both leuco dyes and dye polymers are relatively inexpensive materials, yet research studies have shown that their use as external finishing on tiles or membranes can reduce the experienced mean daily surface temperature range from $28°C-45°C$ to

24°C−38°C (Karlessi et al., 2009). However, these materials are highly susceptible to the phenomenon of photodegradation when used outdoors, which is the main obstacle to their widespread use in the built environment. To address this issue, researchers have been exploring two main categories of non-dye alternatives: materials that undergo a color transition through temperature-induced molecular rearrangements, and those that experience a color transition due to nanoscale effects. The first group includes three main non-dye thermochromic materials: conjugated polymers (CPs), liquid crystals (LCs), and Schiff bases (SBs). Each of these materials works differently and has its own set of advantages and disadvantages. CPs exhibit a heat-induced disorder in the side chains, which causes the main chain to twist, reducing the effective conjugated length of the molecules and resulting in a thermochromic transition (Di Césare et al., 1997). LCs are mesophases with some degree of molecular order (that varies across their nematic, smectic, and cholesteric subclasses) that can switch from one phase to another (Viswanatha & Rajaramb, 2018). SBs—either organic and organometallic—display reversible thermochromic behavior that can be controlled by light, heat, and electric current (Hadjoudis & Mavridis, 2004). The second group of materials, that is, nanoscale thermochromics, includes quantum dots (QDs), plasmonic elements, and photonic crystals. QDs are fascinating inorganic particles that possess the unique ability to absorb and emit light at varying wavelengths (Hines & Kamat, 2014), showing potentially higher cooling capacity as compared to bulk alternatives (Garshasbi et al., 2020). Optimal heat-rejection tuning for urban overheating mitigation applications has been recently investigated (Garshasbi et al., 2021, 2022). On the other hand, plasmonic elements are metals that exhibit surface plasmon resonance when interacting with light. By controlling factors like size, shape, and distance between nanoparticles, thermochromic plasmonic can be produced (Caseri, 2010). Another appealing technology is photonic crystals, which have periodic refractive indices that can be used to tailor the optical response of materials (Gu et al., 2001). They can be utilized to create adaptive structures with temperature-dependent properties. By designing a nanostructured multilayer alternating planes consisting of materials with temperature-dependent thickness and nonthermo-responsive polymers, an adaptive structure with temperature-dependent properties can be created.

2.6.1.3   Materials with tunable and adaptive spectral emittance

Photoluminescent materials emit light when exposed to ultraviolet or visible radiation and can contribute to heat rejection, which makes them a favorable modulator of the surface energy balance equation (Kousis et al., 2020). This effect is particularly useful in areas with high levels of incoming solar radiation, whether direct or diffuse. When the emission of light ceases as soon as the excitation ends, the phenomenon is called fluorescence. On the other hand, if light is emitted for a more extended period, the phenomenon is known as phosphorescence. Fluorescence is a mechanism where electrons excited by light relax from their lowest excited state to the ground state. During this process, a photon is emitted, which usually has a lower energy and longer wavelength than the absorbed photon, as some energy is lost through vibrational relaxation. Fluorescent materials can be categorized into two types: conventional bulk fluorescent and nanoscale materials. Examples of bulk fluorescent materials include ruby (Al_2O_3:Cr), Egyptian blue ($CaCuSi_4O_{10}$), and Han blue ($BaCuSi_4O_{10}$). Phosphorescent materials are a captivating subset of photoluminescent materials that exhibit a unique spin-forbidden process. This process involves excited electrons decaying within different spin multiplicity states, resulting in the emission of radiation in the form of light that can persist for an extended period, ranging from seconds to a couple of days (Mukherjee & Thilagar, 2015). Classic phosphorescent materials emit light within the visible region through photoluminescence. They are composed of inorganic components, such as oxide, silicate, selenide, nitride, oxynitride, and sulfide, and are usually doped with transition metal ions, such as Eu^{3+}, Eu^{2+}, Gd^{3+}, Tb^{3+}, Sm^{3+}, Pr^{3+}, Ho^{3+}, and Yb^{3+}, as well as rare-earth ions, such as Cr^{3+} and Mn^{2+}. Both applications have shown promising results in terms of cooling potential, reaching peak surface temperature reductions of $4°C$ during the hottest hours of a typical summer day (Kousis et al., 2020). Additionally, photoluminescent QDs and perovskite nanocrystals have undergone optimization to enhance their cooling potential and have even been integrated into the upper layer of colored PDRCs, achieving subambient temperature drops of up to $5°C$ under direct sunlight (Wang, Zhang et al., 2022).

2.6.2 Materials improving the latent heat flux

Eq. (2.1) contains a crucial term known as latent heat, which represents the energy transferred through evaporation and condensation processes. To optimize the effectiveness of this term, material-based solutions have been developed for evaporative cooling. These solutions involve the use of permeable, pervious, and porous water-retaining materials that are commonly integrated into pavement applications. By utilizing these materials, water can permeate the surface and enter a lower layer, where it is stored and gradually evaporates as the upper layer experiences increasing thermal loads. This process not only enhances the cooling effect but also addresses the issue of run-off. Permeable pavers—generally made of concrete, plastic, or rubber—have small apertures or spacing lugs that allow water to enter the surface and provide evaporation channels for cooling (Qin, 2015). On the contrary, pervious pavers allow water to pass through them, improving water infiltration and reducing run-off. Recent studies suggest that if water is ponded at or near the surface or brought there through a capillary column, evaporation can decrease the surface temperature by up to 9°C−10°C (Liu et al., 2018), but if the ponding water level is too deep, the cooling effect is limited (Nemirovsky et al., 2013). To address this issue, a new type of surface has been developed: porous water-retaining pavements. These surfaces hold water primarily at the top layer, allowing for evaporative cooling to occur. They are usually asphalt- or cement-based and have porosity comparable to permeable pavements, but their water permeability is about one or two magnitudes lower. Water-retentive fillers such as blast-furnace slag, pervious mortar, bottom ash, peat moss, hydrophilic fibers, and other water-absorbing fillers are embedded in the concrete (Katsunori & Kazuya, 2008; Nakayama & Fujita, 2010). Depending on the filler materials, a water-retentive pavement can retain 0.15−0.27 g/cm^3 rainwater when the surface is sufficiently wet (Yamagata et al., 2008). The pore structure of the water-retentive filler allows it to draw water from the base layer and stay cool for longer than other types of pavements. To further enhance their cooling capacity, researchers are exploring the use of wastewater to replenish them, achieving surface temperature reductions of about 8°C during the day and 3°C at night (Yamagata et al., 2008). Other ideas for enhancing the cooling capacity of water-retentive pavements include the use of high absorptive fillers and novel pore structure designs.

2.6.3   Materials improving the heat storage

The heat storage term in Eq. 2.1 can be modulated by using PCMs and heat-harvesting solutions. PCMs absorb and release significant amounts of heat during a phase change, thereby enhancing the apparent thermal mass of the surface and reducing sensible heating. Because of their better volumetric stability, PCM-impregnated solutions utilize solid−liquid PCMs that melt and solidify at specific temperatures and that can be added directly or microencapsulated into concrete (Fabiani & Pisello, 2018), membranes (Fabiani, Piselli et al., 2020), gypsum (Fabiani et al., 2023), or other matrices to increase their thermal inertia. While commonly used for indoor applications, PCM-impregnated aggregates can be integrated in concrete paving (Sharifi & Sakulich, 2009) to reduce surface temperature during the peak hours by up to 5°C in field conditions (Ryms et al., 2015).

On the other hand, heat-harvesting solutions capture and utilize the excess heat absorbed by built surfaces, mostly pavements, and deliver it to dedicated energy storage systems, thus producing a cooler and more comfortable urban environment while generating useful energy. This is usually achieved by embedding asphalt solar collectors and fluid channels or air-convection channels and photovoltaic solar cell to extract the heat absorbed in the surface layer (Chiarelli et al., 2015; Gao et al., 2010). Results show remarkable surface temperature reductions, for example, up to 19°C (Shaopeng et al., 2011). While heat-harvesting solutions have been tested in experimental settings, further scrutiny and verification are necessary before they can be widely implemented. Pavements that are subjected to high traffic and heavy-vehicle loadings require specific engineered techniques to strengthen the pavement structure (Northmore & Tighe, 2016).

2.6.4 Materials improving the sensible heat flux

Heat conduction plays a critical role in Eq. (2.1) and can be modulated by enhancing the thermal diffusivity of pavement materials, which leads to the effective transfer and dissipation of heat. Incorporating high thermal diffusivity materials like quartzite or graphite into asphalt mixtures can improve their thermal properties, reducing the UHI effect (Vo et al., 2017). Notably, replacing all aggregates with quartzite increases the thermal conductivity of the mixture by 135%, thus lowering the maximum surface temperature by up to 4°C (Dawson et al., 2012). On the other hand, thermal induction technologies can form a thermal conductivity gradient by adding different contents of high thermal conductivity materials to different surface layers. This disperses heat into the subgrade soil, reducing pavement temperature. Heat reflective coatings can be applied to low-conduction asphalt pavements, achieving a maximum temperature reduction of up to 7.1°C (Yinfei et al., 2014). Thermal conductivity channels in the pavement, such as steel rods or combinations of iron powder, graphite, and copper powder, can accelerate the heat transfer downward and reduce the maximum pavement temperature by up to 6.5°C (Yinfei et al., 2018). When the combination of iron powder, graphite, and copper powder is used, the cooling effect is optimized, reducing the pavement temperature by up to 6.1°C (Jiang & Wang, 2020).

2.7 What is the cities' state of the art?

The scientific advancements in research and innovation and the themes most debated in literature concerning urban heat mitigation and adaptation do not necessarily mirror in what local governments are planning and implementing in their climate actions and policies.

Several initiatives have been launched over the latest decades to mobilize subnational players and catalyze climate mitigation and adaptation actions in a bottom-up, yet coordinated and structured fashion. Since 2008, the European Commission endorses and supports cities' efforts all over the world through the Global Covenant of Mayors for Climate and Energy (GCoM) and notably through the provision of capacity building, technical assistance, methodological guidance, sharing of best practices, and peer learning opportunities. Concerning climate mitigation, the overarching aim is to cut emissions by 55% by 2030 as compared to 1990 levels and to reach climate neutrality by 2050 while building resilience and creating jobs and growth (European Commission, 2022; Melica et al., 2022). The initiative helps consolidate practices to monitor and report on energy consumption and greenhouse gas emissions as well as on risks and vulnerabilities at the local level, allowing decision makers to identify priority sectors, set evidence-based targets, and plan relevant measures (Kona et al., 2018; Rivas et al., 2022). Within the GCoM's adaptation pillar, cities (1) identify adaptation goals, (2) conduct a risk and vulnerability assessment (RVA) to identify climate hazards and assign them a degree of current and future intensity and probability, (3) pinpoint vulnerable sectors and groups, (4) define adaptation actions (that could also have an impact on mitigation), and (5) monitor progress. Among all reported climate hazards, extreme heat features at the very top of the ranking (Melica et al., 2022).

As of September 2022, 11,279 cities and local governments joined the GCoM. Most of them (10,052) come from EU-27 and 1703 have already submitted an RVA (Baldi et al., 2023). Based on the data collected through the MyCovenant reporting framework (European Commission, 2023), Table 2.1 shows how the signatories mentioning extreme heat in their RVA consider the intensity and probability of its occurrence in the present and the future. A total of 378 cities indicated that extreme heat is both a high-impact and high-probability hazard already today, and for 97.3% of them (368 cities), the situation is expected to worsen. Fig. 2.4 complements the analysis by mapping the cities in different classes of urgency for action against extreme heat.

TABLE 2.1 Number of GCoM signatories having associated extreme heat to different degrees of current impact, current probability, expected intensity change and expected frequency change in their RVA.

Extreme heat	Current impact [% of total cities]	Current probability [% of total cities]		Expected intensity change [% of total cities]	Expected frequency change [% of total cities]
High	40	36	Increase	92	92
Medium	33	40	No change	5	4
Low	20	24	Decrease	1	0
Not known	7	1	Not known	2	3
Total cities	1301*	1703		1703	1703

The total of cities is lower for current impact, as it was not a compulsory reporting field in the past.

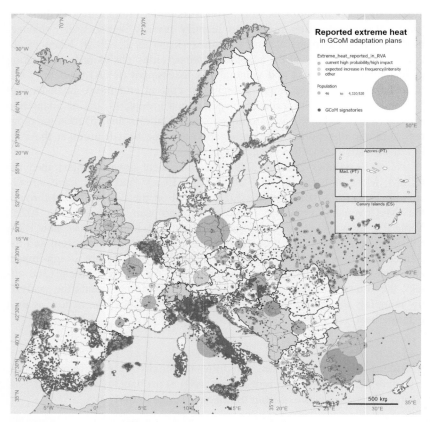

FIGURE 2.4 Map of GCoM signatories with respect to extreme heat. Red halos distinguish cities that declared extreme heat as either a high probability or high impact hazard (currently); other cities are haloed in orange if they foresee a worsening of the heat hazard or in green in any other case where extreme heat is contemplated in the city's RVA. The circles size is proportional to the population. Gray circles without halos represent signatories that have not committed to climate adaptation under the GCoM, not performed an RVA, or not included extreme heat among the hazards.

Against this backdrop, the question arises on whether cities are planning and implementing action to combat local overheating. The analysis of the MyCovenant dataset reveals that 81% (277 over 344) and 64% (1051 out of 1643) of the cities having indicated extreme heat as a high probability/impact or worsening hazard have put in place actions of mitigation/adaptation to urban overheating (hereinafter, UHI actions). In total, 7500 + UHI actions are included in the 2679 plans of 2599 signatories. For 41% of these signatories and for 65.9% of the UHI actions, extreme heat represents a hazard. Other hazards are also addressed through these actions, notably droughts and water scarcity, heavy precipitation, floods and sea level rise, and wildfires (each accounting for over 10% of the UHI actions). More than 65% of the UHI actions deal with greenery and water-based technologies (from tree planting to urban forests, from green roofs/facades to artificial water features), followed by actions on urban surfaces aiming at increased permeability or solar reflection (from biologically active/water-retentive surfaces to cool roofs and pavements) that account for about 35%. Actions dealing with solar shadings, wind-breaking provisions, and stimuli-responsive/smart systems or generic forms of naturalization and heat island counteraction account for less than 10% each (actions can fall under multiple categories). Combinations across different types of strategies are very uncommon. As such, GCoM signatories seem to rely almost completely on nature-based solutions in their strategies against local overheating, with very limited exploration of engineered and adaptive solutions. Vulnerable sectors are typically buildings, environment and biodiversity, land use planning, water management, agriculture and forestry, health, and civil protection/emergency. Further, UHI actions are frequently targeting all vulnerable groups. When specific vulnerable segments are identified, these are typically elderly and children.

Capitalizing on the experience gathered through the Covenant, in 2021, the European Commission raised the bar in terms of climate ambition at the urban level by launching the Mission on 100 Climate-Neutral and Smart Cities by 2030 (European Commission, 2021b). The Cities Mission's primary goal is to deliver about 100 absolute- or net-zero emission cities in less than a decade, by taking a cross-sectoral and demand-led approach. In total, 362 cities from 35 countries including all EU Member States expressed their interest in soon becoming climate-neutral through the mission, by answering a comprehensive questionnaire of 374 questions.

A section of the questionnaire investigated the current policies and revealed that several areas concerning the creation of an enabling urban environment are addressed by cities, notably urban regeneration (61%), UHI mitigation (52%), and mixed-use development and sprawl containment (32%), which may together provide the configurational, compositional, and infrastructural basis to move faster toward climate neutrality. As such, only about half of the mission cities are currently putting mitigation strategies in place against the UHI effect. Despite the limited coverage, mission cities testify on the importance of tackling UHI mitigation in their portfolio of actions by including urban planning measures that aim at UHI mitigation, urban regeneration, and sprawl containment among their "key" (i.e., outstanding in terms of impact, innovation, resource-efficiency, cost-efficiency, time-efficiency, and replicability) measures. Further, municipalities stress the critical role played by universities in the creation of knowledge and awareness for mitigating the heat island effect.

In terms of mitigation strategies, the Cities Mission dataset confirms the trends observed across GCoM signatories. Almost all cities having included

UHI in their policies are planning to increase the local UGI by expanding the tree and vegetative cover. Half of them are considering the installation of water-based features (fountains, pools, ponds, and/or misting systems) to boost evaporative cooling and, interestingly, more than 30% are leveraging "smart growth" practices to combat the onset of localized hot spots for heat and pollutants from a design and planning perspective. These practices promote mixed-use development, sprawl containment, compact building and urban design, walkable neighborhood, and permeable and well ventilated urban fabrics that preserve open space and undeveloped land (EPA, 2022). More than three quarters of the cities indicated that they use a combination of mitigation options, most typically leveraging green and blue infrastructure. The increase of the local solar albedo through the implementation of cool pavements (86 cities, 45.5%) or reflective roofs (24 cities, 12.7%) is comparatively underexplored, as also observed in the Covenant dataset.

In conclusion, data gathered through the Global Covenant of Mayors and the Cities Mission shed some light on the urban heat mitigation strategies cities are pursuing as part of their climate action plans, thus providing a baseline to prioritize research and outreach activities. Scientific studies should aim at fully disclosing the pros and cons of the most popular solutions so that urban overheating can be dealt with conscientiously and diligently. Solutions that come with high and multiple benefits, but with a scientific language that may sound unfamiliar to local authorities, such as cool and supercool materials, should be made more visible and palatable, by, for example, pushing for large-scale pilot and demonstrative projects. The fact that only a portion of the cities perceiving urban heat as a climate hazard are taking active action against it rings an alarm bell and calls for dedicated outreach activities and collaborations between R&I institutions and local governments to confirm the threat, map it, and tackle it through site-specific mitigation design. Cities that historically suffered from no or mild overheating episodes may have to rethink their planning in light of the recent escalation of extreme events, even in colder climates. Finally, integrated and combinatorial approaches in the use of heat mitigation strategies currently receive little attention, even among cities particularly keen on taking up the climate challenge. Guidelines on how to combine fruitfully specific solutions based on their underlying physical process and on contextual factors (e.g., climate and urban layout) should be formulated and disseminated. To channel future efforts in all these directions and create an enabling environment for colearning and cocreation, the European Commission has launched another mission, dedicated to adaptation to climate change (European Commission, 2021a). The Adaptation Mission focuses on supporting EU regions, cities, and local authorities in their efforts to better understand climate risks and build resilience against its impacts by 2030. Selected participants will be frontrunners in climate adaptation and will test and deploy innovative solutions.

Disclaimer

The views expressed here are purely those of the authors and may not, under any circumstances, be regarded as an official position of the European Commission.

References

Abdi, B., Hami, A., & Zarehaghi, D. (2020). Impact of small-scale tree planting patterns on outdoor cooling and thermal comfort. *Sustainable Cities and Society*, *56*, 102085. Available from https://doi.org/10.1016/j.scs.2020.102085.

Abunnasr, Y., & Mhawej, M. (2022). Towards a combined Landsat-8 and Sentinel-2 for 10-m land surface temperature products: The Google Earth Engine monthly Ten-ST-GEE system. *Environmental Modelling & Software*, *155*, 105456. Available from https://doi.org/10.1016/J.ENVSOFT.2022.105456.

Adélaïde, L., Chanel, O., & Pascal, M. (2022). Health effects from heat waves in France: An economic evaluation. *The European Journal of Health Economics*, *23*, 119–131. Available from https://doi.org/10.1007/s10198-021-01357-2.

Adger, W. N. (2006). Vulnerability. *Global Environmental Change*, *16*, 268–281. Available from https://doi.org/10.1016/j.gloenvcha.2006.02.006.

Aflaki, A., Mirnezhad, M., Ghaffarianhoseini, A., Ghaffarianhoseini, A., Omrany, H., Wang, Z.-H., et al. (2017). Urban heat island mitigation strategies: A state-of-the-art review on Kuala Lumpur, Singapore and Hong Kong. *Cities (London, England)*, *62*, 131–145. Available from https://doi.org/10.1016/j.cities.2016.09.003.

Alikhani, S., Nummi, P., & Ojala, A. (2021). Urban wetlands: A review on ecological and cultural values. *Water*, *13*, 3301.

Aljawabra, F., & Nikolopoulou, M. (2018). Thermal comfort in urban spaces: A cross-cultural study in the hot arid climate. *International Journal of Biometeorology*, *62*, 1901–1909. Available from https://doi.org/10.1007/s00484-018-1592-5.

Alvi, U., Suomi, J., & Käyhkö, J. (2022). A cost-effective method for producing spatially continuous high-resolution air temperature information in urban environments. *Urban Climate*, *42*, 101123. Available from https://doi.org/10.1016/J.UCLIM.2022.101123.

Amazirh, A., Merlin, O., & Er-Raki, S. (2019). Including Sentinel-1 radar data to improve the disaggregation of MODIS land surface temperature data. *ISPRS Journal of Photogrammetry and Remote Sensing*, *150*, 11–26. Available from https://doi.org/10.1016/J.ISPRSJPRS.2019.02.004.

Anand, J., Sailor, D. J., & Baniassadi, A. (2021). The relative role of solar reflectance and thermal emittance for passive daytime radiative cooling technologies applied to rooftops. *Sustainable Cities and Society*, *65*, 102612.

Anguelovski, I., Connolly, J. J. T., Cole, H., Garcia-Lamarca, M., Triguero-Mas, M., Baró, F., et al. (2022). Green gentrification in European and North American cities. *Nature Communications*, *13*, 1–13.

Arghavani, S., Malakooti, H., & Bidokhti, A. A. (2019). Numerical evaluation of urban green space scenarios effects on gaseous air pollutants in Tehran Metropolis based on WRF-Chem model. *Atmospheric Environment (Oxford, England: 1994)*, *214*, 116832. Available from https://doi.org/10.1016/j.atmosenv.2019.116832.

Asgarian, A., Amiri, B. J., & Sakieh, Y. (2015). Assessing the effect of green cover spatial patterns on urban land surface temperature using landscape metrics approach. *Urban Ecosystem*, *18*, 209–222. Available from https://doi.org/10.1007/s11252-014-0387-7.

Atwa, S., Ibrahim, M. G., & Murata, R. (2020). Evaluation of plantation design methodology to improve the human thermal comfort in hot-arid climatic responsive open spaces. *Sustainable Cities and Society*, *59*, 102198. Available from https://doi.org/10.1016/j.scs.2020.102198.

Balczó, M. G., & Ruck, B. (2009). Numerical modeling of flow and pollutant dispersion in street canyons with tree planting. *Meteorologische Zeitschrift*, *18*, 197–206. Available from https://doi.org/10.1127/0941-2948/2009/0361.

Baldi, M., Bertoldi, P., Franco de Los Rios, C., Melica, G., & Treville, A. (2023). GCoM - MyCovenant, 3rd release - September 2022. European Commission. *Joint Research Centre (JRC) [Dataset]*, pid. Available from http://data.europa.eu/89h/e97d17e7-c047-42d7-9db5-d10229d073eb.

Bao, J., Li, X., & Yu, C. (2015). The construction and validation of the heat vulnerability index, a review. *International Journal of Environmental Research and Public Health*, *12*, 7220–7234. Available from https://doi.org/10.3390/ijerph120707220.

Bartesaghi-Koc, C., Haddad, S., Pignatta, G., Paolini, R., Prasad, D., & Santamouris, M. (2021). Can urban heat be mitigated in a single urban street? Monitoring, strategies, and performance results from a real scale redevelopment project. *Solar Energy*, *216*, 564–588.

Bartesaghi Koc, C., Osmond, P., & Peters, A. (2018). Evaluating the cooling effects of green infrastructure: A systematic review of methods, indicators and data sources. *Solar Energy*. Available from https://doi.org/10.1016/j.solener.2018.03.008.

Bartesaghi-Koc, C., Osmond, P., & Peters, A. (2019). Spatio-temporal patterns in green infrastructure as driver of land surface temperature variability: The case of Sydney. *International Journal of Applied Earth Observation and Geoinformation*, *83*, 101903. Available from https://doi.org/10.1016/j.jag.2019.101903.

Bartesaghi-Koc, C., Osmond, P., & Peters, A. (2020). Quantifying the seasonal cooling capacity of 'green infrastructure types' (GITs): An approach to assess and mitigate surface urban heat island in Sydney, Australia. *Landscape and Urban Planning*, *203*, 103893. Available from https://doi.org/10.1016/j.landurbplan.2020.103893.

Bartesaghi-Koc, C., Osmond, P., & Peters, A. (2022). Innovative use of spatial regression models to predict the effects of green infrastructure on land surface temperatures. *Energy and Buildings*, *254*, 111564. Available from https://doi.org/10.1016/j.enbuild.2021.111564.

Bassett, R., Cai, X., Chapman, L., Heaviside, C., Thornes, J. E., Muller, C. L., et al. (2016). Observations of urban heat island advection from a high-density monitoring network. *Quarterly Journal of the Royal Meteorological Society*, *142*, 2434–2441. Available from https://doi.org/10.1002/qj.2836.

Bernard, J., Rodler, A., Morille, B., & Zhang, X. (2018). How to design a park and its surrounding urban morphology to optimize the spreading of cool air. *Climate*, *6*, 10. Available from https://doi.org/10.3390/cli6010010.

Bevilacqua, P., Mazzeo, D., Bruno, R., & Arcuri, N. (2016). Experimental investigation of the thermal performances of an extensive green roof in the Mediterranean area. *Energy and Buildings*, *122*, 63–79. Available from https://doi.org/10.1016/j.enbuild.2016.03.062.

Bigras, F. J. (2000). Selection of white spruce families in the context of climate change: Heat tolerance. *Tree Physiology*, *20*, 1227–1234. Available from https://doi.org/10.1093/treephys/20.18.1227.

Binarti, F., Koerniawan, M. D., Triyadi, S., Utami, S. S., & Matzarakis, A. (2020). A review of outdoor thermal comfort indices and neutral ranges for hot-humid regions. *Urban Climate*, *31*, 100531. Available from https://doi.org/10.1016/j.uclim.2019.100531.

Bita, C. E., & Gerats, T. (2013). Plant tolerance to high temperature in a changing environment: Scientific fundamentals and production of heat stress-tolerant crops. *Frontiers in Plant Science*, *4*. Available from https://doi.org/10.3389/fpls.2013.00273.

Bizheva, K. K., Siegel, A. M., & Boas, D. A. (1998). Path-length-resolved dynamic light scattering in highly scattering random media: The transition to diffusing wave spectroscopy. *Physical Review. E, Statistical Physics, Plasmas, Fluids, and Related Interdisciplinary Topics, 58*, 7664−7667. Available from https://doi.org/10.1103/PhysRevE.58.7664.

Bouzarovski, S., Frankowski, J., & Tirado Herrero, S. (2018). Low-carbon gentrification: When climate change encounters residential displacement. *International Journal of Urban and Regional Research, 42*, 845−863.

Bowler, D. E., Buyung-Ali, L., Knight, T. M., & Pullin, A. S. (2010). Urban greening to cool towns and cities: A systematic review of the empirical evidence. *Landscape and Urban Planning*. Available from https://doi.org/10.1016/j.landurbplan.2010.05.006.

Broadbent, A. M., Coutts, A. M., Tapper, N. J., & Demuzere, M. (2018). The cooling effect of irrigation on urban microclimate during heatwave conditions. *Urban Climate*. Available from https://doi.org/10.1016/j.uclim.2017.05.002.

Brousse, O., Martilli, A., Foley, M., Mills, G., & Bechtel, B. (2016). WUDAPT, an efficient land use producing data tool for mesoscale models? Integration of urban LCZ in WRF over Madrid. *Urban Climate, 17*, 116−134. Available from https://doi.org/10.1016/j.uclim.2016.04.001.

Bueno, B., Norford, L., Pigeon, G., & Britter, R. (2011). Combining a detailed building energy model with a physically-based urban canopy model. *Boundary-Layer Meteorology, 140*, 471−489. Available from https://doi.org/10.1007/s10546-011-9620-6.

Cai, Z., Tang, Y., Chen, K., & Han, G. (2019). Assessing the heat vulnerability of different local climate zones in the old areas of a chinese megacity. *Sustainability*, 11. Available from https://doi.org/10.3390/su11072032.

Cao, X., Onishi, A., Chen, J., & Imura, H. (2010). Quantifying the cool island intensity of urban parks using ASTER and IKONOS data. *Landscape and Urban Planning, 96*, 224−231. Available from https://doi.org/10.1016/j.landurbplan.2010.03.008.

Capone, C., Cacciatore, A., Princigallo, A., Berardi, R., & Muscio, A. (2023). Cool cementitious materials for a more sustainable urban environment. *Journal of Architectural Engineering*, 29. Available from https://doi.org/10.1061/jaeied.aeeng-1415.

Carlosena, L., Andueza, Á., Torres, L., Irulegi, O., Hernández-Minguillón, R. J., Sevilla, J., et al. (2021). Experimental development and testing of low-cost scalable radiative cooling materials for building applications. *Solar Energy Materials and Solar Cells, 230*, 111209.

Carlosena, L., Ruiz-Pardo, Á., Feng, J., Irulegi, O., Hernández-Minguillón, R. J., & Santamouris, M. (2020). On the energy potential of daytime radiative cooling for urban heat island mitigation. *Solar Energy, 208*, 430−444. Available from https://doi.org/10.1016/j.solener.2020.08.015.

Carlucci, S., & Pagliano, L. (2012). A review of indices for the long-term evaluation of the general thermal comfort conditions in buildings. *Energy and Buildings, 53*, 194−205. Available from https://doi.org/10.1016/j.enbuild.2012.06.015.

Carolan, M. (2020). "Urban farming is going high tech" digital urban agriculture's links to gentrification and land use. *Journal of the American Planning Association, 86*, 47−59.

Caseri, W. (2010). Color switching in nanocomposites comprising inorganic nanoparticles dispersed in a polymer matrix. *Journal of Materials Chemistry, 20*, 5582−5592. Available from https://doi.org/10.1039/b926280f.

Chae, D., Kim, M., Jung, P.-H., Son, S., Seo, J., Liu, Y., et al. (2020). Spectrally selective inorganic-based multilayer emitter for daytime radiative cooling. *ACS Applied Materials & Interfaces, 12*, 8073−8081. Available from https://doi.org/10.1021/acsami.9b16742.

Chae, D., Son, S., Liu, Y., Lim, H., & Lee, H. (2020). High-performance daytime radiative cooler and near-ideal selective emitter enabled by transparent sapphire substrate, 2001577, 1−9. Available from https://doi.org/10.1002/advs.202001577.

Charoenkit, S., & Yiemwattana, S. (2016). Living walls and their contribution to improved thermal comfort and carbon emission reduction: A review. *Building and Environment, 105,* 82–94. Available from https://doi.org/10.1016/j.buildenv.2016.05.031.

Chatzidimitriou, A., Liveris, P., Bruse, M., & Topli, L. (2013). Urban redevelopment and microclimate improvement: A design project in thessaloniki, greece. *PLEA,* Sustain Archit a Renew Futur 2013.

Chen, L., & Ng, E. (2012). Outdoor thermal comfort and outdoor activities: A review of research in the past decade. *Cities (London, England).* Available from https://doi.org/10.1016/j.cities.2011.08.006.

Chen, M., Pang, D., & Yan, H. (2022). Highly solar reflectance and infrared transparent porous coating for non-contact heat dissipations. *IScience, 25,* 104726. Available from https://doi.org/10.1016/j.isci.2022.104726.

Chen, Y., Mandal, J., Li, W., Smith-Washington, A., Tsai, C.-C., Huang, W., et al. (2020). Colored and paintable bilayer coatings with high solar-infrared reflectance for efficient cooling. *Science Advances, 6,* eaaz5413. Available from https://doi.org/10.1126/sciadv.aaz5413.

Chen, A., Yao, X. A., Sun, R., & Chen, L. (2014). Effect of urban green patterns on surface urban cool islands and its seasonal variations. *Urban Forestry & Urban Greening, 13,* 646–654. Available from https://doi.org/10.1016/j.ufug.2014.07.006.

Cheng, C. (2016). Spatial climate justice and green infrastructure assessment: A case study for the huron river watershed, michigan, USA.

Cheng, W., Li, D., Liu, Z., & Brown, R. D. (2021). Approaches for identifying heat-vulnerable populations and locations: A systematic review. *The Science of the Total Environment, 799,* 149417. Available from https://doi.org/10.1016/j.scitotenv.2021.149417.

Chi, D. A., González, M. E., Valdivia, R., & Gutiérrez, J. E. (2021). Parametric design and comfort optimization of dynamic shading structures. *Sustainability, 13,* 7670.

Chiarelli, A., Dawson, A. R., & García, A. (2015). Parametric analysis of energy harvesting pavements operated by air convection. *Applied Energy, 154,* 951–958. Available from https://doi.org/10.1016/j.apenergy.2015.05.093.

Ching, J., Brown, M., Burian, S., Chen, F., Cionco, R., Hanna, A., et al. (2009). National urban database and access portal tool. *Bulletin of the American Meteorological Society, 90,* 1157–1168. Available from https://doi.org/10.1175/2009BAMS2675.1.

Ching, J., Mills, G., Bechtel, B., See, L., Feddema, J., Wang, X., et al. (2018). WUDAPT: An urban weather, climate, and environmental modeling infrastructure for the anthropocene. *Bulletin of the American Meteorological Society, 99,* 1907–1924. Available from https://doi.org/10.1175/BAMS-D-16-0236.1.

Colaninno, N., & Morello, E. (2022). Towards an operational model for estimating day and night instantaneous near-surface air temperature for urban heat island studies: Outline and assessment. *Urban Climate, 46.* Available from https://doi.org/10.1016/j.uclim.2022.101320.

Conlon, K. C., Mallen, E., Gronlund, C. J., Berrocal, V. J., Larsen, L., & O'Neill, M. S. (2020). Mapping human vulnerability to extreme heat: A critical assessment of heat vulnerability indices created using principal components analysis. *Environmental Health Perspectives, 128,* 97001.

Conrads, L. A., & Van Der Hage, J. C. H. (1971). A new method of air-temperature measurement in urban climatological studies. *Atmospheric Environment (Oxford, England: 1994), 5.* Available from https://doi.org/10.1016/0004-6981(71)90119-3.

Cureau, R. J., Pigliautile, I., & Pisello, A. L. (2022). A new wearable system for sensing outdoor environmental conditions for monitoring hyper-microclimate. *Sensors, 22.* Available from https://doi.org/10.3390/s22020502.

Cutter, S. L., Boruff, B. J., & Shirley, W. L. (2003). Social vulnerability to environmental hazards. *Social Science Quarterly*, *84*, 242−261.

Daemei, A. B., Azmoodeh, M., Zamani, Z., & Khotbehsara, E. M. (2018). Experimental and simulation studies on the thermal behavior of vertical greenery system for temperature mitigation in urban spaces. *Journal of Building Engineering*, *20*, 277−284. Available from https://doi.org/10.1016/j.jobe.2018.07.024.

Davtalab, J., & Heidari, A. (2021). The effect of kharkhona on outdoor thermal comfort in Hot and dry climate: A case study of Sistan Region in Iran. *Sustainable Cities and Society*, *65*, 102607. Available from https://doi.org/10.1016/j.scs.2020.102607.

Dawson, A. R., Dehdezi, P. K., Hall, M. R., Wang, J., & Isola, R. (2012). Enhancing thermal properties of asphalt materials for heat storage and transfer applications. *Road Materials and Pavement Design*, *13*, 784−803. Available from https://doi.org/10.1080/14680629.2012.735791.

de Freitas, C. R., & Grigorieva, E. A. (2015). A comprehensive catalogue and classification of human thermal climate indices. *International Journal of Biometeorology*, *59*, 109−120. Available from https://doi.org/10.1007/s00484-014-0819-3.

Deilami, K., Kamruzzaman, M., & Liu, Y. (2018). Urban heat island effect: A systematic review of spatio-temporal factors, data, methods, and mitigation measures. *International Journal of Applied Earth Observation and Geoinformation*, *67*, 30−42. Available from https://doi.org/10.1016/j.jag.2017.12.009.

Demuzere, M., Kittner, J., & Bechtel, B. (2021). LCZ generator: A web application to create local climate zone maps. Front. *Environmental Sciences: An International Journal of Environmental Physiology and Toxicology*, *9*.

Demuzere, M., Kittner, J., Martilli, A., Mills, G., Moede, C., Stewart, I. D., et al. (2022). A global map of local climate zones to support earth system modelling and urban-scale environmental science. *Earth System Science Data*, *14*, 3835−3873. Available from https://doi.org/10.5194/essd-14-3835-2022.

Demuzere, M., Orru, K., Heidrich, O., Olazabal, E., Geneletti, D., Orru, H., et al. (2014). Mitigating and adapting to climate change: Multi-functional and multi-scale assessment of green urban infrastructure. *Journal of Environmental Management*, *146*, 107−115. Available from https://doi.org/10.1016/j.jenvman.2014.07.025.

Di Césare, N., Belletête, M., Durocher, G., & Leclerc, M. (1997). Towards a theoretical design of thermochromic polythiophenes. *Chemical Physics Letters*, *275*, 533−539. Available from https://doi.org/10.1016/S0009-2614(97)00777-X.

Dong, L., Mitra, C., Greer, S., & Burt, E. (2018). The dynamical linkage of atmospheric blocking to drought, heatwave and urban heat island in Southeastern US: A multiscale case study. *Atmosphere (Basel)*, *9*. Available from https://doi.org/10.3390/atmos9010033.

Dovie, D. B. K., Dzodzomenyo, M., & Ogunseitan, O. A. (2017). Sensitivity of health sector indicators' response to climate change in Ghana. *The Science of the Total Environment*, *574*, 837−846. Available from https://doi.org/10.1016/j.scitotenv.2016.09.066.

Drake, J. E., Tjoelker, M. G., Vårhammar, A., Medlyn, B. E., Reich, P. B., Leigh, A., et al. (2018). Trees tolerate an extreme heatwave via sustained transpirational cooling and increased leaf thermal tolerance. *Global Change Biology*, *24*, 2390−2402. Available from https://doi.org/10.1111/gcb.14037, PM - 29316093.

Du, H., Cai, W., Xu, Y., Wang, Z., Wang, Y., & Cai, Y. (2017). Quantifying the cool island effects of urban green spaces using remote sensing data. *Urban Forestry & Urban Greening*, *27*, 24−31. Available from https://doi.org/10.1016/j.ufug.2017.06.008.

Efthymiou, C., Santamouris, M., Kolokotsa, D., & Koras, A. (2016). Development and testing of photovoltaic pavement for heat island mitigation. *Solar Energy*. Available from https://doi.org/10.1016/j.solener.2016.01.054.

El-Bardisy, W. M., Fahmy, M., & El-Gohary, G. F. (2016). Climatic sensitive landscape design: Towards a better microclimate through plantation in public schools, Cairo, Egypt. *Procedia - Social and Behavioral Sciences*, *216*, 206−216. Available from https://doi.org/10.1016/j.sbspro.2015.12.029.

Ella, I.I., Gaiya, S.N., Gofwen, C.N., & Ola-Adisa, E.O. (2018). An appraisal of simple shading devices to mitigate the effects of urban heat islands on buildings in Nigeria.

Ellis, K. N., Hathaway, J. M., Mason, L. R., Howe, D. A., Epps, T. H., & Brown, V. M. (2017). Summer temperature variability across four urban neighborhoods in Knoxville, Tennessee, USA. *Theoretical and Applied Climatology*, *127*, 701−710. Available from https://doi.org/10.1007/s00704-015-1659-8.

Eludoyin, O. M., & Oluwatumise, O. E. (2021). *Thermal comfort and vulnerability of residents to heat stress in Ikare, Akoko, Ondo State, Nigeria. Handb. Clim. Chang. Manag. Res. Leadership, Transform* (pp. 3273−3297). Springer.

EPA. (2022). Smart growth and heat islands. https://www.epa.gov/heatislands/smart-growth-and-heat-islands Accessed 30.03.23.

Epstein, Y., & Moran, D. S. (2006). Thermal comfort and the heat stress indices. *Industrial Health*, *44*, 388−398.

Esch, T., Heldens, W., Hirner, A., Keil, M., Marconcini, M., Roth, A., et al. (2017). Breaking new ground in mapping human settlements from space − The Global Urban Footprint. *ISPRS Journal of Photogrammetry and Remote Sensing*, *134*, 30−42. Available from https://doi.org/10.1016/j.isprsjprs.2017.10.012.

European Commission. (2021a). EU mission: Adaptation to climate change. https://research-and-innovation.ec.europa.eu/funding/funding-opportunities/funding-programmes-and-open-calls/horizon-europe/eu-missions-horizon-europe/adaptation-climate-change_en Accessed 30.03.23.

European Commission. (2021b). EU mission: Climate-neutral and smart cities. https://ec.europa.eu/info/research-and-innovation/funding/funding-opportunities/funding-programmes-and-open-calls/horizon-europe/eu-missions-horizon-europe/climate-neutral-and-smart-cities_en Accessed 08.02.22.

European Commission. (2022). Global covenant of mayors - Cities taking action against climate change.

European Commission. (2023). Global covenant of mayors - A complete collection of action plans and monitoring reports from MyCovenant reporting platform. https://data.jrc.ec.europa.eu/collection/id-00354.

Fabiani, C., Castaldo, V. L., & Pisello, A. L. (2020). Thermochromic materials for indoor thermal comfort improvement: Finite difference modeling and validation in a real case-study building. *Applied Energy*, *262*, 114147.

Fabiani, C., Piselli, C., & Pisello, A. L. (2020). Thermo-optic durability of cool roof membranes: Effect of shape stabilized phase change material inclusion on building energy efficiency. *Energy and Buildings*, *207*. Available from https://doi.org/10.1016/j.enbuild.2019.109592.

Fabiani, C., Pisello, A. L., & Paksoy, H. (2018). Novel building materials. *Comprehensive Energy Systems*, *2−5*, 980−1017. Available from https://doi.org/10.1016/B978-0-12-809597-3.00257-1.

Fabiani, C., & Pisello, A. L. (2018). Coupling the transient plane source method with a dynamically controlled environment to study PCM-doped building materials. *Energy and Buildings*, *180*, 122−134. Available from https://doi.org/10.1016/j.enbuild.2018.09.008.

Fabiani, C., Santini, C., Barbanera, M., Giannoni, T., Rubino, G., Cotana, F., et al. (2023). Phase change materials-impregnated biomass for energy efficiency in buildings: Innovative material production and multiscale thermophysical characterization. *Journal of Energy Storage*, 58. Available from https://doi.org/10.1016/j.est.2022.106223.

Fan, H., Yu, Z., Yang, G., Liu, T. Y., Liu, T. Y., Hung, C. H., et al. (2019). How to cool hot-humid (Asian) cities with urban trees? An optimal landscape size perspective. *Agricultural and Forest Meteorology*, *265*, 338−348. Available from https://doi.org/10.1016/j.agrformet.2018.11.027.

Fang, Z., Lin, Z., Mak, C. M., Niu, J., & Tse, K.-T. (2018). Investigation into sensitivities of factors in outdoor thermal comfort indices. *Building and Environment*, *128*, 129−142. Available from https://doi.org/10.1016/j.buildenv.2017.11.028.

Feng, J., Haddad, S., Gao, K., Garshasbi, S., Ulpiani, G., Santamouris, M., et al. (2023). *Fighting urban climate change—State of the art of mitigation technologies. Urban Clim. Chang. Heat Islands* (pp. 227−296). Elsevier TS-CrossRef. Available from https://doi.org/10.1016/B978-0-12-818977-1.00006-5.

Feng, J., Khan, A., Doan, Q.-V., Gao, K., & Santamouris, M. (2021). The heat mitigation potential and climatic impact of super-cool broadband radiative coolers on a city scale. *Cell Reports Physical Science*, *2*, 100485.

Feng, J., Saliari, M., Gao, K., & Santamouris, M. (2022). On the cooling energy conservation potential of super cool roofs. *Energy and Buildings*, 112076.

Feng, J., Santamouris, M., & Gao, K. (2020). The radiative cooling efficiency of silica sphere embedded polymethylpentene (TPX) systems. *Solar Energy Materials and Solar Cells*, *215*, 110671. Available from https://doi.org/10.1016/j.solmat.2020.110671.

Fenner, D., Meier, F., Bechtel, B., Otto, M., & Scherer, D. (2017). Intra and inter 'local climate zone' variability of air temperature as observed by crowdsourced citizen weather stations in Berlin. *Germany. Meteorologische Zeitschrift*, *26*, 525−547. Available from https://doi.org/10.1127/metz/2017/0861.

Ferrante, G., Asta, F., Cilluffo, G., Sario, M., Michelozzi, P., & La Grutta, S. (2020). The effect of residential urban greenness on allergic respiratory diseases in youth: A narrative review. *World Allergy Organization Journal*, *13*, 100096. Available from https://doi.org/10.1016/j.waojou.2019.100096, PM - 32071664.

Feyisa, G. L., Dons, K., & Meilby, H. (2014). Efficiency of parks in mitigating urban heat island effect: An example from Addis Ababa. *Landscape and Urban Planning*, *123*, 87−95. Available from https://doi.org/10.1016/j.landurbplan.2013.12.008.

Fischer, E. M., Seneviratne, S. I., Lüthi, D., & Schär, C. (2007). Contribution of land-atmosphere coupling to recent European summer heat waves. *Geophysical Research Letters*, 34.

Flanagan, B.E., Gregory, E.W., Hallisey, E.J., Heitgerd, J.L., & Lewis, B. (2011). A social vulnerability index for disaster management, 8. Available from https://doi.org/10.2202/1547-7355.1792.

Franchini, M., & Mannucci, P. M. (2015). Impact on human health of climate changes. *European Journal of Internal Medicine*, *26*, 1−5. Available from https://doi.org/10.1016/j.ejim.2014.12.008.

Fu, J., Dupre, K., Tavares, S., King, D., & Banhalmi-Zakar, Z. (2022). Optimized greenery configuration to mitigate urban heat: A decade systematic review. *Frontiers of Architectural Research*, *11*, 466−491. Available from https://doi.org/10.1016/j.foar.2021.12.005.

Füssel, H.-M., & Klein, R. J. T. (2006). Climate change vulnerability assessments: An evolution of conceptual thinking. *Climatic Change*, *75*, 301−329. Available from https://doi.org/10.1007/s10584-006-0329-3.

Gaitani, N., Burud, I., Thiis, T., & Santamouris, M. (2017). High-resolution spectral mapping of urban thermal properties with unmanned aerial vehicles. *Building and Environment*, *121*, 215−224. Available from https://doi.org/10.1016/j.buildenv.2017.05.027.

Gao, K., & Santamouris, M. (2019). The use of water irrigation to mitigate ambient overheating in the built environment: Recent progress. *Building and Environment*, *164*, 106346. Available from https://doi.org/10.1016/j.buildenv.2019.106346.

Gao, K., Santamouris, M., & Feng, J. (2020). On the cooling potential of irrigation tó mitigate urban heat island. *The Science of the Total Environment*, *740*, 139754. Available from https://doi.org/10.1016/j.scitotenv.2020.139754.

Gao, Q., Huang, Y., Li, M., Liu, Y., & Yan, Y. Y. (2010). Experimental study of slab solar collection on the hydronic system of road. *Solar Energy*, *84*, 2096−2102. Available from https://doi.org/10.1016/j.solener.2010.09.008.

Garshasbi, S., Huang, S., Valenta, J., & Santamouris, M. (2020). Can quantum dots help to mitigate urban overheating? An experimental and modelling study. *Solar Energy*, *206*, 308−316. Available from https://doi.org/10.1016/j.solener.2020.06.010.

Garshasbi, S., Huang, S., Valenta, J., & Santamouris, M. (2021). Enhancing the cooling potential of photoluminescent materials through evaluation of thermal and transmission loss mechanisms. *Scientific Reports*, *11*, 1−9.

Garshasbi, S., Huang, S., Valenta, J., & Santamouris, M. (2022). Adjusting optical and fluorescent properties of quantum dots: Moving towards best optical heat-rejecting materials. *Solar Energy*, *238*, 272−279. Available from https://doi.org/10.1016/j.solener.2022.04.026.

Garshasbi, S., & Santamouris, M. (2019). Using advanced thermochromic technologies in the built environment: Recent development and potential to decrease the energy consumption and fight urban overheating. *Solar Energy Materials and Solar Cells*, *191*, 21--32.

Geletič, J., Lehnert, M., Savić, S., & Milošević, D. (2018). Modelled spatiotemporal variability of outdoor thermal comfort in local climate zones of the city of Brno, Czech Republic. *The Science of the Total Environment*, *624*, 385−395. Available from https://doi.org/10.1016/j.scitotenv.2017.12.076, PM - 29258039.

Graça, M., Cruz, S., Monteiro, A., & Neset, T.-S. (2022). Designing urban green spaces for climate adaptation: A critical review of research outputs. *Urban Climate*, *42*, 101126. Available from https://doi.org/10.1016/j.uclim.2022.101126.

Grose, M. J. (2014). Gaps and futures in working between ecology and design for constructed ecologies. *Landscape and Urban Planning*, *132*, 69−78. Available from https://doi.org/10.1016/j.landurbplan.2014.08.011.

Gu, Z. Z., Kubo, S., Qian, W., Einaga, Y., Tryk, D. A., Fujishima, A., et al. (2001). Varying the optical stop band of a three-dimensional photonic crystal by refractive index control. *Langmuir: The ACS Journal of Surfaces and Colloids*, *17*, 6751−6753. Available from https://doi.org/10.1021/la0110186.

Guzman-Echavarria, G., Middel, A., & Vanos, J. (2022). Beyond heat exposure—New methods to quantify and link personal heat exposure, stress, and strain in diverse populations and climates: The journal temperature toolbox. *Temperature*, 1−21. Available from https://doi.org/10.1080/23328940.2022.2149024.

Haddad, S., Paolini, R., Ulpiani, G., Synnefa, A., Hatvani-Kovacs, G., Garshasbi, S., et al. (2020). Holistic approach to assess co-benefits of local climate mitigation in a hot humid region of Australia. *Scientific Reports*, *10*, 14216. Available from https://doi.org/10.1038/s41598-020-71148-x.

Haddad, S., Ulpiani, G., Paolini, R., Synnefa, A., & Santamouris, M. (2020). Experimental and theoretical analysis of the urban overheating and its mitigation potential in a hot arid city—Alice Springs. *Architectural Science Review*, 63. Available from https://doi.org/10.1080/00038628.2019.1674128.

Hadjoudis, E., & Mavridis, I. M. (2004). Photochromism and thermochromism of Schiff bases in the solid state: Structural aspects. *Chemical Society Reviews*, *33*, 579−588. Available from https://doi.org/10.1039/b303644h.

Hami, A., Abdi, B., & Zarehaghi, D. (2019). Maulan S Bin. Assessing the thermal comfort effects of green spaces: A systematic review of methods, parameters, and plants' attributes. *Sustainable Cities and Society*, *49*, 101634. Available from https://doi.org/10.1016/j.scs.2019.101634.

Harfouche, A., Meilan, R., & Altman, A. (2011). Tree genetic engineering and applications to sustainable forestry and biomass production. *Trends in Biotechnology*, *29*, 9−17. Available from https://doi.org/10.1016/j.tibtech.2010.09.003.

Harlan, S. L., Brazel, A. J., Prashad, L., Stefanov, W. L., & Larsen, L. (2006). Neighborhood microclimates and vulnerability to heat stress. *Social Science & Medicine*, *63*, 2847−2863. Available from https://doi.org/10.1016/j.socscimed.2006.07.030.

Havenith, G., & Fiala, D. (2011). Thermal indices and thermophysiological modeling for heat stress. *Comprehensive Physiology*, *6*, 255−302.

He, B.-J., Wang, J., Liu, H., & Ulpiani, G. (2021). Localized synergies between heat waves and urban heat islands: Implications on human thermal comfort and urban heat management. *Environmental Research*, *193*, 110584. Available from https://doi.org/10.1016/j.envres.2020.110584.

He, B.-J., Wang, J., Zhu, J., & Qi, J. (2022). Beating the urban heat: Situation, background, impacts and the way forward in China. *Renewable and Sustainable Energy Reviews*, *161*, 112350. Available from https://doi.org/10.1016/j.rser.2022.112350.

He, B.-J., Wang, W., & Sharifi, A. (2023). Xiao L. Progress, knowledge gap and future directions of urban heat mitigation and adaptation research through a bibliometric review of history and evolution. *Energy and Buildings*.

He, B.-J., Zhao, D., Xiong, K., Qi, J., Ulpiani, G., Pignatta, G., et al. (2021). A framework for addressing urban heat challenges and associated adaptive behavior by the public and the issue of willingness to pay for heat resilient infrastructure in Chongqing, China. *Sustainable Cities and Society*, *75*, 103361. Available from https://doi.org/10.1016/j.scs.2021.103361.

He, Y., Yu, H., Ozaki, A., & Dong, N. (2020). Thermal and energy performance of green roof and cool roof: A comparison study in Shanghai area. *Journal of Cleaner Production*, *267*, 122205. Available from https://doi.org/10.1016/j.jclepro.2020.122205.

Hidalgo García, D. (2021). Analysis and precision of the terrestrial surface temperature using landsat 8 and sentinel 3 images: Study applied to the city of Granada (Spain). *Sustainable Cities and Society*, *71*, 102980. Available from https://doi.org/10.1016/J.SCS.2021.102980.

Hilhorst, D., & Bankoff, G. (2013). *Introduction: Mapping vulnerability. Mapp. vulnerability* (pp. 20−28). Routledge.

Hines, D. A., & Kamat, P. V. (2014). Recent advances in quantum dot surface chemistry. *ACS Applied Materials & Interfaces*, *6*, 3041−3057. Available from https://doi.org/10.1021/am405196u.

Hirsch, A. L., Evans, J. P., Thomas, C., Conroy, B., Hart, M. A., Lipson, M., et al. (2021). Resolving the influence of local flows on urban heat amplification during heatwaves. *Environmental Research Letters*, *16*, 64066.

Hossain, M. M., Jia, B., & Gu, M. (2015). A metamaterial emitter for highly efficient radiative cooling. *Advanced Optical Materials*, *3*, 1047−1051. Available from https://doi.org/10.1002/adom.201500119.

Howe, D. A., Hathaway, J. M., Ellis, K. N., & Mason, L. R. (2017). Spatial and temporal variability of air temperature across urban neighborhoods with varying amounts of tree canopy.

Urban Forestry & Urban Greening, *27*, 109−116. Available from https://doi.org/10.1016/j. ufug.2017.07.001.

Imran, H. M., Kala, J., Ng, A. W. M., & Muthukumaran, S. (2019). Effectiveness of vegetated patches as Green Infrastructure in mitigating Urban Heat Island effects during a heatwave event in the city of Melbourne. *Weather and Climate Extremes*, *25*, 100217. Available from https://doi.org/10.1016/j.wace.2019.100217.

Inostroza, L., Palme, M., & De La Barrera, F. (2016). A heat vulnerability index: Spatial patterns of exposure, sensitivity and adaptive capacity for Santiago de Chile. *PLoS One*, *11*, e0162464.

Jendritzky, G., & Tinz, B. (2009). The thermal environment of the human being on the global scale. *Global Health Action*, *2*, 2005. Available from https://doi.org/10.3402/gha.v2i0.2005.

Jeon, S., Son, S., Lee, S. Y., Chae, D., Bae, J. H., Lee, H., et al. (2020). Multifunctional daytime radiative cooling devices with simultaneous light-emitting and radiative cooling functional layers. *ACS Applied Materials & Interfaces*, *12*, 54763−54772. Available from https://doi.org/10.1021/acsami.0c16241.

Jerzy, Z., Łukasz, S., Anna, Z., Kornelia, K., & Maksym, B. (2020). Water retention in nature-based solutions—Assessment of potential economic effects for local social groups. *Water*, 12. Available from https://doi.org/10.3390/w12123347.

Jiang, L., & Wang, S. (2020). Enhancing heat release of asphalt pavement by a gradient heat conduction channel. *Construction and Building Materials*, *230*. Available from https://doi.org/10.1016/j.conbuildmat.2019.117018.

Jiang, W., Xiao, J., Yuan, D., Lu, H., Xu, S., & Huang, Y. (2018). Design and experiment of thermoelectric asphalt pavements with power-generation and temperature-reduction functions. *Energy and Buildings*, *169*, 39−47. Available from https://doi.org/10.1016/J. ENBUILD.2018.03.049.

Jing, W., Zhang, S., Zhang, W., Chen, Z., Zhang, C., Wu, D., et al. (2021). Scalable and flexible electrospun film for daytime subambient radiative cooling. *ACS Applied Materials & Interfaces*. Available from https://doi.org/10.1021/acsami.1c05364.

Jo, H.-K., & McPherson, G. E. (1995). Carbon storage and flux in urban residential greenspace. *Journal of Environmental Management*, *45*, 109−133. Available from https://doi.org/10.1006/jema.1995.0062.

Johnson, D. P., Stanforth, A., Lulla, V., & Luber, G. (2012). Developing an applied extreme heat vulnerability index utilizing socioeconomic and environmental data. *Applied Geography (Sevenoaks, England)*, *35*, 23−31. Available from https://doi.org/10.1016/j. apgeog.2012.04.006.

Kala, J., Kauwe, M. G., Pitman, A. J., Medlyn, B. E., Wang, Y.-P., Lorenz, R., et al. (2016). Impact of the representation of stomatal conductance on model projections of heatwave intensity. *Scientific Reports*, *6*, 23418. Available from https://doi.org/10.1038/srep23418, PM - 26996244.

Kamath, H. G., Singh, M., Magruder, L. A., Yang, Z.-L., & Niyogi, D. (2022). GLOBUS: GLObal building heights for. *Urban Studies*. Available from https://doi.org/10.48550/ARXIV.2205.12224.

Kantamaneni, K. (2019). Evaluation of social vulnerability to natural hazards: A case of Barton on Sea, England. *Arabian Journal of Geosciences*, *12*, 628. Available from https://doi.org/10.1007/s12517-019-4819-9.

Karanja, J., & Kiage, L. (2021). Perspectives on spatial representation of urban heat vulnerability. *The Science of the Total Environment*, *774*, 145634. Available from https://doi.org/10.1016/j.scitotenv.2021.145634.

Karlessi, T., Santamouris, M., Apostolakis, K., Synnefa, A., & Livada, I. (2009). Development and testing of thermochromic coatings for buildings and urban structures. *Solar Energy*, *83*, 538−551. Available from https://doi.org/10.1016/j.solener.2008.10.005.

Katsunori, T., & Kazuya, Y. (2008). Road temperature mitigation effect of road cool, a water-retentive material using blast furnace slag. *JEF GIHO*, *19*, 28−32.

Kawamoto, Y. (2017). Effect of land-use change on the urban heat island in the Fukuoka−Kitakyushu metropolitan area, Japan. *Sustainability*, 9. Available from https://doi.org/10.3390/su9091521.

Kershaw, S. E., & Millward, A. A. (2012). A spatio-temporal index for heat vulnerability assessment. *Environmental Monitoring and Assessment*, *184*, 7329−7342. Available from https://doi.org/10.1007/s10661-011-2502-z.

Khan, A., Carlosena, L., Feng, J., Khorat, S., Khatun, R., Doan, Q.-V., et al. (2022). Optically modulated passive broadband daytime radiative cooling materials can cool cities in summer and heat cities in winter. *Sustainability*, *14*, 1110.

Khan, A., Carlosena, L., Khorat, S., Khatun, R., Doan, Q.-V., Feng, J., et al. (2021). On the winter overcooling penalty of super cool photonic materials in cities. *Solar Energy Advances*, *1*, 100009.

Kikegawa, Y., Genchi, Y., Yoshikado, H., & Kondo, H. (2003). Development of a numerical simulation system toward comprehensive assessments of urban warming countermeasures including their impacts upon the urban buildings' energy-demands. *Applied Energy*, *76*, 449−466. Available from https://doi.org/10.1016/S0306-2619(03)00009-6.

Kim, S., & Ryu, Y. (2015). Describing the spatial patterns of heat vulnerability from urban design perspectives. *International Journal of Sustainable Development & World Ecology*, *22*, 189−200. Available from https://doi.org/10.1080/13504509.2014.1003202.

Kim, S. W., & Brown, R. D. (2021). Urban heat island (UHI) intensity and magnitude estimations: A systematic literature review. *The Science of the Total Environment*, 779. Available from https://doi.org/10.1016/j.scitotenv.2021.146389.

Kolbe, K. (2019). Mitigating urban heat island effect and carbon dioxide emissions through different mobility concepts: Comparison of conventional vehicles with electric vehicles, hydrogen vehicles and public transportation. *Transport Policy*, *80*, 1−11. Available from https://doi.org/10.1016/j.tranpol.2019.05.007.

Kona, A., Bertoldi, P., Monforti-Ferrario, F., Rivas, S., & Dallemand, J. F. (2018). Covenant of mayors signatories leading the way towards 1.5 degree global warming pathway. *Sustainable Cities and Society*, *41*, 568−575. Available from https://doi.org/10.1016/j.scs.2018.05.017.

Kong, F., Yin, H., James, P., Hutyra, L. R., & He, H. S. (2014). Effects of spatial pattern of green-space on urban cooling in a large metropolitan area of eastern China. *Landscape and Urban Planning*, *128*, 35−47. Available from https://doi.org/10.1016/j.landurbplan.2014.04.018.

Kotthaus, S., & Grimmond, C. S. B. (2014). Energy exchange in a dense urban environment − Part I: Temporal variability of long-term observations in central London. *Urban Climate*, *10*, 261−280. Available from https://doi.org/10.1016/j.uclim.2013.10.002.

Kousis, I., Fabiani, C., Gobbi, L., & Pisello, A. L. (2020). Phosphorescent-based pavements for counteracting urban overheating − A proof of concept. *Solar Energy*, *202*, 540−552. Available from https://doi.org/10.1016/j.solener.2020.03.092.

Kousis, I., Manni, M., & Pisello, A. L. (2022). Environmental mobile monitoring of urban microclimates: A review. *Renewable and Sustainable Energy Reviews*, *169*. Available from https://doi.org/10.1016/j.rser.2022.112847.

Kousis, I., Pigliautile, I., & Pisello, A. L. (2021). Intra-urban microclimate investigation in urban heat island through a novel mobile monitoring system. *Scientific Reports*, *11*. Available from https://doi.org/10.1038/s41598-021-88344-y.

Krayenhoff, E. S., Jiang, T., Christen, A., Martilli, A., Oke, T. R., Bailey, B. N., et al. (2020). A multi-layer urban canopy meteorological model with trees (BEP-Tree): Street tree impacts on pedestrian-level climate. *Urban Climate*, *32*, 100590. Available from https://doi.org/10.1016//j.uclim.2020.100590.

Krpo, A., Salamanca, F., Martilli, A., & Clappier, A. (2010). On the impact of anthropogenic heat fluxes on the urban boundary layer: A two-dimensional numerical study. *Boundary-Layer Meteorology*, *136*, 105–127. Available from https://doi.org/10.1007/s10546-010-9491-2.

Kusaka, H., Kondo, H., Kikegawa, Y., & Kimura, F. (2001). A simple single-layer urban canopy model for atmospheric models: Comparison with multi-layer and slab models. *Boundary-Layer Meteorology*, *101*, 329–358. Available from https://doi.org/10.1023/A:1019207923078.

Kuznetsov, S., & Tomitsch, M. (2018). *A study of urban heat: Understanding the challenges and opportunities for addressing wicked problems in HCI. Proc. 2018 CHI Conf. Hum. Factors Comput. Syst* (pp. 1–13). New York, NY, USA: Association for Computing Machinery. Available from https://doi.org/10.1145/3173574.3174137.

Kyriakodis, G.-E., & Santamouris, M. (2018). Using reflective pavements to mitigate urban heat island in warm climates - Results from a large scale urban mitigation project. *Urban Climate*, *24*, 326–339. Available from https://doi.org/10.1016/j.uclim.2017.02.002.

Ladan, T. A., Ibrahim, M. H., Ali, S. S. B. S., & Saputra, A. (2022). A geographical review of urban farming and urban heat island in developing countries. *IOP Conference Series: Earth and Environmental Science*, *986*, 12071. Available from https://doi.org/10.1088/1755-1315/986/1/012071.

Leal Filho, W., Echevarria Icaza, L., Neht, A., Klavins, M., & Morgan, E. A. (2018). Coping with the impacts of urban heat islands. A literature based study on understanding urban heat vulnerability and the need for resilience in cities in a global climate change context. *Journal of Cleaner Production*, *171*, 1140–1149. Available from https://doi.org/10.1016/j.jclepro.2017.10.086.

Lee, J., Jung, Y., Lee, M., Hwang, J. S., Guo, J., Shin, W., et al. (2022). Biomimetic reconstruction of butterfly wing scale nanostructures for radiative cooling and structural coloration. *Nanoscale Horizons*, *7*, 1054–1064. Available from https://doi.org/10.1039/D2NH00166G.

Lee, S.-H., Lee, H., Park, S.-B., Woo, J.-W., Lee, D.-I., & Baik, J.-J. (2016). Impacts of in-canyon vegetation and canyon aspect ratio on the thermal environment of street canyons: Numerical investigation using a coupled WRF-VUCM model. *Quarterly Journal of the Royal Meteorological Society*, *142*, 2562–2578.

Leuzinger, S., Vogt, R., & Körner, C. (2010). Tree surface temperature in an urban environment. *Agricultural and Forest Meteorology*, *150*, 56–62. Available from https://doi.org/10.1016/j.agrformet.2009.08.006.

Levinson, R., Chen, S., Slack, J., Goudey, H., Harima, T., & Berdahl, P. (2020). Design, characterization, and fabrication of solar-retroreflective cool-wall materials. *Solar Energy Materials and Solar Cells*, *206*. Available from https://doi.org/10.1016/j.solmat.2019.110117.

Li, D., & Bou-Zeid, E. (2013). Synergistic interactions between urban heat islands and heat waves: The impact in cities is larger than the sum of its parts. *Journal of Applied Meteorology and Climatology*, *52*, 2051–2064. Available from https://doi.org/10.1175/JAMC-D-13-02.1.

Li, X., Sun, B., Sui, C., Nandi, A., Fang, H., Peng, Y., et al. (2020). Integration of daytime radiative cooling and solar heating for year-round energy saving in buildings. *Nature Communications*, *11*, 1–9.

Li, D., Yuan, J., & Kopp, R. E. (2020). Escalating global exposure to compound heat-humidity extremes with warming. *Environmental Research Letters*, *15*, 64003.

Li, F., Yigitcanlar, T., Nepal, M., Thanh, K. N., & Dur, F. (2022). Understanding urban heat vulnerability assessment methods: A PRISMA review. *Energies*, 15. Available from https://doi.org/10.3390/en15196998.

Lin, Y.-S., Medlyn, B. E., Duursma, R. A., Prentice, I. C., Wang, H., Baig, S., et al. (2015). Optimal stomatal behaviour around the world. *Nature Climate Change*, *5*, 459−464. Available from https://doi.org/10.1038/nclimate2550.

Lin, W., Yu, T., Chang, X., Wu, W., & Zhang, Y. (2015). Calculating cooling extents of green parks using remote sensing: Method and test. *Landscape and Urban Planning*, *134*, 66−75. Available from https://doi.org/10.1016/j.landurbplan.2014.10.012.

Liu, W., Li, Z., Cheng, H., & Chen, S. (2020). Dielectric resonance-based optical metasurfaces: From fundamentals to applications. *IScience*, 23. Available from https://doi.org/10.1016/j.isci.2020.101868.

Liu, Y., Li, T., & Peng, H. (2018). A new structure of permeable pavement for mitigating urban heat island. *The Science of the Total Environment*, *634*, 1119−1125. Available from https://doi.org/10.1016/j.scitotenv.2018.04.041.

Liu, X., Yue, W., Yang, X., Hu, K., Zhang, W., & Huang, M. (2020). Mapping urban heat vulnerability of extreme heat in hangzhou via comparing two approaches. *Complexity*, *2020*, 9717658. Available from https://doi.org/10.1155/2020/9717658.

Luber, G., & McGeehin, M. (2008). Climate change and extreme heat events. *American Journal of Preventive Medicine*, *35*, 429−435. Available from https://doi.org/10.1016/j.amepre.2008.08.021.

Mandal, J., Fu, Y., Overvig, A. C., Jia, M., Sun, K., Shi, N. N., et al. (2018). Hierarchically porous polymer coatings for highly efficient passive daytime radiative cooling. *Science*, *362*, 315−319. Available from https://doi.org/10.1126/science.aat9513.

Manni, M., Cardinali, M., Lobaccaro, G., Goia, F., Nicolini, A., & Rossi, F. (2020). Effects of retro-reflective and angular-selective retro-reflective materials on solar energy in urban canyons. *Solar Energy*, *209*, 662−673. Available from https://doi.org/10.1016/j.solener.2020.08.085.

Maronga, B., Gross, G., Raasch, S., Banzhaf, S., Forkel, R., Heldens, W., et al. (2019). Development of a new urban climate model based on the model PALM-project overview, planned work, and first achievements. *Meteorologische Zeitschrift*, 1−15.

Martilli, A., Clappier, A., & Rotach, M. W. (2002). An urban surface exchange parameterisation for mesoscale models. *Boundary-Layer Meteorology*, *104*, 261−304. Available from https://doi.org/10.1023/A:1016099921195.

Martin, M., Chong, A., Biljecki, F., & Miller, C. (2022). Infrared thermography in the built environment: A multi-scale review. *Renewable and Sustainable Energy Reviews*, *165*, 112540. Available from https://doi.org/10.1016/J.RSER.2022.112540.

Martinez-Juarez, P., Chiabai, A., Suárez, C., & Quiroga, S. (2019). Insights on urban and periurban adaptation strategies based on stakeholders' perceptions on hard and soft responses to climate change. *Sustainability*, 11. Available from https://doi.org/10.3390/su11030647.

Masson, V. (2000). A physically-based scheme for the urban energy budget in atmospheric models. *Boundary-Layer Meteorology*, *94*, 357−397. Available from https://doi.org/10.1023/A:1002463829265.

McMichael, A. J., Woodruff, R. E., & Hales, S. (2006). Climate change and human health: Present and future risks. *Lancet*, *367*, 859−869. Available from https://doi.org/10.1016/S0140-6736(06)68079-3.

Meier, F., Fenner, D., Grassmann, T., Otto, M., & Scherer, D. (2017). Crowdsourcing air temperature from citizen weather stations for urban climate research. *Urban Climate*, *19*, 170−191. Available from https://doi.org/10.1016/j.uclim.2017.01.006.

Meili, N., Manoli, G., Burlando, P., Bou-Zeid, E., Chow, W. T. L., Coutts, A. M., et al. (2020). An urban ecohydrological model to quantify the effect of vegetation on urban climate and hydrology (UT&C v1.0). *Geoscientific Model Development, 13*, 335−362. Available from https://doi.org/10.5194/gmd-13-335-2020.

Melica, G., Treville, A., Franco De Los Rios, C., Baldi, M., Monforti-Ferrario, F., Palermo, V., et al. (2022). Covenant of Mayors: 2021 assessment. *Luxembourg (Luxembourg): Publications Office of the European Union.* Available from https://doi.org/10.2760/58412.

Meng, X., Meng, L., Gao, Y., & Li, H. (2022). A comprehensive review on the spray cooling system employed to improve the summer thermal environment: Application efficiency, impact factors, and performance improvement. *Building and Environment*, 109065.

Miller, V. S., & Bates, G. P. (2007). The thermal work limit is a simple reliable heat index for the protection of workers in thermally stressful environments. *The Annals of Occupational Hygiene, 51*, 553−561. Available from https://doi.org/10.1093/annhyg/mem035.

Milošević, D. D., Bajšanski, I. V., & Savić, S. M. (2017). Influence of changing trees locations on thermal comfort on street parking lot and footways. *Urban Forestry & Urban Greening, 23*, 113−124. Available from https://doi.org/10.1016/j.ufug.2017.03.011.

Min, S., Jeon, S., Yun, K., & Shin, J. (2022). All-color sub-ambient radiative cooling based on photoluminescence. *ACS Photonics, 9*, 1196−1205. Available from https://doi.org/10.1021/acsphotonics.1c01648.

Mokhtari, R., Ulpiani, G., & Ghasempour, R. (2022). The cooling station: Combining hydronic radiant cooling and daytime radiative cooling for urban shelters. *Applied Thermal Engineering, 211*, 118493. Available from https://doi.org/10.1016/j.applthermaleng.2022.118493.

Morakinyo, T. E., Kong, L., Lau, K. K.-L., Yuan, C., & Ng, E. (2017). A study on the impact of shadow-cast and tree species on in-canyon and neighborhood's thermal comfort. *Building and Environment, 115*, 1−17. Available from https://doi.org/10.1016/j.buildenv.2017.01.005.

Morakinyo, T. E., Lai, A., Lau, K. K.-L., & Ng, E. (2019). Thermal benefits of vertical greening in a high-density city: Case study of Hong Kong. *Urban Forestry & Urban Greening, 37*, 42−55. Available from https://doi.org/10.1016/j.ufug.2017.11.010.

Morakinyo, T. E., & Lam, Y. F. (2016). Simulation study on the impact of tree-configuration, planting pattern and wind condition on street-canyon's micro-climate and thermal comfort. *Building and Environment, 103*, 262−275. Available from https://doi.org/10.1016/j.buildenv.2016.04.025.

Morakinyo, T. E., Ouyang, W., Lau, K. K.-L., Ren, C., & Ng, E. (2020). Right tree, right place (urban canyon): Tree species selection approach for optimum urban heat mitigation - development and evaluation. *The Science of the Total Environment, 719*, 137461. Available from https://doi.org/10.1016/j.scitotenv.2020.137461, PM - 32114235.

Mourou, C., Zamorano, M., Ruiz, D. P., & Martín-Morales, M. (2022). Cool surface strategies with an emphasis on the materials. *dimension: A review. Applied Sciences, 12*, 1893.

Mukherjee, S., & Thilagar, P. (2015). Recent advances in purely organic phosphorescent materials. *Chemical Communications, 51*, 10988−11003. Available from https://doi.org/10.1039/c5cc03114a.

Mussetti, G. (2019). Urban climate modelling with explicit representation of street trees. *ETH Zurich*.

Nag, P. K., Dutta, P., Nag, A., & Kjellstrom, T. (2013). Extreme heat events: Perceived thermal response of indoor and outdoor workers. *International Journal of Current Research and Review, 5*, 65.

Nagase, A., & Dunnett, N. (2010). Drought tolerance in different vegetation types for extensive green roofs: Effects of watering and diversity. *Landscape and Urban Planning, 97*, 318−327. Available from https://doi.org/10.1016/j.landurbplan.2010.07.005.

Nakayama, T., & Fujita, T. (2010). Cooling effect of water-holding pavements made of new materials on water and heat budgets in urban areas. *Landscape and Urban Planning, 96,* 57−67. Available from https://doi.org/10.1016/j.landurbplan.2010.02.003.

Napoly, A., Grassmann, T., Meier, F., & Fenner, D. (2018). Development and application of a statistically-based quality control for crowdsourced air temperature data. *Frontiers in Earth Science, 6,* 118. Available from https://doi.org/10.3389/FEART.2018.00118/BIBTEX.

Nath, B., Ni-Meister, W., & Özdoğan, M. (2021). Fine-scale urban heat patterns in New York city measured by ASTER satellite—The role of complex spatial structures. *Remote Sensing, 13.* Available from https://doi.org/10.3390/rs13193797.

Nayak, S. G., Shrestha, S., Kinney, P. L., Ross, Z., Sheridan, S. C., Pantea, C. I., et al. (2018). Development of a heat vulnerability index for New York State. *Public Health, 161,* 127−137. Available from https://doi.org/10.1016/j.puhe.2017.09.006.

Nemirovsky, E. M., Welker, A. L., & Lee, R. (2013). Quantifying evaporation from pervious concrete systems: Methodology and hydrologic perspective. *Journal of Irrigation and Drainage Engineering, 139,* 271−277. Available from https://doi.org/10.1061/(asce)ir.1943-4774.0000541.

Northmore, A. B., & Tighe, S. L. (2016). Performance modelling of a solar road panel prototype using finite element analysis. *International Journal of Pavement Engineering, 17,* 449−457. Available from https://doi.org/10.1080/10298436.2014.993203.

Núñez Peiró, M., Sánchez-Guevara Sánchez, C., & Neila González, F. J. (2019). Source area definition for local climate zones studies. A systematic review. *Building and Environment, 148,* 258−285. Available from https://doi.org/10.1016/j.buildenv.2018.10.050.

Oke, T. R. (1988). The urban energy balance. *Progress in Physical Geography: Earth and Environment, 12,* 471−508. Available from https://doi.org/10.1177/030913338801200401.

Oke, T. R., Mills, G., Christen, A., & Voogt, J. A. (2017). *Urban climates.* Cambridge University Press.

Oliveira, S., Andrade, H., & Vaz, T. (2011). The cooling effect of green spaces as a contribution to the mitigation of urban heat: A case study in Lisbon. *Building and Environment, 46,* 2186−2194. Available from https://doi.org/10.1016/j.buildenv.2011.04.034.

Pardyjak, E. R., & Brown, M. (2003). QUIC-URB v. 1.1: Theory and user's guide. *Los Alamos Natl Lab Los Alamos, NM.*

Parsons, K. (2014). *Human thermal environments: The effects of hot, moderate, and cold environments on human health, comfort, and performance.* CRC press.

Pauleit, S., Zölch, T., Erlwein, S., Reischl, A., Rahman, M., Pretzsch, H., et al. (2021). Urban green infrastructures for climate change adaptation: A multiscale approach. In V. Costanzo, G. Evola, & L. Marletta (Eds.), *Urban heat stress mitig. Solut* (pp. 301−321). London: Routledge, TS-CrossRef. Available from https://doi.org/10.1201/9781003045922-15-18.

Peretti, C., Zarrella, A., De Carli, M., & Zecchin, R. (2013). The design and environmental evaluation of earth-to-air heat exchangers (EAHE). A literature review. *Renewable and Sustainable Energy Reviews, 28,* 107−116. Available from https://doi.org/10.1016/j.rser.2013.07.057.

Pérez, G., Coma, J., Martorell, I., & Cabeza, L. F. (2014). Vertical greenery systems (VGS) for energy saving in buildings: A review. *Renewable and Sustainable Energy Reviews, 39,* 139−165. Available from https://doi.org/10.1016/j.rser.2014.07.055.

Pigliautile, I., D'Eramo, S., & Pisello, A. L. (2021). Intra-urban microclimate mapping for citizens' wellbeing: Novel wearable sensing techniques and automatized data-processing. *Journal of Cleaner Production, 279.* Available from https://doi.org/10.1016/j.jclepro.2020.123748.

Pigliautile, I., & Pisello, A. L. (2020). Environmental data clustering analysis through wearable sensing techniques: New bottom-up process aimed to identify intra-urban granular morphologies from pedestrian transects. *Building and Environment*, 171. Available from https://doi.org/10.1016/j.buildenv.2019.106641.

Pioppi, B., Pigliautile, I., Piselli, C., & Pisello, A. L. (2020). Cultural heritage microclimate change: Human-centric approach to experimentally investigate intra-urban overheating and numerically assess foreseen future scenarios impact. *The Science of the Total Environment*, 703. Available from https://doi.org/10.1016/j.scitotenv.2019.134448.

Pisello, A. L., Castaldo, V. L., Piselli, C., Fabiani, C., & Cotana, F. (2017). Thermal performance of coupled cool roof and cool façade: Experimental monitoring and analytical optimization procedure. *Energy and Buildings*, *157*, 35−52. Available from https://doi.org/10.1016/j.enbuild.2017.04.054.

Potgieter, J., Nazarian, N., Lipson, M. J., Hart, M. A., Ulpiani, G., Morrison, W., et al. (2021). Combining high-resolution land use data with crowdsourced air temperature to investigate intra-urban microclimate. *Frontiers in Environmental Science*, *9*, 385. Available from https://doi.org/10.3389/FENVS.2021.720323/BIBTEX.

Qi, J., Ding, L., & Lim, S. (2022). A decision-making framework to support urban heat mitigation by local governments. *Resources, Conservation and Recycling*, *184*, 106420. Available from https://doi.org/10.1016/j.resconrec.2022.106420.

Qi, J., Ding, L., & Lim, S. (2020). Planning for cooler cities: A framework to support the selection of urban heat mitigation techniques. *Journal of Cleaner Production*, *275*, 122903.

Qi, Y., Xiang, B., & Zhang, J. (2017). Effect of titanium dioxide (TiO_2) with different crystal forms and surface modifications on cooling property and surface wettability of cool roofing materials. *Solar Energy Materials and Solar Cells*, *172*, 34−43.

Qin, Y. (2015). A review on the development of cool pavements to mitigate urban heat island effect. *Renewable and Sustainable Energy Reviews*. Available from https://doi.org/10.1016/j.rser.2015.07.177.

Quaglia, S. Climate crisis study finds heatwaves have cost global economy $16tn. Guard 2022:. https://www.theguardian.com/environment/2022/oct/28/climate-crisis-heatwaves-cost-global-economy#:~:text = Heatwaves brought on by human, human health and other areas. Accessed 18.03.23.

Raman, A. P., Anoma, M. A., Zhu, L., Rephaeli, E., & Fan, S. (2014). Passive radiative cooling below ambient air temperature under direct sunlight. *Nature*, *515*, 540−544. Available from https://doi.org/10.1038/nature13883.

Rawat, M., & Singh, R. N. (2021). A study on the comparative review of cool roof thermal performance in various regions. *Energy and Built Environment*.

Reid, C. E., O'neill, M. S., Gronlund, C. J., Brines, S. J., Brown, D. G., Diez-Roux, A. V., et al. (2009). Mapping community determinants of heat vulnerability. *Environmental Health Perspectives*, *117*, 1730−1736.

Rinner, C., Patychuk, D., Bassil, K., Nasr, S., Gower, S., & Campbell, M. (2010). The role of maps in neighborhood-level heat vulnerability assessment for the city of Toronto. *Cartography and Geographic Information Science*, *37*, 31−44. Available from https://doi.org/10.1559/152304010790588089.

Rivas, S., Urraca, R., & Bertoldi, P. (2022). Covenant of mayors 2020 achievements: A two-speed climate action process. *Sustainability*, *14*. Available from https://doi.org/10.3390/su142215081.

Rodríguez Algeciras, J. A., Coch, H., De la Paz Pérez, G., Chaos Yeras, M., & Matzarakis, A. (2016). Human thermal comfort conditions and urban planning in hot-humid climates—The case of Cuba. *International Journal of Biometeorology*, *60*, 1151−1164. Available from https://doi.org/10.1007/s00484-015-1109-4.

Rodriguez, L. R., Ramos, J. S., de la Flor, F. J. S., & Dominguez, S. A. (2020). Analyzing the urban heat Island: Comprehensive methodology for data gathering and optimal design of mobile transects. *Sustainable Cities and Society*, 55.

Rogers, A., Medlyn, B. E., Dukes, J. S., Bonan, G., Caemmerer, S., Dietze, M. C., et al. (2017). A roadmap for improving the representation of photosynthesis in Earth system models. *The New Phytologist*, *213*, 22−42. Available from https://doi.org/10.1111/nph.14283, PM - 27891647.

Romero-Lankao, P., Qin, H., & Dickinson, K. (2012). Urban vulnerability to temperature-related hazards: A meta-analysis and meta-knowledge approach. *Global Environmental Change*, *22*, 670−683. Available from https://doi.org/10.1016/j.gloenvcha.2012.04.002.

Rotach, M. W. (1993). Turbulence close to a rough urban surface part I: Reynolds stress. *Boundary-Layer Meteorology*, *65*, 1−28. Available from https://doi.org/10.1007/BF00708816.

Roxon, J., Ulm, F.-J., & Pellenq, R. J.-M. (2020). Urban heat island impact on state residential energy cost and CO_2 emissions in the United States. *Urban Climate*, *31*, 100546. Available from https://doi.org/10.1016/j.uclim.2019.100546.

Ryms, M., Lewandowski, W. M., Klugmann-Radziemska, E., Denda, H., & Wcisło, P. (2015). The use of lightweight aggregate saturated with PCM as a temperature stabilizing material for road surfaces. *Applied Thermal Engineering*, *81*, 313−324. Available from https://doi.org/10.1016/j.applthermaleng.2015.02.036.

Saaroni, H., Amorim, J. H., Hiemstra, J. A., & Pearlmutter, D. (2018). Urban green infrastructure as a tool for urban heat mitigation: Survey of research methodologies and findings across different climatic regions. *Urban Climate*, *24*, 94−110. Available from https://doi.org/10.1016/j.uclim.2018.02.001.

Salamanca, F., Krpo, A., Martilli, A., & Clappier, A. (2010). A new building energy model coupled with an urban canopy parameterization for urban climate simulations—Part I. formulation, verification, and sensitivity analysis of the model. *Theoretical and Applied Climatology*, *99*, 331−344.

Santamouris, M. (2013). Using cool pavements as a mitigation strategy to fight urban heat island - A review of the actual developments. *Renewable and Sustainable Energy Reviews*. Available from https://doi.org/10.1016/j.rser.2013.05.047.

Santamouris, M., Ban-Weiss, G., Osmond, P., Paolini, R., Synnefa, A., Cartalis, C., et al. (2018). Progress in urban greenery mitigation science−assessment methodologies advanced technologies and impact on cities. *Journal of Civil Engineering and Management*, *24*, 638−671.

Santamouris, M., & Feng, J. (2018). Recent progress in daytime radiative cooling: Is it the air conditioner of the future. *Buildings*, *8*. Available from https://doi.org/10.3390/buildings8120168.

Santamouris, M., & Osmond, P. (2020). Increasing green infrastructure in cities: Impact on ambient temperature, air quality and heat-related mortality and morbidity. *Buildings*, *10*, 233.

Santamouris, M., Paolini, R., Haddad, S., Synnefa, A., Garshasbi, S., Hatvani-Kovacs, G., et al. (2020). Heat mitigation technologies can improve sustainability in cities. An holistic experimental and numerical impact assessment of urban overheating and related heat mitigation strategies on energy consumption, indoor comfort, vulnerability and heat-related m. *Energy and Buildings*, *217*, 110002. Available from https://doi.org/10.1016/j.enbuild.2020.110002.

Santamouris, M., & Yun, G. Y. (2020). Recent development and research priorities on cool and super cool materials to mitigate urban heat island. *Renewable Energy*, *161*, 792−807. Available from https://doi.org/10.1016/j.renene.2020.07.109.

Santiago, J.-L., Buccolieri, R., Rivas, E., Calvete-Sogo, H., Sanchez, B., Martilli, A., et al. (2019). CFD modelling of vegetation barrier effects on the reduction of traffic-related pollutant concentration in an avenue of Pamplona, Spain. *Sustainable Cities and Society*, *48*, 101559. Available from https://doi.org/10.1016/j.scs.2019.101559.

Santiago, J.-L., Martilli, A., & Martin, F. (2017). On dry deposition modelling of atmospheric pollutants on vegetation at the microscale: Application to the impact of street vegetation on air quality. *Boundary-Layer Meteorology*, *162*, 451−474. Available from https://doi.org/10.1007/s10546-016-0210-5.

Sauer, J. A., & Muñoz-Esparza, D. (2020). The FastEddy®resident-GPU accelerated large-eddy simulation framework: Model formulation, dynamical-core validation and performance benchmarks. *Journal of Advances in Modeling Earth Systems*, *12*, e2020MS002100.

Savva, Y., Denneler, B., Koubaa, A., Tremblay, F., Bergeron, Y., & Tjoelker, M. G. (2007). Seed transfer and climate change effects on radial growth of jack pine populations in a common garden in Petawawa, Ontario, Canada. *Forest Ecology and Management*, *242*, 636−647. Available from https://doi.org/10.1016/j.foreco.2007.01.073.

Shahidan, M. F., Jones, P. J., Gwilliam, J., & Salleh, E. (2012). An evaluation of outdoor and building environment cooling achieved through combination modification of trees with ground materials. *Building and Environment*, *58*, 245−257. Available from https://doi.org/10.1016/j.buildenv.2012.07.012.

Shao, H., & Kim, G. (2022). A comprehensive review of different types of green infrastructure to mitigate urban heat islands: Progress, functions, and benefits. *Land*, *11*, 1792. Available from https://doi.org/10.3390/land11101792.

Shaopeng, W., Mingyu, C., & Jizhe, Z. (2011). Laboratory investigation into thermal response of asphalt pavements as solar collector by application of small-scale slabs. *Applied Thermal Engineering*, *31*, 1582−1587. Available from https://doi.org/10.1016/j.applthermaleng.2011.01.028.

Sharifi, A. (2021). Co-benefits and synergies between urban climate change mitigation and adaptation measures: A literature review. *The Science of the Total Environment*, *750*, 141642. Available from https://doi.org/10.1016/j.scitotenv.2020.141642.

Sharifi, N.P., & Sakulich, A. (2009). Application of phase change materials in structures and pavements, 82−88.

Shen, H., Jiang, Y., Li, T., Cheng, Q., Zeng, C., & Zhang, L. (2020). Deep learning-based air temperature mapping by fusing remote sensing, station, simulation and socioeconomic data. *Remote Sensing of Environment*, *240*, 111692. Available from https://doi.org/10.1016/J.RSE.2020.111692.

Sheng, C., An, Y., Du, J., & Li, X. (2019). Colored radiative cooler under optical tamm resonance. *ACS Photonics*, *6*, 2545−2552. Available from https://doi.org/10.1021/acsphotonics.9b01005.

Shih, W.-Y., & Mabon, L. (2021). Understanding heat vulnerability in the subtropics: Insights from expert judgements. *International Journal of Disaster Risk Reduction*, *63*, 102463. Available from https://doi.org/10.1016/j.ijdrr.2021.102463.

Shokry, G., Anguelovski, I., Connolly, J. J. T., Maroko, A., & Pearsall, H. (2022). "They didn't see it coming": Green resilience planning and vulnerability to future climate gentrification. *Housing Policy Debate*, *32*, 211−245. Available from https://doi.org/10.1080/10511482.2021.1944269.

Sillmann, J., & Roeckner, E. (2008). Indices for extreme events in projections of anthropogenic climate change. *Climatic Change*, *86*, 83−104. Available from https://doi.org/10.1007/s10584-007-9308-6.

Skoulika, F., Santamouris, M., Kolokotsa, D., & Boemi, N. (2014). On the thermal characteristics and the mitigation potential of a medium size urban park in Athens, Greece. *Landscape and Urban Planning*, *123*, 73−86. Available from https://doi.org/10.1016/j.landurbplan.2013.11.002.

Smith, J. P., Meerow, S., & Turner, B. L. (2021). Planning urban community gardens strategically through multicriteria decision analysis. *Urban Forestry & Urban Greening*, *58*, 126897. Available from https://doi.org/10.1016/j.ufug.2020.126897.

Son, S., Jeon, S., Chae, D., Lee, S. Y., Liu, Y., Lim, H., et al. (2021). Colored emitters with silica-embedded perovskite nanocrystals for efficient daytime radiative cooling. *Nano Energy*, *79*. Available from https://doi.org/10.1016/j.nanoen.2020.105461.

Song, Q., Tran, T., Herrmann, K., Lauster, T., Breitenbach, M., & Retsch, M. (2022). A tailored indoor setup for reproducible passive daytime cooling characterization. *Cell Reports Physical Science*, *3*, 100986. Available from https://doi.org/10.1016/j.xcrp.2022.100986.

Spronken-Smith, R. A., & Oke, T. R. (1998). The thermal regime of urban parks in two cities with different summer climates. *International Journal of Remote Sensing*, *19*, 2085−2104. Available from https://doi.org/10.1080/014311698214884.

Stewart, I. D. (2011). A systematic review and scientific critique of methodology in modern urban heat island literature. *International Journal of Climatology*, *31*, 200−217. Available from https://doi.org/10.1002/joc.2141.

Stewart, I. D., & Oke, T. R. (2012). Local climate zones for urban temperature studies. *Bulletin of the American Meteorological Society*, *93*, 1879−1900. Available from https://doi.org/10.1175/BAMS-D-11-00019.1.

Sun, Y., Ji, Y., Javed, M., Li, X., Fan, Z., Wang, Y., et al. (2022). Preparation of passive daytime cooling fabric with the synergistic effect of radiative cooling and evaporative cooling. *Advanced Materials Technologies*, *7*, 2100803.

Sun, S., Xu, X., Lao, Z., Liu, W., Li, Z., Higueras García, E., et al. (2017). Evaluating the impact of urban green space and landscape design parameters on thermal comfort in hot summer by numerical simulation. *Building and Environment*, *123*, 277−288. Available from https://doi.org/10.1016/j.buildenv.2017.07.010.

Taha, H. (1997). Modeling the impacts of large-scale albedo changes on ozone air quality in the South Coast Air Basin. *Atmospheric Environment (Oxford, England: 1994)*, *31*, 1667−1676. Available from https://doi.org/10.1016/S1352-2310(96)00336-6.

Taha, H., Levinson, R., Mohegh, A., Gilbert, H., Ban-Weiss, G., & Chen, S. (2018). Air-temperature response to neighborhood-scale variations in albedo and canopy cover in the real world: Fine-resolution meteorological modeling and mobile temperature observations in the Los Angeles climate archipelago. *Climate*, *6*, 53. Available from https://doi.org/10.3390/cli6020053.

Takebayashi, H. (2017). Influence of urban green area on air temperature of surrounding built-up area. *Climate*, *5*, 60. Available from https://doi.org/10.3390/cli5030060.

Taleghani, M., Sailor, D. J., Tenpierik, M., & van den Dobbelsteen, A. (2014). Thermal assessment of heat mitigation strategies: The case of Portland State University, Oregon, USA. *Building and Environment*, *73*, 138−150. Available from https://doi.org/10.1016/J.BUILDENV.2013.12.006.

Tan, Z., Lau, K. K.-L., & Ng, E. (2016). Urban tree design approaches for mitigating daytime urban heat island effects in a high-density urban environment. *Energy and Buildings*, *114*, 265−274. Available from https://doi.org/10.1016/j.enbuild.2015.06.031.

Tapia, C., Abajo, B., Feliu, E., Mendizabal, M., Martinez, J. A., Fernández, J. G., et al. (2017). Profiling urban vulnerabilities to climate change: An indicator-based vulnerability assessment for European cities. *Ecological Indicators*, *78*, 142−155. Available from https://doi.org/10.1016/j.ecolind.2017.02.040.

Teixeira, C. P., Fernandes, C. O., & Ahern, J. (2022). Adaptive planting design and management framework for urban climate change adaptation and mitigation. *Urban Forestry & Urban Greening*, *70*, 127548. Available from https://doi.org/10.1016/j.ufug.2022.127548.

Thom, J. K., Coutts, A. M., Broadbent, A. M., & Tapper, N. J. (2016). The influence of increasing tree cover on mean radiant temperature across a mixed development suburb in Adelaide, Australia. *Urban Forestry & Urban Greening, 20*, 233−242. Available from https://doi.org/10.1016/j.ufug.2016.08.016.

Thornton, P. K., Ericksen, P. J., Herrero, M., & Challinor, A. J. (2014). Climate variability and vulnerability to climate change: A review. *Global Change Biology, 20*, 3313−3328.

Touati, N., Gardes, T., & Hidalgo, J. (2020). A GIS plugin to model the near surface air temperature from urban meteorological networks. *Urban Climate, 34*. Available from https://doi.org/10.1016/j.uclim.2020.100692.

Tsoka, S., Tsikaloudaki, A., & Theodosiou, T. (2018). Analyzing the ENVI-met microclimate model's performance and assessing cool materials and urban vegetation applications−A review. *Sustainable Cities and Society, 43*, 55−76. Available from https://doi.org/10.1016/j.scs.2018.08.009.

Ulpiani, G. (2019). Water mist spray for outdoor cooling: A systematic review of technologies, methods and impacts. *Applied Energy, 254*, 113647. Available from https://doi.org/10.1016/j.apenergy.2019.113647.

Ulpiani, G. (2021). On the linkage between urban heat island and urban pollution island: Three-decade literature review towards a conceptual framework. *The Science of the Total Environment, 751*, 141727. Available from https://doi.org/10.1016/j.scitotenv.2020.141727.

Ulpiani, G., Hart, M. A., Di Virgilio, G., Maharaj, A. M., Lipson, M. J., & Potgieter, J. (2022). A citizen centred urban network for weather and air quality in Australian schools. *Science Data, 9*, 129. Available from https://doi.org/10.1038/s41597-022-01205-9.

Ulpiani, G., Ranzi, G., Feng, J., & Santamouris, M. (2021). Expanding the applicability of daytime radiative cooling: Technological developments and limitations. *Energy and Buildings, 243*, 110990. Available from https://doi.org/10.1016/j.enbuild.2021.110990.

Ulpiani, G., Ranzi, G., & Santamouris, M. (2020). Experimental evidence of the multiple microclimatic impacts of bushfires in affected urban areas: The case of Sydney during the 2019/2020 Australian season. *Environmental Research Communications, 2*, 65005. Available from https://doi.org/10.1088/2515-7620/ab9e1a.

Ulpiani, G., Ranzi, G., Shah, K. W., Feng, J., & Santamouris, M. (2020). On the energy modulation of daytime radiative coolers: A review on infrared emissivity dynamic switch against overcooling. *Solar Energy, 209*, 278−301. Available from https://doi.org/10.1016/j.solener.2020.08.077.

Urban, J., Ingwers, M. W., McGuire, M. A., & Teskey, R. O. (2017). Increase in leaf temperature opens stomata and decouples net photosynthesis from stomatal conductance in Pinus taeda and Populus deltoides x nigra. *Journal of Experimental Botany, 68*, 1757−1767. Available from https://doi.org/10.1093/jxb/erx052, PM - 28338959.

Vahmani, P., & Ban-Weiss, G. (2016). Climatic consequences of adopting drought-tolerant vegetation over Los Angeles as a response to California drought. *Geophysical Research Letters, 43*, 8240−8249. Available from https://doi.org/10.1002/2016GL069658.

Van Renterghem, T., Forssén, J., Attenborough, K., Jean, P., Defrance, J., Hornikx, M., et al. (2015). Using natural means to reduce surface transport noise during propagation outdoors. *Applied Acoustics, 92*, 86−101.

Vanos, J. K., Warland, J. S., Gillespie, T. J., & Kenny, N. A. (2010). Review of the physiology of human thermal comfort while exercising in urban landscapes and implications for bioclimatic design. *International Journal of Biometeorology, 54*, 319−334. Available from https://doi.org/10.1007/s00484-010-0301-9.

Vaz Monteiro, M., Doick, K. J., Handley, P., & Peace, A. (2016). The impact of greenspace size on the extent of local nocturnal air temperature cooling in London. *Urban Forestry & Urban Greening, 16*, 160−169. Available from https://doi.org/10.1016/j.ufug.2016.02.008.

Venter, Z. S., Chakraborty, T., & Lee, X. (2021). Crowdsourced air temperatures contrast satellite measures of the urban heat island and its mechanisms. *Science Advances*, 7, eabb9569.

Vetters, N., Ulpiani, G., Della Valle, N., Treville, A., Palermo, V., & Melica, G., et al. (2021). European Missions − 100 climate-neutral - and smart cities by 2030 − Info Kit for Cities.

Vincent, K., & Cull, T. (2010). A household social vulnerability index (HSVI) for evaluating adaptation projects in developing countries. *PEGNet Conference*, 2−3.

Vinocur, B., & Altman, A. (2005). Recent advances in engineering plant tolerance to abiotic stress: Achievements and limitations. *Current Opinion in Biotechnology*, 16, 123−132. Available from https://doi.org/10.1016/j.copbio.2005.02.001.

Viswanatha, V., & Rajaramb, C. (2018). priyad SRF| DB. Brief review of liquid crystals. *International Journal of Trend in Scientific Research and Development*, 2, 956−961. Available from https://doi.org/10.31142/ijtsrd18770.

Vo, H. V., Park, D.-W., Seo, W.-J., & Yoo, B.-S. (2017). Evaluation of asphalt mixture modified with graphite and carbon fibers for winter adaptation: Thermal conductivity improvement. *Journal of Materials in Civil Engineering*, 29. Available from https://doi.org/10.1061/(asce)mt.1943-5533.0001675.

Voelkel, J., Hellman, D., Sakuma, R., & Shandas, V. (2018). Assessing vulnerability to urban heat: A study of disproportionate heat exposure and access to refuge by socio-demographic status in Portland, Oregon. *International Journal of Environmental Research and Public Health*, 15. Available from https://doi.org/10.3390/ijerph15040640.

Vujovic, S., Haddad, B., Karaky, H., Sebaibi, N., & Boutouil, M. (2021). Urban heat island: Causes, consequences, and mitigation measures with emphasis on reflective and permeable pavements. *CivilEng*, 2, 459−484.

Wahid, A., Gelani, S., Ashraf, M., & Foolad, M. (2007). Heat tolerance in plants: An overview. *Environmental and Experimental Botany*, 61, 199−223. Available from https://doi.org/10.1016/j.envexpbot.2007.05.011.

Wang, S., Jiang, T., Meng, Y., Yang, R., Tan, G., & Long, Y. (2021). Scalable thermochromic smart windows with passive radiative cooling regulation. *Science*, 374, 1501−1504. Available from https://doi.org/10.1126/science.abg0291.

Wang, X., Liu, X., Li, Z., Zhang, H., Yang, Z., Zhou, H., et al. (2020). Scalable flexible hybrid membranes with photonic structures for daytime radiative cooling. *Advanced Functional Materials*, 30. Available from https://doi.org/10.1002/adfm.201907562.

Wang, J., Liu, S., Meng, X., Gao, W., & Yuan, J. (2021). Application of retro-reflective materials in urban buildings: A comprehensive review. *Energy and Buildings*, 247, 111137.

Wang, J., Meng, Q., Zou, Y., Qi, Q., Tan, K., Santamouris, M., et al. (2022). Performance synergism of pervious pavement on stormwater management and urban heat island mitigation: A review of its benefits, key parameters, and co-benefits approach. *Water Research*, 221, 118755. Available from https://doi.org/10.1016/j.watres.2022.118755.

Wang, T., Wu, Y., Shi, L., Hu, X., Chen, M., & Wu, L. (2021). A structural polymer for highly efficient all-day passive radiative cooling. *Nature Communications*, 12, 365. Available from https://doi.org/10.1038/s41467-020-20646-7.

Wang, Z., Xie, Y., Mu, M., Feng, L., Xie, N., & Cui, N. (2022). Materials to mitigate the urban heat island effect for cool pavement: A brief review. *Buildings*, 12. Available from https://doi.org/10.3390/buildings12081221.

Wang, X., Zhang, Q., Wang, S., Jin, C., Zhu, B., Su, Y., et al. (2022). Sub-ambient full-color passive radiative cooling under sunlight based on efficient quantum-dot photoluminescence. *Science Bulletin*, 67, 1874−1881. Available from https://doi.org/10.1016/j.scib.2022.08.028.

Warren, E. L., Young, D. T., Chapman, L., Muller, C., Grimmond, C. S. B., & Cai, X. M. (2016). The birmingham urban climate laboratory-a high density, urban meteorological dataset, from 2012-2014. *Science Data*, *3*, 1−8. Available from https://doi.org/10.1038/sdata.2016.38.

Whiteoak, K., & Saigar, J. (2019). Estimating the economic benefits of urban heat island mitigation—Economic analysis. *Melbourne, Aust Coop Res Cent Water Sensitive Cities*.

Wilhelmi, O. V., & Hayden, M. H. (2010). Connecting people and place: A new framework for reducing urban vulnerability to extreme heat. *Environmental Research Letters*, *5*, 14021. Available from https://doi.org/10.1088/1748-9326/5/1/014021.

Williams, S., Nitschke, M., Sullivan, T., Tucker, G. R., Weinstein, P., Pisaniello, D. L., et al. (2012). Heat and health in Adelaide, South Australia: Assessment of heat thresholds and temperature relationships. *The Science of the Total Environment*, *414*, 126−133. Available from https://doi.org/10.1016/j.scitotenv.2011.11.038.

Wolf, T., & McGregor, G. (2013). The development of a heat wave vulnerability index for London, United Kingdom. *Weather and Climate Extremes*, *1*, 59−68. Available from https://doi.org/10.1016/j.wace.2013.07.004.

Wong, N. H., Kwang Tan, A. Y., Chen, Y., Sekar, K., Tan, P. Y., Chan, D., et al. (2010). Thermal evaluation of vertical greenery systems for building walls. *Building and Environment*, *45*, 663−672. Available from https://doi.org/10.1016/j.buildenv.2009.08.005.

Wong, N. H., Tan, C. L., Kolokotsa, D. D., & Takebayashi, H. (2021). Greenery as a mitigation and adaptation strategy to urban heat. *Nature Reviews Earth & Environment*, *2*, 166−181.

Xie, N., Akin, M., & Shi, X. (2019). Permeable concrete pavements: A review of environmental benefits and durability. *Journal of Cleaner Production*, *210*, 1605−1621. Available from https://doi.org/10.1016/j.jclepro.2018.11.134.

Xue, X., Qiu, M., Li, Y., Zhang, Q. M., Li, S., Yang, Z., et al. (2020). Creating an eco-friendly building coating with smart subambient radiative cooling. *Advanced Materials*, *32*, 1−8. Available from https://doi.org/10.1002/adma.201906751.

Yamagata, H., Nasu, M., Yoshizawa, M., Miyamoto, A., & Minamiyama, M. (2008). Heat island mitigation using water retentive pavement sprinkled with reclaimed wastewater. *Water Science and Technology: A Journal of the International Association on Water Pollution Research*, *57*, 763−771. Available from https://doi.org/10.2166/wst.2008.187.

Yan, H., Wang, K., Lin, T., Zhang, G., Sun, C., Hu, X., et al. (2021). The challenge of the urban compact form: Three-dimensional index construction and urban land surface temperature impacts. *Remote Sensing*, *13*, 1067.

Yang, L., Liu, X., & Qian, F. (2020). Research on water thermal effect on surrounding environment in summer. *Energy and Buildings*, *207*, 109613.

Yang, X., Peng, L. L. H., Jiang, Z., Chen, Y., Yao, L., He, Y., et al. (2020). Impact of urban heat island on energy demand in buildings: Local climate zones in Nanjing. *Applied Energy*, *260*. Available from https://doi.org/10.1016/j.apenergy.2019.114279.

Yang, J., Wang, Z.-H., Chen, F., Miao, S., Tewari, M., Voogt, J. A., et al. (2015). Enhancing hydrologic modelling in the coupled weather research and forecasting−urban modelling system. *Boundary-Layer Meteorology*, *155*, 87−109. Available from https://doi.org/10.1007/s10546-014-9991-6.

Yang, J., Yang, Y., Sun, D., Jin, C., & Xiao, X. (2021). Influence of urban morphological characteristics on thermal environment. *Sustainable Cities and Society*, *72*. Available from https://doi.org/10.1016/j.scs.2021.103045.

Yang, Z., & Zhang, J. (2021). Bioinspired radiative cooling structure with randomly stacked fibers for efficient all-day passive cooling. *ACS Applied Materials & Interfaces*, *13*, 43387−43395.

Yang, C., Zhu, W., Sun, J., Xu, X., Wang, R., Lu, Y., et al. (2021). Assessing the effects of 2D/3D urban morphology on the 3D urban thermal environment by using multi-source remote sensing data and UAV measurements: A case study of the snow-climate city of Changchun, China. *Journal of Cleaner Production, 321*, 128956. Available from https://doi.org/10.1016/j.jclepro.2021.128956.

Yao, Y., Luo, Y., Huang, J., & Zhao, Z. (2013). Comparison of monthly temperature extremes simulated by CMIP3 and CMIP5 models. *Journa of Climate n.d., 26*, 7692−7707. Available from https://doi.org/10.1175/JCLI-D-12-00560.1.

Yasmeen, S., & Liu, H. (2019). Evaluation of thermal comfort and heat stress indices in different countries and regions−A review. *IOP Conference Series: Materials Science and Engineering, 609*, 52037.

Yenneti, K., Ding, L., Prasad, D., Ulpiani, G., Paolini, R., Haddad, S., et al. (2020). Urban overheating and cooling potential in Australia: An evidence-based review. *Climate, 8*. Available from https://doi.org/10.3390/cli8110126.

Yenneti, K., Tripathi, S., Wei, Y. D., Chen, W., & Joshi, G. (2016). The truly disadvantaged? Assessing social vulnerability to climate change in urban India. *Habitat International, 56*, 124−135. Available from https://doi.org/10.1016/j.habitatint.2016.05.001.

Yinfei, D., Qin, S., & Shengyue, W. (2014). Highly oriented heat-induced structure of asphalt pavement for reducing pavement temperature. *Energy and Buildings, 85*, 23−31. Available from https://doi.org/10.1016/j.enbuild.2014.09.035.

Yinfei, D., Qin, S., & Shengyue, W. (2015). Bidirectional heat induced structure of asphalt pavement for reducing pavement temperature. *Applied Thermal Engineering, 75*. Available from https://doi.org/10.1016/j.applthermaleng.2014.10.011.

Yinfei, D., Zheng, H., Jiaqi, C., & Weizheng, L. (2018). A novel strategy of inducing solar absorption and accelerating heat release for cooling asphalt pavement. *Solar Energy, 159*, 125−133. Available from https://doi.org/10.1016/j.solener.2017.10.086.

Yu, W., Vaneckova, P., Mengersen, K., Pan, X., & Tong, S. (2010). Is the association between temperature and mortality modified by age, gender and socio-economic status. *The Science of the Total Environment, 408*, 3513−3518. Available from https://doi.org/10.1016/j.scitotenv.2010.04.058.

Yu, Z., Guo, X., Jørgensen, G., & Vejre, H. (2017). How can urban green spaces be planned for climate adaptation in subtropical cities. *Ecological Indicators, 82*, 152−162. Available from https://doi.org/10.1016/j.ecolind.2017.07.002.

Yu, Z., Xu, S., Zhang, Y., Jørgensen, G., & Vejre, H. (2018). Strong contributions of local background climate to the cooling effect of urban green vegetation. *Scientific Reports, 8*, 1−9.

Yu, Z., Yang, G., Zuo, S., Jørgensen, G., Koga, M., & Vejre, H. (2020). Critical review on the cooling effect of urban blue-green space: A threshold-size perspective. *Urban Forestry & Urban Greening, 49*, 126630. Available from https://doi.org/10.1016/j.ufug.2020.126630.

Yuan, C., Zhu, R., Tong, S., Mei, S., & Zhu, W. (2022). Impact of anthropogenic heat from air-conditioning on air temperature of naturally ventilated apartments at high-density tropical cities. *Energy and Buildings, 268*. Available from https://doi.org/10.1016/j.enbuild.2022.112171.

Yun, G. Y., Ngarambe, J., Duhirwe, P. N., Ulpiani, G., Paolini, R., Haddad, S., et al. (2020). Predicting the magnitude and the characteristics of the urban heat island in coastal cities in the proximity of desert landforms. The case of Sydney. *The Science of the Total Environment, 709*, 136068. Available from https://doi.org/10.1016/j.scitotenv.2019.136068.

Zhai, Y., Ma, Y., David, S. N., Zhao, D., Lou, R., Tan, G., et al. (2017). Scalable-manufactured randomized glass-polymer hybrid metamaterial for daytime radiative cooling. *Science (New York, N.Y.) (80-), 355*, 1062−1066. Available from https://doi.org/10.1126/science.aai7899.

Zhang, W., McManus, P., & Duncan, E. (2018). A raster-based subdividing indicator to map urban heat vulnerability: A case study in Sydney, Australia. *International Journal of Environmental Research and Public Health*, 15. Available from https://doi.org/10.3390/ijerph15112516.

Zhao, J., Zhao, X., Liang, S., Zhou, T., Du, X., Xu, P., et al. (2020). Assessing the thermal contributions of urban land cover types. *Landscape and Urban Planning*, *204*, 103927. Available from https://doi.org/10.1016/j.landurbplan.2020.103927.

Zheng, Y., Ren, H., Guo, J., Ghent, D., Tansey, K., Hu, X., et al. (2019). Land surface temperature retrieval from sentinel-3A sea and land surface temperature radiometer, using a split-window algorithm. *Remote Sensing*, *11*, 650. Available from https://doi.org/10.3390/rs11060650.

Zhu, Z., Wulder, M. A., Roy, D. P., Woodcock, C. E., Hansen, M. C., Radeloff, V. C., et al. (2019). Benefits of the free and open Landsat data policy. *Remote Sensing of Environment*, *224*, 382−385. Available from https://doi.org/10.1016/j.rse.2019.02.016.

Zölch, T., Maderspacher, J., Wamsler, C., & Pauleit, S. (2016). Using green infrastructure for urban climate-proofing: An evaluation of heat mitigation measures at the micro-scale. *Urban Forestry & Urban Greening*, *20*, 305−316. Available from https://doi.org/10.1016/j.ufug.2016.09.011.

Zonato, A., Martilli, A., Gutierrez, E., Chen, F., He, C., Barlage, M., et al. (2021). Exploring the effects of rooftop mitigation strategies on urban temperatures and energy consumption. *Journal of Geophysical Research: Atmospheres*, *126*, e2021JD035002.

Zoras, S., & Dimoudi, A. (2016). Exploiting earth cooling to mitigate heat on cities' scale. In M. Santamouris, & D. D. Kolokotsa (Eds.), *Urban Clim. Mitig. Tech.* Routledge.

Chapter 3

The impact of heat mitigation and adaptation technologies and urban climate

Hideki Takebayashi
Department of Architecture, Graduate School of Engineering, Kobe University, Kobe, Japan

3.1 Introduction

Based on the experience of the extreme high temperature (heatwave) of the summer in recent years, Kobe city has been studying and implementing several adaptation measures for extreme high temperatures. As a demonstration of cool spots in outdoor spaces, the fractal sunshades with fine mist spray were provided in the plaza in front of a famous department store and on the north—south street in front of a central train station in the summer of 2019. Two vehicles with water tanks sprinkled 32 tons of water on 25.8 ha of downtown streets every day except rainy days in the summer of 2020 to 2022. Paved surfaces in parks and sidewalks in downtown areas have been watered and mist spraying has been implemented in parks, open spaces, bus stops, and so forth since 2019. By using the thermal environment index of the human body as a common index, it is possible to compare countermeasure technologies by different mechanisms, such as watering on road, sunshade with mist spray, water surface, watering on pavement, and mist spray in a park, in an open space, and at a bus stop. In this chapter, the effects of these adaptation measures on improving the thermal environment of pedestrians and visitors are presented using thermal environment index standard new effective temperature (SET*) based on the demonstration experiments. In addition, more appropriate measures to improve the thermal environment of pedestrians and visitors are presented as a possibility to utilize the combined roadway and sidewalk space in the street canyon as a human-centered space.

3.2 Demonstration of cool spots in outdoor spaces

Kobe city is located on Osaka Bay. The climate is classified as a warm temperate climate; according to Köppen and Geiger, this climate is classified as

Mitigation and Adaptation of Urban Overheating. DOI: https://doi.org/10.1016/B978-0-443-13502-6.00003-8
91

Cfa. The average annual temperature is 16.7°C. The average annual rainfall is 1216 mm. Kobe city is one of the typical core cities in Japan, facing the sea and adjacent to mountains rich in nature, similar to many other cities in Japan. The Mayor of Kobe has been considering and implementing measures to deal with extreme heat since around 2018.

3.2.1 Watering on roadway

Based on discussions between Kobe City engineers and experts, including the author, watering on roadways was selected as a measure to mitigate extreme heat. The discussion was as follows. The introduction of highly reflective or water-retentive pavement in a certain urban area requires a larger expenditure and may be implemented over a period of several years. Furthermore, high reflectance is not acceptable on roadways from a visual quality perspective, and reflectance may decrease over time due to vehicle travel. Continuous water supply is necessary to ensure a stable evaporative cooling effect in water-retentive pavements. Although continuous water supply technologies are available on the market (Akagawa & Komiya, 2000), they require a substantial cost and management effort. In contrast, watering on roadways is controllable according to citizen need and expenditure and allows for continuous improvement in the location and timing of watering.

As shown in Fig. 3.1, two vehicles with water tanks sprinkled 32 tons of water on 25.8 ha of downtown streets every day except on rainy days in the summer of 2020 to 2022. In order to achieve effective measures for extreme

FIGURE 3.1 Watering on roadway.

heat, it is necessary to accumulate evidence and discuss specific methods of roadway watering based on the evidence. Except on rainy days, two 4-ton trucks watered approximately 32 tons per day (4 cups for each car per day) at 9:00, 11:00, 12:30, 14:00, and 15:30. The total watering area is 258,000 [m^2], and the amount of each watering per unit area is 24.8 [g/(m^{2}times)] (=32,000,000 [g] / 258,000 [m^2] / 5 [times]). If all the water evaporated in 1 hour, the latent heat of evaporation would be 62,000 [J/(m^{2}h)] (=2500 [J/g] × 24.8 [g/(m^{2}h)]), which is 17.2 [W/m^2].

Measurements were taken on July 30 and August 5, 2020, from 9:00 to 11:00 and from 13:00 to 15:00. Air temperature, relative humidity, wind direction and velocity, and surface temperature on the watered roadway were measured immediately before and after the water supply, and every 5 minutes until 30 minutes later. Since only one of the multiple lanes was watered, no decrease in air temperature or increase in relative humidity was observed in the measurements at a height of 1.2 m. Wind direction tended to follow the road, and wind velocity tended to be higher on the wider road, which was not affected by the water supply.

The SET* difference between the roadway with and without watering, just after watering, and 30 minutes after watering is shown in Fig. 3.2. SET* was calculated using the MRT and the measured values of air temperature, relative humidity, and wind velocity at each measurement point. MRT was calculated from the surface temperature measured on road and other surfaces and the view factor of each surface to the human body. The amounts of

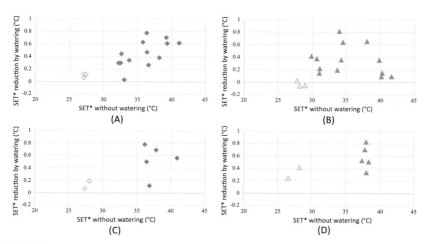

FIGURE 3.2 Standard new effective temperature difference between the roadway with and without watering just after watering and 30 minutes after watering White symbols are in the shade and color symbols are in the sun. (A) On north–south road immediately after watering, (B) on north–south road 30 minutes after watering, (C) on east–west road immediately after watering, and (D) on east–west road 30 minutes after watering.

clothing and metabolic rate were assumed to be 0.6 clo (1 clo = 0.155 Km^2/W) and 1.4 met (1 met = 58.2 W/m^2), respectively. The horizontal axis is SET* without watering, and the vertical axis is the difference between SET* without and with watering. The left sides are immediately after watering, and the right sides are 30 minutes after watering. The upper sides are on north−south road, and the lower sides are on east−west road. The maximum difference in SET* is about 0.8°C. It can be seen that the solar radiation directly incident on the human body is quite dominant to the SET* via MRT. Since the pedestrians being evaluated were assumed to be located at the boundary between the roadway and the sidewalk, the view factor of watered roadway is almost 0.2. Considering a maximum surface temperature reduction of 10°C and a view factor of 0.2 for a watered roadway, the maximum MRT and SET* reduction would be about 2°C and 0.8°C, respectively.

Fig. 3.3 shows the measured surface temperature decrease immediately after watering at 12:30 p.m. on the north and south lanes of the east−west road, excluding rainy days from July 22 to September 4, 2022, separately for sunny and cloudy days. Surface temperatures were measured continuously by a thermal camera set on the roadside. The north side roadway is shown in orange, and the south side roadway is shown in blue. Sunny and cloudy are classified according to the hourly weather conditions by the Kobe meteorological observatory. Because noise (e.g., influence of vehicles) was observed when the time interval was short, the thermal images were extracted every minute after confirming measurements taken at 15 seconds intervals.

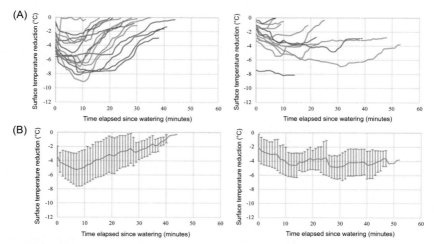

FIGURE 3.3 Measured surface temperature decrease immediately after watering at 12:30 p.m. on the north and south lanes of the east−west road, excluding rainy days from July 22 to September 4, 2022: (A) sunny days and cloudy days and (B) mean and standard deviation on sunny days and cloudy days.

The surface temperature immediately before watering (one minute before) was used as the reference, and surface temperatures were shown up to the time when they became higher than the reference temperature. The horizontal axis represents the time elapsed from watering time, since the watering time varies depending on the sprinkler vehicle's timing.

The mean and standard deviation of the measured surface temperature decrease immediately after watering are also shown in Fig. 3.3. The number of measurements varied by time of day and the standard deviation was large, but a general trend was observed. Although the surface temperature may decrease by 2°C to 4°C immediately after watering, the maximum decrease in surface temperature is not observed immediately after watering, but gradually decreases for about 10 minutes after watering. It is considered that evaporation of the water is the cause of this effect. Even on sunny days, the maximum surface temperature reduction does not reach 10°C.

Table 3.1 shows the amount and duration of the surface temperature decrease at each watering time. The maximum reduction of surface temperature is about 10°C. The duration of the surface temperature decrease is the time it returns to the surface temperature before watering, but it is difficult to determine them in many cases. As can be seen from Fig. 3.3, surface temperature may not rise to the level before watering under cloudy or shaded conditions.

3.2.2   Sunshade with mist spray

Because of the administrative effort and cost involved in continuous irrigation and periodic watering, more strategic research is needed to realize cool pavements to improve the outdoor human thermal environment. Several studies have already examined the improvement of the outdoor human thermal environment by cool pavements, but more practical studies are needed. Djekic et al. (2018) discussed the relationship between color, roughness, and shade of several materials and surface temperature and argued that the effects of sunshine and shade are dominant in human thermal comfort declaration based on the analysis of thermal comfort declaration survey. Wang et al. (2018) reported that water sprinkling can reduce surface temperature and wet bulb globe temperature (WBGT) at a height of 0.5 m above the cold pavement by up to 10°C and 2°C, respectively. The author also reported that the most dominant factor on the thermal environment of outdoor spaces is solar radiation shielding, followed by surface cover improvement (Takebayashi & Kyogoku, 2018). They also state that solar radiation shielding by trees is necessary at least 10 m from buildings on the south side and at least 6 m from buildings on the east or west side (Takebayashi et al., 2017). In addition, based on the results of measurements and calculations in a large-scale redevelopment plaza in front of a major railway station, it has been discussed that the thermal environment design of outdoor spaces should be considered in the following order: shading by buildings, shading by trees, and improvement of surface materials (Takebayashi, 2019).

TABLE 3.1 Amount and duration of the surface temperature decrease at each watering time. – means it could not be determined.

	Sunny condition				Cloudy condition			
	Decrease (°C)		Duration (minutes)		Decrease (°C)		Duration (minutes)	
	Max	Average	Max	Average	Max	Average	Max	Average
9:00	1.6	1.0	–	–	2.4	1.7	–	–
11:00	3.8	3.5	–	–	2.0	1.7	–	–
12:30	9.1	5.7	37	26	8.2	3.9	–	–
14:00	9.7	5.9	31	24	8.2	4.0	40	34
15:30	10.4	6.7	–	–	7.3	4.6	–	–

(A)   (B)

FIGURE 3.4 Sunshade with mist spray: (A) sunshade-1 on the north—south road and (B) sunshade-2 in an open space.

As shown in Fig. 3.4, the sunshade was fractal-shaped (Sakai et al., 2012), 2×4 m in size, 2.5 m in height, and mist sprays were installed on four square columns at a height of 1.5 m. Sunshades-1 and -2 in the plaza in front of a famous department store and on the north—south street in front of a central train station were the same. Measurements were taken on July 30 and August 5, 2019, from 10:00 to 17:00. Mist was sprayed from 11:00 to 15:00. Air temperature, relative humidity, globe temperature, and solar radiation at the center under the sunshade were measured every 5 seconds using a data logger. In addition, the wind direction and velocity under the sunshade and the surface temperature of representative surface materials were measured hourly by the measurer. The same measurements were carried out at reference points set in the immediate vicinity of each sunshade.

SET* under the sunshade-1 and -2 and nearby reference points without sunshade on July 30, 2019, are shown in Fig. 3.5. Blue lines are sunshade-1, and red lines are sunshade-2. Dashed lines are under the sunshade, and dotted lines are the reference points without sunshade. SET* was calculated using the measured air temperature, relative humidity, wind velocity, and MRT calculated from the measured surface temperature and solar radiation. The amount of clothing and metabolic rate were assumed to be 0.6 clo and 1.4 met, respectively. Air temperature under the sunshade was lower than the nearby reference point without the sunshade, but the absolute humidity showed little difference between the two points with and without the sunshade. The sprayed mist did not directly affect the air temperature and humidity sensors, but the decrease in surface temperature and solar radiation shielding affected the decrease in air temperature.

Sunshade-1 was located on the north—south road, so it was shaded by the surrounding buildings before 10:00 and after 14:00. Sunshade-2 was located in an open space, so it was shaded by the surrounding buildings only after 16:00. The large decrease in SET* from 11:00 to 13:00 was due to the fractal sunshade at the top blocking solar radiation to the central measurement point

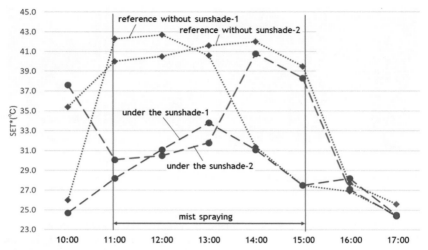

FIGURE 3.5 Standard new effective temperature under the sunshade-1 and -2 and nearby reference points without sunshade on July 30, 2019.

under the sunshade. Outside of this time period, there was little difference between SET*s with and without the sunshade. The shielding of solar radiation to the measurement point is quite dominant for the SET*.

Time variation of shade by the sunshade is important for improving thermal sensation. The effect of the mist spray was hardly observed. Changes with time in solar radiation shade by sunshades and roadside trees are shown in Fig. 3.6. The red line shows the shade by time. Upper sides are shaded by sunshades that are 2 × 4 m, 2.5 m high, and lower sides are shaded by roadside trees that are 10 m high. Left sides are on north−south road, and right sides are on east−west road. In the case of sunshades on the north−south road sidewalk, the sunshade is effective on the western sidewalk in the morning and on the eastern sidewalk in the afternoon. However, the shadow directly under the sunshade is limited to before and after noon. Vertical louvers, blinds, and so forth are required to ensure the effect directly under the sunshade. In the case of sunshades on the east−west road sidewalk, the sunshade is effective on the northern sidewalk, but it is not effective on the southern sidewalk. In the case of roadside trees on the north−south road sidewalk, the sunshade is effective on a relatively wide area, but it is limited in morning or evening on either sidewalk. In the case of roadside trees on the east−west road sidewalk, the sunshade is effective along the sidewalk. From this, it is necessary to consider the distance in the east−west direction when arranging sunshades and trees in order to obtain the effect of solar radiation shielding over a wide area. On the other hand, if sunshades and trees are arranged in the east−west direction, a space against extreme heat will be formed.

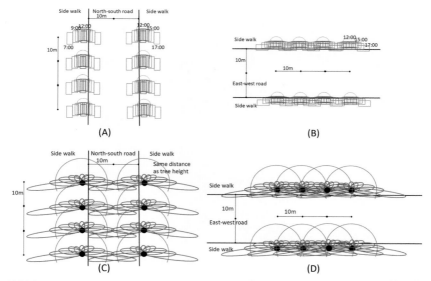

FIGURE 3.6 Changes with time in solar radiation shade by sunshades and roadside trees. (A) Sunshade on the north−south road sidewalk, (B) sunshade on the east−west road sidewalk, (C) roadside tree on the east−west road sidewalk, and (D) roadside tree on the north−south road sidewalk.

3.2.3   Water surface and watering on pavement

The effect of a cool pavement is mainly obtained by reflection or evaporation, but the effect of evaporation is relatively promising because it is necessary to avoid the solar radiation reflected by the pavement from incident on the human body. Various studies have been conducted on a cool pavement using evaporation. For example, the author performed comparative surface heat budget measurements on several types of asphalt, concrete, and ceramic pavements developed in Japan, as well as soil and turf surfaces (Djekic et al., 2018). Cool pavements based on evaporation are classified as porous pavements, permeable pavements, and pervious pavements. In all cool pavements based on evaporation, the evaporation rate gradually decreases, and the cooling effect lasts only a few days after precipitation or watering. Relatively continuous evaporation can be expected in pavements with water-retaining materials incorporated between porous aggregates, which are called water-retaining pavements (Takebayashi & Moriyama, 2012). In addition, Akagawa and Komiya (2000) proposed the use of water-retaining sheets using capillary action under the surface of the water-retaining material in order to continue more continuous evaporation. Watering on an as-needed basis or periodically is another possible alternative for continuous evaporation.

Running water is artificially circulated in a canal in the park. The location of the measurement points on the water surface is shown in Fig. 3.7.

FIGURE 3.7 Location of the measurement points on the water surface.

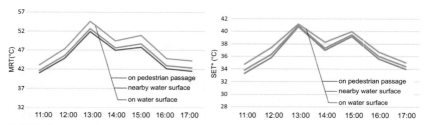

FIGURE 3.8 MRT and standard new effective temperature on the water surface and its surroundings on August 13, 2020.

The measurement point for pedestrians was set in the center of a 7.5 m pedestrian passage. The measurements were taken from 11:00 to 17:00 on August 13, 2020. Air temperature, relative humidity, wind direction and velocity, globe temperature, solar radiation, and surface temperature on water surface, pavements, walls, and plants were measured hourly. MRT and SET* on the water surface and its surroundings on August 13, 2020, are shown in Fig. 3.8. The left side is MRT, and the right side is SET*. The blue line is on the water surface, the red line is near the water surface, and the gray line is on the pedestrian passage. MRT was calculated from the view factors based on the fisheye photographs and surface temperature measurements. SET* was calculated using the calculated MRT and air temperature, relative humidity, and wind velocity measured at each measurement point. The amount of clothing and metabolic rate were assumed to be 0.6 clo and 1.4 met, respectively. At all measurement points, MRT and SET* were higher during the day when solar radiation was higher, but MRT and SET* on the water surface were about 1.8°C and 1.1°C lower on average than those on the pedestrian passage. This is because the surface temperature difference between the water surface and the pedestrian passage is relatively large, about 15°C. As the pedestrians approached the water surface, MRT and SET* were almost the same as on the water surface.

The watering experiment was conducted by watering 2 m width on the same pedestrian pavement in the park. As with watering on the roadway, a sufficient amount of water was supplied to completely wet the surface. Measurements were taken on August 12, 2020, from 8:00 to 8:30 and from 14:00 to 14:30. Air temperature, relative humidity, wind direction and velocity, and surface temperature on the watered pavement surface and its surroundings were measured immediately before and after watering and every 5 minutes until 30 minutes later. MRT and SET* on the pedestrian passage with and without watering before watering to 30 minutes after watering are shown in Fig. 3.9. The blue line is on the pavement with watering, and the red line is on the pavement without watering. Upper sides are MRT, and lower sides are SET*. Left sides are at 8:00, and right sides are at 14:00. MRT and SET* were calculated in the same way as mentioned above. MRT and SET* on the pedestrian passage with watering were about 1.3°C and 4.0°C and 0.9°C and 2.5°C lower on average at 8:00 and 14:00, respectively,

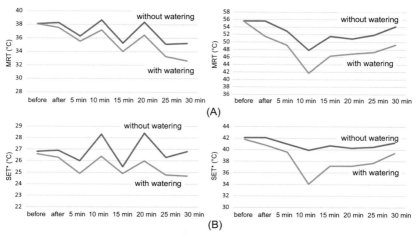

FIGURE 3.9   MRT and standard new effective temperature on pedestrian passage with and without watering before watering to 30 minutes after watering. (A) MRT before watering to 30 minutes after watering (left: watering at 8:00, right: watering at 14:00, on August 14, 2020) and (B) SET* before watering to 30 minutes after watering (left: watering at 8:00, right: watering at 14:00, on August 14, 2020).

compared to those on the pedestrian passage without watering. These effects were obtained because the view factor of the 2 m wide watered pavement was larger than that of the 1 m wide water surface, although the maximum reduction on pavement surface temperature due to watering was 10°C, which was smaller than on water surface (15°C).

3.2.4 Mist spray in a park, in an open space, and at a bus stop

Mist spraying is a technique for locally lowering air temperature by spraying fine mist into the air and using the latent heat of evaporation immediately after spraying. There are two main methods of spraying: spraying water under pressure (one fluid) and releasing water along with pressurized air (two fluids). The particle size of the sprayed mist varies depending on the manufacturers, but it is as fine as 10 to 30 μm and can vaporize in a short time to relieve the heat without making people in the vicinity feel wet. Since the sight of fine mist being sprayed is cool, there are examples of its introduction in private and public facilities. Some cases have been combined with blower fans. For example, in Japan, mist sprayers have been installed in shopping streets, as well as in front of train stations, event venues, shopping centers, amusement facilities, parking areas, city halls, and semioutdoor spaces at building entrances.

The decrease in air temperature and increase in humidity due to spraying mist is an isoenthalpy change. The effect of spraying mist on air temperature decrease depends on the amount of evaporation. The change in isoenthalpy due to spraying mist is evaluated as having no effect on the thermal sensation index, such as SET* or WBGT, which consider the decrease in air temperature and increase in humidity. For example, SET*, which is defined as air temperature when it is equal to the thermal sensation of a hypothetical environment with 50% relative humidity, is evaluated by changing air temperature, which has been lowered by spraying mist, to a higher level in order to reduce the increased humidity to the equivalent of 50%. WBGT is calculated using the black bulb, wet bulb, and dry bulb temperature, but the wet bulb temperature, which has a weight of 70%, does not change due to mist spraying, so it only reflects the effect of the dry bulb temperature, which has a weight of 10%.

Wooden flowerbeds and benches were placed in the park, in which the mist spray outlets and associated pumps and pipes are hidden, as shown in Fig. 3.10A. The mist spray outlet was installed at 1.3 m above the ground. Enough water is supplied to wet the tile surface under shaded conditions. The measurement points were set near and slightly away from the mist outlets. The measurements were carried out from 10:00 to 11:00 and from 16:00 to 17:00 on August 12, 2020. At each measurement point, air temperature, relative humidity, wind direction and velocity, globe temperature, and surface temperature on surrounding materials were measured in 15 minutes.

FIGURE 3.10   Mist spray in a park, in an open space, and at a bus stop: (A) in a park, (B) in an open space, and (C) at a bus stop.

Air temperature decreased by about 1°C, and relative humidity increased by about 1% at 30 cm from the mist outlet. It was observed that there was a large difference in globe temperature. The globe ball that the temperature sensor put inside at 30 cm from the mist outlet got wet, but the globe balls at 130 and 230 cm from the mist outlets did not. This means that when the human body is wetted by mist spray, the effect of improving the thermal sensation is significant. MRT and SET* in front of mist spray are shown in Table 3.2. MRT takes into account the effect of wet globe balls. SET* is calculated using measured air temperature, relative humidity, wind velocity, MRT calculated above, and solar radiation. The reduction in MRT due to

TABLE 3.2 MRT and standard new effective temperature in front of mist sprays on August 12.

	10:00−11:00		16:00−17:00	
	MRT [°C]	SET* [°C]	MRT [°C]	SET* [°C]
30 cm from the mist outlet on the tile surface	34.9	28.5	28.2	24.1
130 cm from the mist outlet on the tile surface	40.1	31.3	31.1	26.2
30 cm from the mist outlet on the grass surface	27.6	23.5	31.8	29.0
230 cm from the mist outlet on the grass surface	47.0	39.5	42.9	38.2

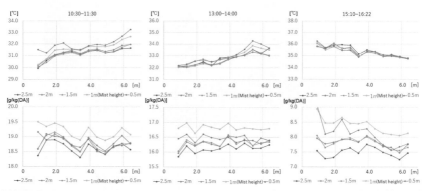

FIGURE 3.11 Measurement results of air temperature and humidity in front of mist spraying in an open space in a park between 10:30−11:30, 13:00−14:00, and 15:10−16:22 on August 27, 2021.

wetting of the globe ball at 30 cm from the mist outlet had a greater effect on the reduction in SET* than the reduction in air temperature due to evaporation of the mist.

Fig. 3.11 shows measurement results of air temperature and humidity in front of mist spraying in an open space in a park between 10:30−11:30, 13:00−14:00, and 15:10−16:22 on August 27, 2021. The horizontal distribution perpendicular to the spray axis was not confirmed significantly, so the values averaged for each height are shown. In each measurement period, the frame with the thermohygrometer and thermocouples was moved, so the data measured at fixed points were used for time-correctness. The measurement data were corrected to the start time of each measurement period. Therefore Fig. 3.11 only shows the spatial distribution of air temperature and humidity with distance from the mist spray. Depending on the meteorological conditions of the day, the air temperature was low in the morning and continuously increased in the afternoon and evening. Conversely, humidity was high in the morning and continuously decreased in the afternoon and evening. At all times, the humidity at the height of the mist spray, 1 m high, was the highest, followed by 0.5 m high, and the humidity at the upper part was slightly lower. The relationship between distance from the mist spray and humidity is not clearly confirmed. In the morning and afternoon, heating from the ground surface receiving solar radiation was observed, especially at a distance of more than 5 m from the mist spray. In the evening, the entire area was heated and hot. Air temperature decrease of about 0.5°C to 1°C due to mist spray on an open space is limited to within 2 m of the mist spray, due to the effect of heating from the ground surface receiving solar radiation. In contrast, an increase in humidity of about 0.5 to 1 g/kg(DA) at the height of the mist spray is not spatially limited. In this case, the ratio of the decrease in air temperature to the increase in humidity is $dT/dX = (0.5 \sim 1.0) < 2.5$ [K/(g/kg(DA))], so people should not feel cool.

Fig. 3.12 shows measurement results of air temperature, humidity, and wind velocity at semiopen bus stops on September 21, 2021. This is a

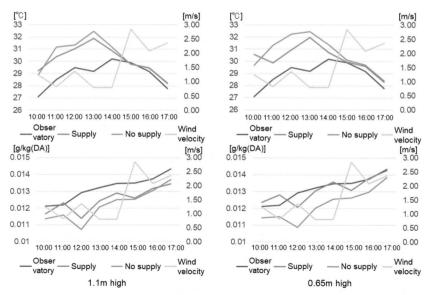

FIGURE 3.12 Measurement results of air temperature, humidity, and wind velocity at semiopen bus stops on September 21, 2021.

comparison of the results of hourly measurements taken at "Supply" and "No supply" bus stops. The meteorological observatory is located about 1.7 km east of the objective site. The wind velocity was measured at a height of 1 m at the objective site. The wind direction was south along the street direction throughout the day. By 13:00, both "supply" and "no supply" air temperatures were higher than the meteorological observatory, and "supply" air temperature was slightly lower than "no supply" air temperature. Air temperature in the semiopen bus stop tends to increase during the hours of sunlight, when ventilation is suppressed. After 15:00, the wind velocity increased and air temperature at the semiopen bus stop was almost the same as the meteorological observatory. The humidity at the bus stop was lower than the meteorological observatory all day, and "supply" humidity was slightly higher than "no supply" humidity.

Fig. 3.13 shows air temperature and humidity changes by mist spraying and iso-SET* lines at semiopen bus stops at 11:00, 12:00, and 13:00 on September 21, 2021. The SET* was calculated assuming a typical summer shade condition; MRT is 30°C, wind velocity is 1.0 m/s, clothing volume is 0.6 clo, and metabolic rate is 2 met. Air temperature decreased slightly to 1°C. The humidity increased slightly to 1 g/kg(DA). In this case, the ratio of the decrease in air temperature to the increase in humidity is $dT/dX = (0.2 \sim 1.0) < 2.5$ [K/(g/kg(DA))], so people will not feel cold. However, since the change is roughly along the iso-SET* line, it will not feel hot.

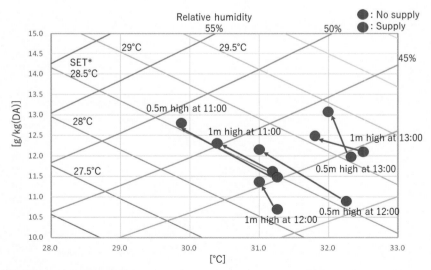

FIGURE 3.13 Air temperature and humidity changes by mist spraying and iso-standard new effective temperature lines at semiopen bus stops at 11:00, 12:00, and 13:00 on September 21, 2021.

3.2.5 Possibility of implementing adaptation measures for extreme heat based on improving effects of human thermal sensation

Table 3.3 shows a summary of the effects on the improvement of human thermal sensation.

When the roadway is watered at a high surface temperature (over 40°C), the surface temperature reduction is large (about 10°C) and, accordingly, MRT and SET* on the nearby sidewalks also decrease. However, even if the surface temperature reduction is the same, the MRT and SET* reductions are significantly different when water is supplied to nearby and distant lanes.

When incident solar radiation to the human body is shielded by sunshades, the reductions in MRT and SET* are very large (about 15°C and 7°C, respectively). The sun shading effect by street trees is similar to that of sunshades. However, the effect of sunshades and street trees does not always improve the thermal sensation of their lower and surrounding spaces

TABLE 3.3 Summary of the effects on the improvement of human thermal sensation.

	Surface temp. reduction	MRT reduction	SET* reduction	Condition
Watering on road	10°C	1.9°C	0.8°C	Watering on the first lane and pedestrian is beside it
		0.25°C	0.07°C	Watering on Second lane and pedestrian is beside it
Sunshade	–	15°C	7°C	When the human body is shaded
Water surface	15°C	0.2°C	0.1°C	On a passage 3.75 m away
		1.6°C	1.2°C	On a 1 m wide water surface
Watering on the pavement	5°C–10°C	1.3°C–4.0°C	0.9°C–2.5°C	On a passage of 2 m width watered
Mist spray	–	2.9°C–19.4°C	1.2°C–8.2°C	When the human body gets wet
Street tree	17.4°C	16.1°C	6.8°C	When the human body and ground surface is shaded

throughout the day. It is important to design the layout of sunshades and street trees, taking into account the temporal changes in solar radiation shading caused by these shades and trees.

When water is running in a park or other locations, the water-supplied temperature is lower than the surface temperature that is reduced by water sprinkling. However, when pedestrians and visitors are on the sidewalk away from the water surface, the MRT and SET* reductions are smaller than when they are on the watered pavement.

Air temperature decreases slightly to 1°C, and humidity increases slightly 1 g/kg(DA) in the vicinity of the mist outlets under low-wind conditions. When the human body gets wet, MRT and SET* decrease significantly. In situations where it is acceptable for the human body to be wetted by the mist spray, such as in recreational areas, the effect of improving the thermal sensation can be expected.

As shown in Table 3.3, the improvement of human thermal sensation varies depending on the distance from the countermeasure technologies to the human body. Based on the characteristics of the target space, each technology should be selected appropriately in consideration of the following points. The author discusses these issues with Kobe City government officers.

- Regarding mist spray, is it acceptable for mist to wet the human body?
- Regarding the water surface, is it acceptable to have a walking space on the water surface?
- Regarding the watering pavement, is it acceptable that the pavement underfoot is wet?
- Regarding watering road, is it possible to supply water up to the edge of the road?
- Regarding street trees, is it possible to maintain the trees, such as pruning?
- Regarding sunshade, is it possible to match the shade area with the location of visitors?

3.3   Strategies to implement adaptation measures for extreme high temperatures into street canyon

Using GIS building data, a more detailed calculation was used to derive a thermal environment map in the street canyon and to examine strategies for implementing adaptation measures for extreme high temperatures. Incident solar radiation distribution was calculated by the Arc-GIS tool (Takebayashi et al., 2015). In this method, direct and sky solar radiation influenced by surrounding buildings were considered, but reflected solar radiation was not considered to avoid the calculation of complex interreflection of each surface. Since the urban blocks were not configured with surface materials with high reflectivity, the error due to this was considered to be small. Surface temperature was calculated on each ground and wall surface by the surface heat budget equation, which balances radiation, convection, evaporation, and conduction heat flux by the given physical properties of each surface (Takebayashi, 2019). Radiation heat flux was calculated by incident solar radiation and mutual infrared radiation by using building shape data. Convection and evaporation heat flux were calculated by air temperature, relative humidity, wind velocity, and surface temperature on each ground and wall surface, by using the heat and moisture transfer coefficients between the air and each surface. Conduction heat flux was calculated by the one-dimensional transient heat conduction equation with internal temperature as the boundary condition on each surface material (Takebayashi, 2019). Since long-term calculations were not carried out, the effect of the stored heat of solar radiation in the morning on the surface temperature from after-noon to evening was considered, but the previous day's effect on the surface temperature in the morning was not considered.

Building shape and land cover distribution in the objective area are shown in Fig. 3.14. The objective area is the economic, administrative, and cultural center of Kobe city, where people gather from inside and outside the city. Sannomiya station is located in the northeastern end, and Motomachi station is located in the northwestern end of this area. The south end of this area is connected to the port area. The east—west shopping streets are located on the north side of this area, but because these pedestrian streets have arcades, they are excluded from the calculation in this study. Relatively large-scale buildings such as offices, department stores, banks, museums,

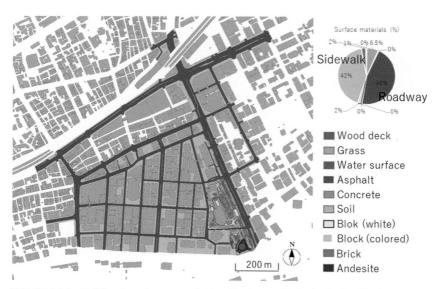

FIGURE 3.14 Building shape (gray—top view) and land cover distribution in the objective area.

hotels, city halls, and condominiums are located from the center to the south side of this area. The objective area was divided into 2 m grid, and surface materials were set for each grid. Although asphalt, block, and grass were set for the surfaces in the street canyons, wood deck, grass, water surface, asphalt, concrete, soil, block (white), block (colored), brick, and andesite were set for the surfaces only in the central park (Higashi-Yuenchi). The crown width and tree height of each street tree and park tree were set by a field survey and Google Earth. The ratio of the tree canopy area in the entire objective area was 9.7%, which was almost the same in the east−west road and the north−south road; it was large at 31.9% in the central park and small in intersection and open space.

Time change of the ground surface temperature, MRT, and SET* are calculated on a typical summer sunny day, August 5, 2019. Air temperature and relative humidity are given by the measurement data obtained in the Kobe local meteorological observatory located near the objective area. It is assumed that a solar absorption rate, a clothing amount, and a metabolic rate of the human body are 0.5, 0.6 clo, and 1.0 met, respectively, and transmittance of solar radiation of the tree is 0.06. The metabolic rate of the human body should be set according to the activity, for example, walking on roads, standing at intersections, sitting in open spaces, walking, sitting, and exercising in parks. If these values are respectively set according to each place, recognition of the SET* distribution becomes complicated, so a constant value of 1.0 met when the human body is at rest is given throughout the objective area.

Distribution of SET* at 1.5 m height at 1:00 p.m. on August 5, 2019 is shown in Fig. 3.15. Ground surface temperature is low in the area shaded by the surrounding buildings and trees. Since daily integral incident solar radiation is small in the streets shaded by street trees, surface temperature is low in both east−west and north−south roads. MRT is low in the median strip in some streets and the central park, where incident solar radiation and surrounding surface temperature are low. SET* is more affected by MRT than wind velocity.

Diurnal variation of the spatial distribution frequency of SET* at 1.5 m height in each study area on August 5, 2019 is shown in Fig. 3.16. SET* is low in the shaded area since incident solar radiation is dominant on SET*. Although the proportion of the shaded area is steady at 30% to 40% from 9:00 a.m. to 4:00 p.m. on the east−west road and in the central park, it becomes minimum around noon in other areas. Even around the noon, the pedestrians may find the shaded places on the east−west road due to the southern building's shade and in the central park due to the tree's shade. On the other hand, pedestrians may find the shaded places on the north−south road until 10:00 a.m. and after 3:00 p.m., due to the eastern building's shade in the morning and the western building's shade in the afternoon. Overall, pedestrians should look for places to avoid extreme heat on the north−south road in the morning and afternoon and on the east−west road and in the central park around the noon. It is difficult to find shaded places at intersections, open spaces, and boulevards (southwest−northeast road).

Without street trees   With street trees

22 (°C)   43 (°C)

FIGURE 3.15   Distribution of standard effective temperature at 1.5 m height at 1:00 p.m. on August 5, 2019 (Left: Without street trees, Right: With street trees).

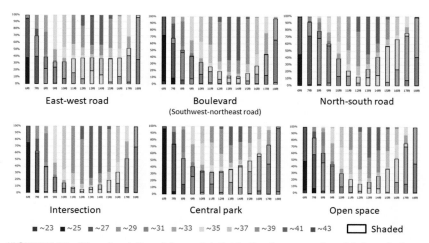

East-west road   Boulevard
(Southwest-northeast road)   North-south road

Intersection   Central park   Open space

■ ~23 ■ ~25 ■ ~27 ■ ~29 ■ ~31 ■ ~33 ■ ~35 ■ ~37 ■ ~39 ■ ~41 ■ ~43   ☐ Shaded

FIGURE 3.16   Diurnal variation of the spatial distribution frequency of modeled standard new effective temperature at 1.5 m height in each study area on August 5, 2019.

3.3.1   Spatiotemporal distribution of standard new effective temperature on roadway and sidewalk

Diurnal variations of spatial distribution frequency of SET* at 1.5 m height on east—west road and north—south road on August 5, 2020 are shown in Fig. 3.17. Based on the relationship between SET* and thermal sensation by Ishii et al. (1988), it is uncomfortable if SET* exceeds 30°C. It is uncomfortable from 9:00 to 16:00 on east—west roads, while it is limited from

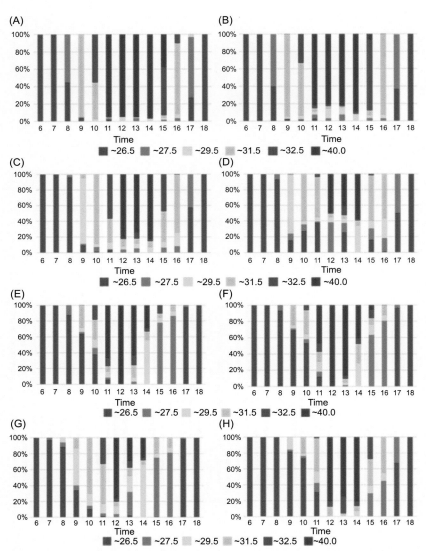

FIGURE 3.17 Diurnal variation of spatial distribution frequency of standard new effective temperature at 1.5 m height on east–west road and north–south road on August 5, 2020. (A) North side roadway on east–west road, (B) south side roadway on east–west road, (C) north side sidewalk on east–west road, (D) south side sidewalk on east–west road, (E) west side road-way on north–south road, (F) east side roadway on north–south road, (G) west side sidewalk on north–south road, and (H) east side sidewalk on north–south road.

11:00 to 14:00 on north—south roads. It is uncomfortable on all roadways and sidewalks from 11:00 to 14:00 on the north—south road, while there are locations where it is not uncomfortable even from 9:00 to 16:00 on the south side sidewalks on the east—west road. It is possible to find shaded areas on the south side sidewalk of the east—west road even from 9:00 to 16:00, due to the buildings on the south side of the road. However, on the north side sidewalk of the east—west road, it is a severe thermal environment similar to that on the roadway. The overall trend is the same for wide roads with many lanes, with only an increase in the number of lanes with severe thermal environments. The same trend is observed on boulevards, where the orientation slightly shifted from east to west, as on the east—west road. The same trend is observed at the intersection as on the east—west road, where the buildings on the south side are low. As for the possibility of human-centered use of road space, a north—south road with a limited width is a candidate, and a sidewalk on the south side of an east—west road is also envisioned.

3.3.2 Effects of watering on standard new effective temperature on roadway

When water was sprinkled on the roads, diurnal variations of spatial distribution frequency of SET* at 1.5 m height on east—west road and north—south road on August 5, 2020 were calculated and are shown in Fig. 3.18. Based on the experimental results with sprinkler vehicles, the calculations are carried out with an evaporation efficiency set at 0.15. When the surface temperature is high before water sprinkling, a surface temperature reduction is confirmed greater than 10°C, and SET* is reduced up to 2°C. Although the SET* reduction around noon is large, the water sprinkling does not lead to comfortable conditions because the conditions before water sprinkling are quite uncomfortable. Sprinkling water in the evening may increase the number of comfortable spaces.

3.3.3 Effects of the water surface on standard new effective temperature on roadway

When the road surface is covered by a water layer, diurnal variations of spatial distribution frequency of SET* at 1.5 m height on the east—west road and north—south road on August 5, 2020 were calculated and are shown in Fig. 3.19. Based on the experimental results with a water surface in a park, it is assumed that the water is supplied at a constant temperature 32°C. If people are assumed to walk or stay on the water surface, SET* is less than 31.5°C at any time, due to lower MRT. If a continuous water supply can be guaranteed and people can approach the water surface, the water surface can be expected to have a significant effect at any time and place. Water sprinkling is the effect

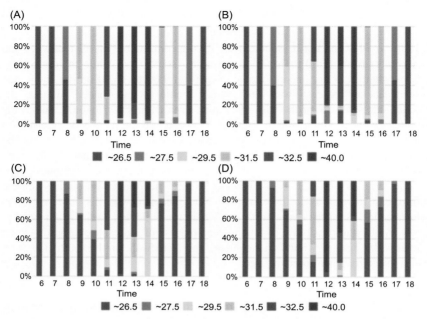

FIGURE 3.18 Diurnal variation of spatial distribution frequency of standard new effective temperature at 1.5 m height on east−west road and north−south road on August 5, 2020, when water is sprinkled on the roads. (A) North side roadway, (B) south side roadway, (C) west side roadway, and (D) east side roadway.

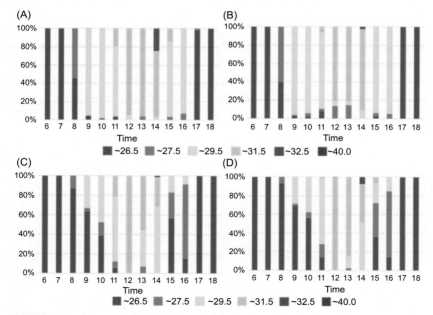

FIGURE 3.19 Diurnal variation of spatial distribution frequency of standard new effective temperature at 1.5 m height on east−west road and north−south road on August 5, 2020, when the road surface is covered by a water layer. (A) North side roadway, (B) south side roadway, (C) west side roadway, and (D) east side roadway.

due to evaporative cooling, whereas the water surface is the effect due to the supply of cooler water.

3.3.4 Effects of sunshade on standard new effective temperature on roadway

When sunshades were installed, diurnal variations of spatial distribution frequency of SET* at 1.5 m height on August 5, 2020 on the east−west road and north−south road were calculated and are shown in Fig. 3.20. 2 × 4 m sunshades are installed at 10 m intervals along a 3.5 m wide lane. The ratio of sunshade area to the lane area is 23%. SET* decreases by up to 6°C

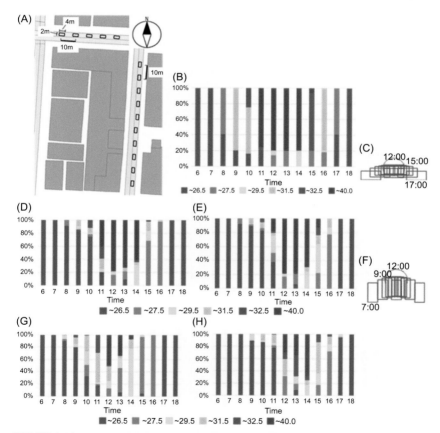

FIGURE 3.20 Diurnal variation of spatial distribution frequency of standard new effective temperature at 1.5 m height on east−west road and north−south road on August 5, 2020, when sunshades are installed. (A) Sunshade installation condition, (B) north side roadway, (C) diurnal shade change, (D) west side roadway, (E) east side roadway, (F) diurnal shade change, (G) west side sidewalk, and (H) east side sidewalk.

around noon. On the east—west road, shade occurs along the lane at any time, so that pedestrians moving along the lane can periodically pass through the shade. On the north—south road, the time required for countermeasures is limited to around noon, so shade is effective even if it does not occur in the target lane only around noon. Shade also occurs on the sidewalks of the north—south road at the time required for countermeasures.

3.3.5 Effects of street trees on standard new effective temperature on roadway

When street trees are installed, diurnal variation of spatial distribution frequency of SET* at 1.5 m height on August 5, 2020 on east—west road and north—south road were calculated and are shown in Fig. 3.21. A cylindrical canopy with a radius of 2 m and a height of 7 m is set at 3 to 10 m above the ground surface. When trees are installed at 15 m intervals, the ratio of the canopy area to the lane area is 23%, same as for the sunshade. SET*

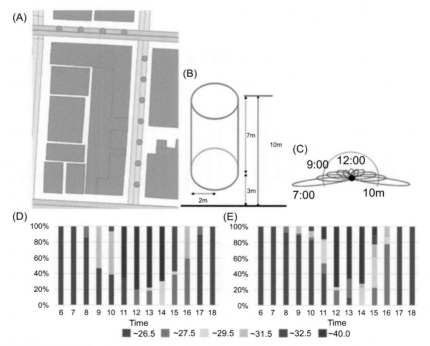

FIGURE 3.21 Diurnal variation of spatial distribution frequency of standard new effective temperature at 1.5 m height on east—west road and north—south road on August 5, 2020, when street trees are installed. (A) Street tree installation condition, (B) tree conditions, (C) diurnal shade change, (D) north side roadway, and (E) east side roadway.

decreases by up to 8.5°C around noon. On east–west roads, tree shade also occurs along the lane at any time, so that pedestrians moving along the lane can periodically pass through the shade. On the north–south road, the time required for countermeasures is limited to around noon, so shade is effective even if it does not occur in the target lane only around noon.

3.3.6 Strategies for watering time based on standard new effective temperature improvement effects on sidewalks

Diurnal variations of spatial distribution frequency of SET* at 1.5 m height on the northern and southern sidewalk on east–west road on August 5, 2020 are shown in Fig. 3.22. The right bar at each time indicates the frequency at which the road was watered. It is uncomfortable from 9:00 to 16:00 on the northern sidewalk, while there are locations where it is not uncomfortable even from 9:00 to 16:00 on the southern sidewalks. Based on the experimental results

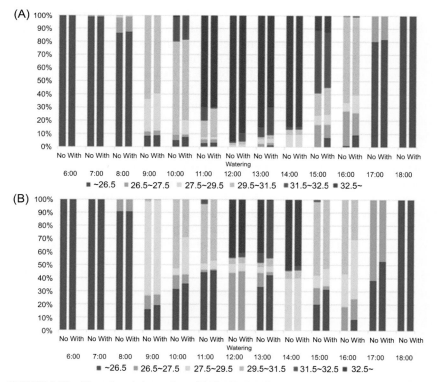

FIGURE 3.22 Diurnal variations of spatial distribution frequency of standard new effective temperature at 1.5 m height on east–west road on August 5, 2020. The left bar is without watering, and the right bar is with watering. (A) On the northern sidewalk and (B) on the southern sidewalk.

TABLE 3.4 Change in percentage of comfortable standard new effective temperature (below 29.5°C) due to roadway watering on east−west road and north−south road.

	10:00	16:00
Northern sidewalk	10% to 20% (+10%)	34% to 39% (+5%)
Southern sidewalk	47% to 71% (+24%)	43% to 69% (+26%)
Eastern sidewalk	72% to 94% (+22%)	85% to 91% (+6%)
Western sidewalk	53% to 62% (+9%)	97% to 99% (+2%)

with sprinkler vehicles, the calculations are carried out with an evaporation efficiency set at 0.15 (Akagawa & Komiya, 2000). When the surface temperature is high before watering, a surface temperature reduction is confirmed to be greater than 10°C, and SET* is reduced up to 2°C. Although the SET* reduction around noon is large, the watering does not lead to comfortable conditions because the conditions before watering are quite uncomfortable. Watering in the evening may increase the number of comfortable spaces.

Change in percentage of comfortable SET* (below 29.5°C) due to roadway watering on east−west road is shown in Table 3.4. While watering around noon does not increase the percentage of comfort, watering at 10:00 and 16:00 increases the percentage of comfort by about 70% on the southern sidewalk, but only by 20% to 40% on the northern sidewalk. The increase of percentage of comfort was as large as 25% on the southern sidewalk. Although surface temperature reduction by watering is almost the same on the northern and southern roadways, SET* tends to be comfortable on the southern sidewalk, where incident solar radiation on the human body is more likely to be shaded. Watering the lanes near the southern sidewalk is considered to be more effective in mitigating the thermal environment for pedestrians.

Diurnal variations of spatial distribution frequency of SET* at 1.5 m height on the eastern and western sidewalk on north−south road on August 5, 2020 are shown in Fig. 3.23. The right bar at each time indicates the frequency at which the road was watered. On the eastern sidewalk, there are comparatively many places that are comfortable (below 29.5°C) until 11:00 a.m. On the western sidewalk, there are relatively many places that are comfortable after 1:00 p.m. Use of the eastern sidewalk in the morning and the western sidewalk in the afternoon is recommended.

While watering around noon does not increase the percentage of comfort, watering in the morning or afternoon increases the percentage of comfort on the eastern and western sidewalk. Change in percentage of comfortable SET* (below 29.5°C) due to roadway watering on north−south road is shown in Table 3.4. Compared to the east−west road (especially on the northern

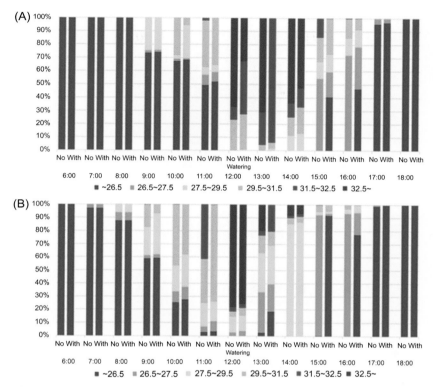

FIGURE 3.23 Diurnal variations of spatial distribution frequency of standard new effective temperature at 1.5 m height on north–south road on August 5, 2020. The left bar is without watering, and the right bar is with watering. (A) On the eastern sidewalk and (B) on the western sidewalk.

sidewalk), the thermal environment on the sidewalks of the north–south road tends to be more mitigated as a whole. Watering the lanes near the eastern sidewalk in the morning and near the western sidewalk in the afternoon is considered to be more effective in mitigating the thermal environment for pedestrians.

3.3.7   Strategies for watering based on standard new effective temperature improvement effects on sidewalks and roadways

Since the form factor of the road surface for the human body on the sidewalk is small, watering the road has a limited effect on the thermal environment for pedestrians on the sidewalk. Measures to provide road space for pedestrians are also considered by revising the traffic network (e.g., Midosuji, the central street in Osaka). Fig. 3.24 shows the change in SET* on the sidewalk and on the roadway due to roadway watering at 10:00 and 16:00. At each

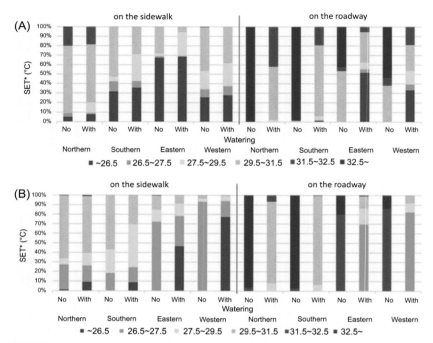

FIGURE 3.24 The change in standard new effective temperature on the sidewalk and on the roadway due to roadway watering. (A) At 10:00 and (B) at 16:00.

TABLE 3.5 Change in percentage of comfortable standard new effective temperature (below 29.5°C) due to roadway watering on north–south road at 10:00 and 16:00.

	10:00		16:00	
	Sidewalk	Roadway	Sidewalk	Roadway
Northern	10% to 20% (+10%)	0% to 2% (+2%)	34% to 39% (+5%)	0% to 8% (+8%)
Southern	47% to 71% (+24%)	0% to 6% (+6%)	43% to 69% (+26%)	0% to 6% (+6%)
Eastern	72% to 94% (+22%)	0% to 62% (+62%)	85% to 91% (+6%)	0% to 87% (+87%)
Western	53% to 62% (+9%)	0% to 54% (+54%)	97% to 99% (+2%)	0% to 92% (+92%)

time, the thermal environment on the sidewalk, which is more affected by shading by surrounding buildings, is mitigated than that on the roadway. In addition, roadway watering increases the percentage of comfortable SET* (below 29.5°C) for pedestrians on the sidewalk. Change in percentage of comfortable SET* (below 29.5°C) due to roadway watering on north−south road at 10:00 and 16:00 is shown in Table 3.5. Watering on the east−west roadway does not increase comfort because of the sun, but approaches comfort on the shaded south sidewalk. Watering on the north−south roadway at the same time approaches comfort both on the roadway and on the sidewalk. Therefore, a more effective strategy for roadway watering would be to guide pedestrians to shaded sidewalks and then water nearby roadways in the sun to reduce surface temperatures.

3.4 Summary

From the measurement results of the cool spots demonstration in outdoor spaces, the human thermal environmental mitigation effects by watering on road, sunshade, water surface, watering on pavement, mist spray, and street tree are compared with each other by using the thermal environmental index SET*. In addition, the issues that should be considered in implementing each heat mitigation technology are identified. The author discusses these issues with Kobe City government officers. Through the thermal environment simulation, strategies for implementing heat mitigation measures are described for pedestrians and visitors on the roadways and on the sidewalks. Furthermore, the optimal times for watering the roadway are identified through the calculation results of the temporal changes in the thermal environmental index of pedestrians on the sidewalks.

References

Akagawa, H., & Komiya, H. (2000). Experimental study on pavement system with continuous wet surface. *Journal of Architecture and Planning (Transactions of AIJ)*, 65(530), 79−85. Available from https://doi.org/10.3130/aija.65.79_1.

Djekic, J., Djukic, A., Vukmirovic, M., Djekic, P., & Dinic Brankovic, M. (2018). Thermal comfort of pedestrian spaces and the influence of pavement materials on warming up during summer. *Energy and Buildings*, *159*, 474−485. Available from https://doi.org/10.1016/j.enbuild.2017.11.004.

Ishii, A., Katayama, T., Shiotsuki, Y., Yoshimizu, H., & Abe, Y. (1988). Experimental stud on comfort sensation of people in the outdoor environment. *Journal of Architecture, Planning and Environmental Engineering (Transactions of AIJ)*, *386*(0), 28−37. Available from https://doi.org/10.3130/aijax.386.0_28.

Sakai, S., Nakamura, M., Furuya, K., Amemura, N., Onishi, M., Iizawa, I., Nakata, J., Yamaji, K., Asano, R., & Tamotsu, K. (2012). Sierpinski's forest: New technology of cool roof with fractal shapes. *Energy and Buildings*, *55*, 28−34. Available from https://doi.org/10.1016/j.enbuild.2011.11.052.

Takebayashi, H., Ishii, E., Moriyama, M., Sakaki, A., Nakajima, S., & Ueda, H. (2015). Study to examine the potential for solar energy utilization based on the relationship between urban morphology and solar radiation gain on building rooftops and wall surfaces. *Solar Energy*, *119*, 362–369. Available from https://doi.org/10.1016/j.solener.2015.05.039, http://www.elsevier.com/inca/publications/store/3/2/9/index.htt.

Takebayashi, H., & Moriyama, M. (2012). Study on surface heat budget of various pavements for urban heat island mitigation. *Advances in Materials Science and Engineering*, *2012*. Available from https://doi.org/10.1155/2012/523051.

Takebayashi, H. (2019). Thermal environment design of outdoor spaces by examining redevelopment buildings opposite central Osaka station. *Climate*, *7*(12), 143. Available from https://doi.org/10.3390/cli7120143.

Takebayashi, H., Kasahara, M., Tanabe, S., & Kouyama, M. (2017). Analysis of solar radiation shading effects by trees in the open space around buildings. *Sustainability*, *9*(8), 1398. Available from https://doi.org/10.3390/su9081398.

Takebayashi, H., & Kyogoku, S. (2018). Thermal environmental design in outdoor space focusing on radiation environment influenced by ground cover material and solar shading, through the examination on the redevelopment buildings in front of central Osaka station. *Sustainability*, *10*(2), 337. Available from https://doi.org/10.3390/su10020337.

Wang, J., Meng, Q., Tan, K., Zhang, L., & Zhang, Y. (2018). Experimental investigation on the influence of evaporative cooling of permeable pavements on outdoor thermal environment. *Building and Environment*, *140*, 184–193. Available from https://doi.org/10.1016/j.buildenv.2018.05.033, http://www.elsevier.com/inca/publications/store/2/9/6/index.htt.

Chapter 4

The impact of heat mitigation on low-income population

Fabrizio Ascione[1], Nicola Bianco[1], Giacomo Manniti[1], Margherita Mastellone[2], Francesco Tariello[3] and Giuseppe Peter Vanoli[4]

[1]Department of Industrial Engineering, Piazzale Tecchio 80, Università degli Studi di Napoli Federico II DII, Napoli, Italy, [2]Department of Architecture, Via Forno Vecchio 36, Università degli Studi di Napoli Federico II DIARC, Napoli, Italy, [3]Department of Agricultural, Environmental and Food Sciences, Via Francesco De Sanctis 1, Università degli Studi del Molise DiAAA, Campobasso, Italy, [4]Department of Medicine and Health Sciences, Via Francesco De Sanctis 1, Università degli Studi del Molise DiMeS, Campobasso, Italy

4.1 Urban growth and development of megacities

Urbanization involves and will involve an increasing percentage of people in the next years. Such population growth in cities is very consistent, and, as indicated by an MIT report (Karagianis, 2014), the share of people living today in urban centers is set to increase and will reach two-thirds by 2050.

Demographic changes, including net population change, rural−urban migration, immigration, and changing age structures, are important drivers of urbanization, economic growth (GDP, R&D investment, and innovation), and socioeconomic development (high quality of life and service provision in cities).

Today, reducing the heat produced in cities by population growth and industrialization is one of the most complex challenges that institutions and governments face. The sources of anthropogenic heat, which are often huge in overcrowded urban settings, together with the large amounts of heat generated by city structures, cause an increase in the urban areas' temperature, determining the urban heat island (UHI) phenomenon. The first observations of UHIs started more than 100 years ago when increased temperatures in the city of London were evaluated for the backcountry. Then, several studies in many cities have been conducted. More recently, the investigations focused on the causes of UHIs' formation and how these affect the thermal comfort and health of users within urban centers.

Mitigation and Adaptation of Urban Overheating. DOI: https://doi.org/10.1016/B978-0-443-13502-6.00004-X

The temperature increase in city centers is related to different factors that depend on the urban density and the city's organization and design. Each urban center has a different size, area occupied, population density, height development, and number of inhabitants, and these variables define the city structure and even future expansion. In the coming years, we will most likely see the development of megacities, which will change the way we consider urban centers. Still today, for example, the city of Chóngqìng in China has a surface area of 82,401 km^2 with a population of more than 32 million, corresponding to about 54% of the entire population of Italy (The World Bank, 2023a). Of course, many other examples can be provided, and thus, just to cite some, the urban areas of Tokyo (>37 million), Shanghai (≈ 30 millions), New York, Mexico City, and Karachi (>24 millions). Even in Europe, Moscow (>18 millions), London (>14 millions), and Paris (>12 millions) are examples of megacities.

According to the "Demographia World Urban Areas" report (Demographia World Urban Areas 18th Annual, 2022), more than 2.3 billion people in the world (about 29% of the total population) live in 990 large urban areas (areas characterized by more than 500 thousand inhabitants), and, among them, 524 are located in Asia, of which about 200 are in China. In the next 15 years, it is estimated that precisely in China, about 300 million people will move to urban centers; in India, on the other hand, the number is slightly less, 250 million people, and in Africa, it is 380 million people, which is challenging (Karagianis, 2014).

In general, today, the data evidence that urbanization is a phenomenon mainly involving developing countries, but, even if with smaller numbers, all other areas of the globe are included. Indeed, in the developed areas of the world, this phenomenon was anticipated in the previous century.

According to the report provided by Euromonitor International and published in 2018, "Megacities: Developing Country Domination," the most populous megacities in 2030 will be Jakarta, Tokyo, Karachi, Manila, and Cairo (Ferrara, 2018). These large urban agglomerations will be the object of attention, as they will be the largest contributors to the world's population growth, emissions, and energy consumption, and they will be the cities with the highest percentage of wealth. In particular, Asian GDP (at market exchange rates) is projected to constitute about half of the world's GDP by 2050 (A. Development Bank, 2011).

4.2 Opportunities and vulnerability of cities

On the one hand, cities offer many opportunities for those who live or will live in them; the challenge is to make them livable. Crime, poverty, disease, crowding, and pollution may be, in the next future, and partly already are the negative effects of intense urbanization, in addition to the rising temperatures on a global and local scale. The urbanization process has the potential to create new prospects for citizens, but these are achieved if urban development is properly planned. Ensuring new job opportunities, reducing poverty and inequality, and enabling access to basic urban infrastructure, energy, and services, especially for the most vulnerable and low-income segments of the citizens, are goals that can be achieved through a well-structured city plan. Often, diversification in the use of urban spaces means that disadvantage is concentrated in certain areas, and this unavoidably leads to the formation of residential spaces, in which more affluent residents have access to the most modern facilities and services. On the other hand, poorer groups are relegated to slum neighborhoods, in which the frequently wrong management of public sources or scarce availability of resources increases the already evident differences.

The COVID-19 pandemic amplified the existing urban inequalities between citizens living in the most served, connected and rich districts, compared to those living in suburbs, as shown in Fig. 4.1.

In addition, political and socioeconomic instabilities, the climate crisis, and sometimes conflicts delay the construction of well-organized and inclusive urban centers. The differences that are generated within urban settings are even one of the main causes of difficulties in implementing measures to reduce the urban heating phenomenon. Specific studies and deepening of the increasing overheating, and its effect, in urban areas are provided by many sources. Ascione et al. (2022) provided a general overview, focusing on European cities, while a deep study of connections between the rising urban heat and the mental healthiness is inferred by Aghamohammadi et al. (2022). Page et al. (2012) demonstrated the increase in heat-related mortality for

FIGURE 4.1   Suburbs in Europe suburbs in Europe: (1) Russia, (2) Germany, (3), and (4) Italy. *From https://www.google.it/maps/preview*

fragile persons, suffering from psychosis, dementia, or dependence on alcohol and drugs.

Santamouris, with other coauthors, reported several examples of relations among mortalities, morbidity, and pathologies due to the increased heat in cities, also proposing mitigation approaches (Adilkhanova et al., 2023; Falasca et al., 2022; Santamouris, 2022; Wang et al., 2022), with some very interesting specifications for low-income buildings (Garshasbi et al., 2023; Haddad et al., 2022), suggesting, in the case of houses on New South Wales in Australia, the need of mitigation strategies to face the winter underheating and the summer indoor overheating. Of course, energy poverty, as it will be discussed in deep in the next sections of this chapter, makes such a phenomenon even worse, because low-income populations cannot mitigate the uncomfortable conditions by using air conditioning equipment. It is a very disheartening, but unfortunately sometimes necessary measure; during heat waves in European cities, institutions often advise the fragile population (often elders and low-income families) to take refuge in air-conditioned environments, such as, for example, subway stations and shopping centers. Of course, the future cannot be this, but the only possible chance for all of us is a serious, immediate, and strong contrast to urban overheating.

Moreover, the overcrowded housing conditions, or the precarious conditions in which a part of the citizenry is often forced to live, create disparities in socioeconomic and political assets and in the availability of services, mainly, educational and health ones.

Poverty and urban inequality remain two of the most hard and complex issues to face in cities. The notorious overcrowding of metropolitan slums, the growing number of homeless in big cities, and the increasing number of people in poverty should generate a political consciousness that wisely addresses these issues. Reducing urban disparities should be a key priority in the design of an inclusive and equitable urban future.

4.3   The impact of overheating on energy poverty

4.3.1   Introduction to energy poverty

Energy poverty is the term referring to a multidimensional problem involving all the causes (low-income, socioeconomic restrictions, political conditions, extreme weather conditions, economic crisis, labor market, welfare state configuration, low-efficient building, energy supplies, etc.) which determine limited access to energy services that would allow for materially and socially sufficient living conditions. It usually represent the issue related to the incapability/impossibility to have adequate indoor thermal conditions during cold months (also as a condition in which people can fall in temporarily). This is the idea for which energy poverty was introduced in 1991 (Boardman & Poverty, 1991), but it is a more general problem and can, in some circumstances, be named energy vulnerability (Castaño-Rosa et al., 2019). Further and equivalently or alternatively used terms to outline the same concept are "fuel poverty," "energy deprivation," "energy precariousness," and "consumer vulnerability." The topic of energy poverty in the interpretative nuance linked to building heating has been extensively investigated in the last 2 decades, and its effects on physical and mental health have been a topic of particular interest (Pan et al., 2021), especially when households with elderly or vulnerable components are considered. Energy poverty is usually exacerbated by economic crises (Halkos & Gkampoura, 2021; Oliveras et al., 2021) and energy crises (Hussain et al., 2023; Simshauser, 2023) because people have less chance of coping with energy costs. The last COVID-19 pandemic and the consequent lockdown, imposed to limit the infection diffusion, have had an indirect effect on energy poverty. COVID-19 worsened the social, economic, environmental, and energy conditions of the population and consequently enhanced energy poverty (Carfora et al., 2022).

In the following subsection, special attention will be given to energy poverty, providing a second key to understanding the phenomenon in a cooling demand-predominant scenario (climate change, urban overheating, and heat waves).

4.3.2   Effects of overheating on summer energy poverty

Climate changes, intense use of energy for anthropogenic activities, urbanization, and overbuilding of inhabited areas make the urban environment overheated, determining implications on public health and vulnerability of the population. This happens especially for those who have scarce economic possibilities and limited access to energy supplies due to economic (prices too high compared to wages) and structural shortcomings (lack of energy infrastructure) and live in buildings of poor quality and low energy efficiency, so much so that they are unable to maintain their homes at an appropriate temperature during the summer period. These people are experiencing what can be termed "summer energy poverty" (SEP, in the following lines).

The urban microclimate, that is the climatic conditions prevailing in the urban area (Tsilini et al., 2015), is one of the contributing factors influencing SEP. An aspect that has a direct implication on the possibility of maintaining indoor conditions to properly cool through an air conditioning system is the possibility of accessing the electricity supply in certain urban areas (Fig. 4.2 shows the urban population for different income economies). The growth of the environmental temperature can increase energy poverty with respect to three aspects: it amplifies the use of electricity-driven cooling devices, increasing the energy expenditure and so the poverty; it forces people to spend most of their time at home also performing leisure activities indoors, increasing energy demands; it worsens health conditions and labor

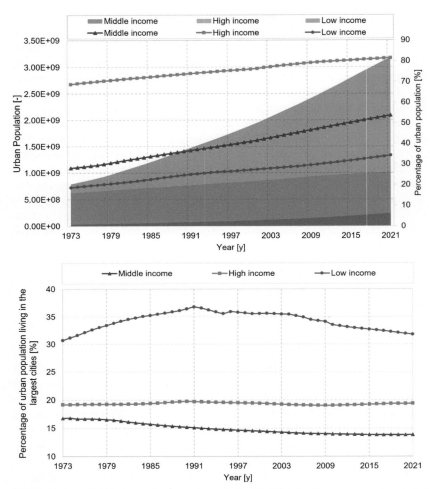

FIGURE 4.2   Population and living conditions: (A) Urban population in absolute terms (colored area) and in percentage (line) for low-income economies (gross national income, GNI ≤ 1085 $ per capita), middle-income economies (1085 $ < GNI < 13'205 $), and high-income economies (GNI per capita ≥ 13'205 $). (B) Percentage of urban population living in the country's largest metropolitan areas in low-, middle-, and high-income economies (The World Bank, 2023b). *Data from The World Bank. (2023b).* World Development Indicators. *https://databank.world-bank.org/source/world-development-indicators.*

productivity, reducing the potential earnings. On the other hand, the poorest people suffer more from heat stress because they have less access or possibility of using energy; this widens the disparities between households and reduces welfare in an area where similar climatic conditions take place (Li et al., 2023). Fig. 4.3 shows the percentage of the urban population with

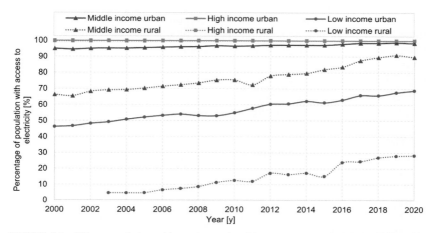

FIGURE 4.3 Urban population with access to electricity: percentage of the urban population with access to electricity in low-, middle-, and high-income economies. *Data from The World Bank. (2023b).* World Development Indicators. *https://databank.worldbank.org/source/world-development-indicators.*

access to electricity considering countries all over the world with low-, middle-, and high-income economies in the period 2000—20. Evidently, the largest percentage of the population that does not have access to electricity is found in those countries with low incomes, decreasing from 53.8% to 31.1% in the first 20 years of the 21st century, whereas it is higher than 99% in high-income economies.

SEP shows its greatest manifestations mainly in urban areas under the impulse of increasingly recurring heat waves and urban overheating. This phenomenon can even be found in rural areas, where it is mainly driven by limited access to energy supplies and climate change. In both cases, the poorest segments of the population are also the most affected. Rural energy poverty has, as an indirect consequence, resulted in reduced agricultural production (Feeny et al., 2021) or in the lack of essential energy services that limit the possibility of keeping buildings sufficiently cool (Ugembe et al., 2022).

The scientific literature that has dealt with the effects of urban overheating on SEP is limited, and this correlation is analyzed in only a few works. For the first time, a methodology to consider energy cooling needs in a new and broader definition of energy poverty was proposed in Spain by Sánchez-Guevara Sánchez et al. (2017).

A neighborhood-level study (Thomson et al., 2019), performed among 521 households in Budapest (Hungary), 620 in Prague (Czech Republic), and 598 in Gdańsk (Poland) and Skopje (North Macedonia), with 55 interviews in each, reviewed the factors that mostly impact SEP vulnerability. The exposure factors that were found to be worthy of interest were UHI (obviously), sunlight exposure through windows, building ventilation, and materials. Concerning sensibility, they recorded that main concerns derive from households with preexisting health problems and/or with older people, while regarding the adaptive capacity, interventions on the building envelope, the use of the rooms, the accessibility/availability of cool outdoor spaces (private or public garden), economic resources, restrictive tenancy relations, and technical and physical limitation in buildings retrofit appeared to be more effective factors.

The authors reported in their review the data concerning "dwelling comfortably cool during summertime" collected in the European Union countries (European Commission, 2012). Eastern and southern countries achieved the highest percentages: Bulgaria 49.7%, Greece 34%, Malta 35.4%, and Portugal 35.7%. These data become significantly worse when considering the low-income population: Bulgaria 70.7%, Greece 48.9%, Malta 40.1%, and Portugal 41.4%.

The study reported in Sanchez-Guevara et al. (2019) explores the areas of Madrid and London where there is greater exposure to SEP by mixing information relating to UHIs, building characteristics, and other aspects of vulnerability (age, salary, health conditions, etc.). Because of the different climatic, socioeconomic, and building conditions, it was found that the issue is not very alarming in London, due to the lower summer temperatures. About 1% of the population lives in areas characterized by low-quality houses, low income, and high temperatures during daytime and nighttime, while 4% of elderly people live in the hottest zones in the daytime. In Madrid, the last percentage increases to 18%, while the percentage of people living in inefficient buildings and with low income and experiencing high temperatures all day long is 2%.

Continuing to refer to the city of Madrid, in the work of Sánchez-Guevara Sánchez et al. (2020), the feminization of the problem of energy poverty is studied. Among the various factors that contribute to identifying energy poverty, aspects purely related to the SEP are considered, such as the availability of cooling plants and UHI. The results show strong gender inequality in energy poverty distribution. In about 23% of households, there is a risk of energy poverty, and in half of them, women are the breadwinners. The condition worsens in the case of women over the age of 65, who are without employment or with a part-time contract, and in the case of women who take care of family members because they spend more time at home.

In Li et al. (2023), the connection between high temperatures and energy poverty in China is assessed. In detail, the authors analyzed the impact of temperature on energy expenditure and income considering data from the period 2014−18. They found that the growth of one standard deviation in cooling degree days (54.4) increases the probability of being in energy poverty by 51.36%−80.83%. This is dramatic, given that cooling degrees days are increasing, as shown in Fig. 4.4, and therefore, this phenomenon will be exacerbated in the next years.

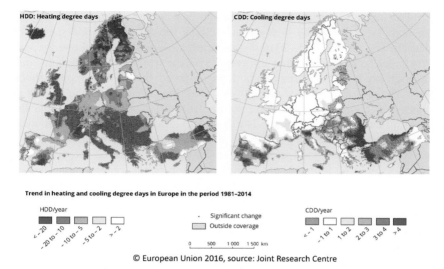

Trend in heating and cooling degree days in Europe in the period 1981–2014

© European Union 2016, source: Joint Research Centre

FIGURE 4.4 Variation of heating and cooling degrees day (1981−2014) in Europe. How the heating (A) and cooling (B) degrees day (1981−2014) in Europe are varying, according to the European Environment Agency (European Environment Agency, 2016). *From https://www.eea. europa.eu/legal/copyright*

The paper of Tabata and Tsai (2020) focused on the Japanese context. Through the low-income high costs indicator and considering a threshold for it equal to 10% (it was calculated based on the expenditure on electricity and the income during the summer period), it is estimated that the percentage of households potentially in a fuel poverty condition is 3.2% within a scenario (data considered), which highlights that 0.93% of the surveyed households report fuel poverty; however, 5.53% of poor households can be considered to be in fuel poverty. Particularly susceptible to this risk are elderly couples, people living in old and/or big homes with old air conditioning systems, and single-parent households.

The fuel poverty for families living in 51 social dwellings in Cadiz (Spain) was examined in Bienvenido-Huertas et al. (2020). Cadiz is in the South of Spain, where energy consumption is mainly due to the cooling of buildings. In particular, it analyzed the potential reduction of energy poverty through the use of natural ventilation strategies despite conventional electricity-driven air conditioning systems in a scenario in which adaptive comfort is assumed. Results show that the use of air conditioners determines the risk of energy poverty in all 51 households if families with a Public Income Indicator of Multiple Effects (IPREM) not sufficiently higher than 1 are considered. Instead, the natural ventilation allows reducing this number to 13 and 16 in the case of families with monthly incomes between 1 and 1.2 times the IPREM.

The impact of rising temperatures and heat waves on energy poverty in Vietnam is evaluated by Feeny et al. (2021), referring to a wide sample of data for the period 2010−2016. Considering a multidimensional energy poverty index, it was found that in perspective, the overall energy poverty rate of Vietnam is 19%. Furthermore, it is observed that a temperature shock (two or more standard deviations from the average) determines the growth in energy poverty of 2.5 percentage points. In addition, it is observed that the most affected areas are in the North and Central Coast regions, and most prone to temperature changes are the households located in rural areas.

In a study performed in Jaffna, Sri Lanka, that involved internally displaced people of four clusters of rehabilitation residence programs in four regions (Kumareswaran et al., 2021), aspects emerging from the thermal comfort survey suggest that the population experiences a condition of energy poverty. About 36% of subjects judged as unacceptable the indoor thermal environments, as the mean thermal preference was found to be -0.6, and 64% of the interviewees preferred a cooled indoor environment. At the same time, it was observed that the use of fans remains low. One fan on average is usually available in each household, and it is switched on for less than 4 hours during sleeping hours. This can seem to be the behavior of energy conservation-inclined people, but it hides a condition of energy poverty if the electricity costs and expenditures are even considered.

Opposite conclusions were found by Awaworyi Churchill et al. (2022). These authors concluded that, for a country like Australia with a mild climate, especially in short-term projection, climate change can reduce energy poverty, as there is a predominant sensitivity to cold days. However, the same authors predict that the persistence of the increase in temperatures could reverse the trend and make the summer issue prevalent. However, in the opinion of the authors of this chapter, the increase in temperature will lead to an increase in energy poverty, given that a large part of the population will not be able to control the environmental conditions in the houses during the warm season, finally exacerbating the scarcity to the access to energy and systems. In addition, the condenser heat of cooling systems and the increase of electricity at the urban scale will exacerbate the local climate, affecting, once again, the population not able to control the indoor microclimate.

There are some final considerations, taking into account the overlays between UHI effects and COVID-19 pandemic aspects in the last 3 years. In recent work (Wang & He, 2023), the effects of UHIs are evaluated by combining them with those of the COVID-19 pandemic. Contrasting factors and aspects are combined. The COVID-19 pandemic has reduced anthropogenic heat released in urban areas due to the lockdown, which, for example, almost canceled travel. On the other hand, the consequent restrictions had negative effects on the economy and aggravated the situation of poverty, mainly among the population living in less energy-efficient and modern buildings, who could not afford an air conditioning system. The heat rejected by the air conditioners moved from the administrative districts, housing most of the offices, and from the school buildings to the residential areas. The contribution of direct solar radiation increased because it is less shielded from airborne pollutants. Overall, the pandemic improved the climatic conditions of cities and had hidden effects on overheating risks. The use of masks, gloves, and special and waterproof clothing for healthcare personnel worsens heat exchanges with the surrounding environment and limits the acceptability of high temperatures. Similarly, the less use of public transport and the increased use of private cars aggravated the risk of heat waves, and the higher physical activity level for walking and cycling trips, added further risks of exposure.

Moreover, fear of infection, isolation, and loneliness have had psychological repercussions on people, making them more susceptible to thermal stress. The impossibility of administering or accessing periodic treatments has worsened the health condition of already weaker subjects, exposing them to greater thermal risk. More directly, the closure of many activities and the loss of many jobs have increased the poor population groups and consequently the SET. The phenomenon of energy poverty, as well as urban climate data, has been affected by the pandemic and by the measures to limit the infection and will be partly altered in the future due to the change in habits. Furthermore, COVID-19, unfortunately, had a greater impact in terms of deaths on the weakest population, the elderly, and the sick, who are, and would also have been, those most exposed to the risk of UHIs and heat waves.

Therefore, even more in recent years, the need to study UHIs emerged, in order to understand their evolution and the causes that make them more intense and frequent over the years. In the next section (Section 4), the causes of UHIs in cities are investigated, with a comparison with the peripheral areas, and the analytical approaches adopted to study them are evaluated.

4.4 The impact of urban heat islands between cities and peripheries

Geographic location, city organization, building structures, and anthropogenic activities largely affect the city microclimate and the formation of UHIs. The detection of temperature changes within urban centers requires careful assessments because temperature differences are influenced by various factors. The accuracy of satellite measurements of air temperature with modern measuring instruments in recent years has allowed for the improvement of the analysis quality and temperature variations mapping with high accuracy. In addition, it is possible to assess daytime and nighttime temperature fluctuations, the frequency of urban precipitation, how urban areas affect such weather events, or which contribution aerosols might have. Thanks to higher-resolution images, information regarding air humidity during the day can be obtained, providing a better understanding of the heat island phenomenon and its effects on different areas of interest.

The urban microclimate is largely affected by the city configuration and by the materials characterizing it. Building walls, roofs, or urban pavement absorb radiation and retain and emit heat differently compared to vegetation, trees, or water bodies. Plants absorb water from the soil through their roots, and by the process of evapotranspiration, the water is transformed into water vapor released into the air. Water bodies, on the other hand, achieve the same result through the process of evaporation that occurs on their surface: evaporation subtracts sensible energy from the air that becomes cooler (according to the so-called "evaporative cooling" effect). In this way, vegetation and water bodies help to cool the air, even if an increase of latent heat should be evidenced, being, finally, an almost isenthalpic effect. This is specified because latent heat must be considered too when outdoor comfort is investigated.

At the same time, the tree shade creates a natural barrier that partially prevents solar radiation from reaching the soil. Frequently in cities, there is less shade and less moisture than in rural landscapes, as structures are generally made of materials such as brick, asphalt, concrete, and steel, often with high absorption coefficients that cause absorption of solar radiation, gradually released as heat, contributing to the rising local and urban temperatures.

Dark artificial materials absorb radiation and emit more heat than natural surfaces because, as they reach higher temperatures due to the slow release of accumulated heat, they tend to generate a rise in ambient temperatures, particularly in the hours after sunset. In addition, building materials are not permeable to water, and prevent the penetration and evaporation cycle of water avoiding rapid cooling.

The urban layout, however, even contributes to the formation of UHIs, because the small distance between buildings and their sizes, affect the wind flow. Indeed, even if buildings are divided by roadways, the high density of urban settings creates conditions for the stagnation of air in certain areas where ventilation is limited. This causes the permanence of "dirty and hot" air, where physical, chemical, and biological agents modifying the natural characteristics of the atmosphere are concentrated. This is related to the issue of poor air quality and the increase in temperatures that is favored by the lack of air exchange, with the penetration of cool and fresh airflows in urban canyons. As described above, buildings release heat, and this rate, if not removed by convective surface—air heat exchange, remains trapped in the dense urban structures, preventing these large thermal masses from cooling effectively.

The formation of UHIs, however, is not exclusively due to the presence of buildings that act as obstacles to the proper ventilation of cities or "thermal flywheels," but this is even largely affected by human activities. Industrial plants, air conditioning systems inside buildings, and cars release heat into urban environments as anthropogenic heat sources. Finally, weather conditions and the geomorphology of the area can contribute to rising temperatures, even more during sunny days, when precipitations and ventilation are poor, or when there are natural barriers, such as mountain ranges that obstruct the wind flows.

Extreme heat in cities is a serious public health problem as heat-related illnesses are rising worldwide, as well as deaths from the most extreme events. Heat and drought are more frequent causes of death than other weather phenomena, and their incidence in urban areas is more pronounced, particularly in low-income communities and older and/or more vulnerable segments of the population, and on fragile children for whom heat waves and extreme heat can lead to premature deaths. Finally, climate change and rising temperatures present their greatest effects on the urban population as opposed to those living in rural areas.

Several studies focus on the impact of rising temperatures and urban heat on mortality rates. According to the data from ECMWF (2023), there were about 70,000 deaths in Europe caused by the 2003 heat wave, and, between 1991 and 2018 in 43 countries, 37% of deaths were due to heat. As reported in a study conducted by Iungman et al. (2023), based on data from 93 European cities, where 57 million citizens over the age of 20 years live, the estimation of premature deaths due to UHI amounts to about 6700 during the summer period (June 1 to August 31, 2015). This study then analyzes the benefits in the reduction of deaths that can be achieved by increasing urban green areas in cities. Based on data obtained by Imhoff et al. (2010) in the United States, for land surface temperatures, considering an annual average of temperature differences between cities and rural areas, the increase in urban microclimates is about $+2.9°C$, except in urban areas in biomes where the climate is arid or semiarid. During the summer period, the highest average value is $+8°C$, recorded in urban areas standing in biomes where there is a large presence of mixed temperate and broadleaf forests.

As is clear from the data presented, the phenomenon of UHI is becoming increasingly important, and, considering the serious consequences it also has on citizen health, many researchers have proposed progressively accurate methods of analysis to understand this phenomenon, as summarized in Table 4.1.

Park et al. (2023), based on a 25-year period of observations, from 1997 to 2021, investigated the interactions between UHI and temperature data in Seoul (South Korea). The study specifically analyzes the strong influence of an anomalous anticyclonic and the expanded Tibetan high during heat waves in South Korea. From the data obtained, based on the difference between the average of urban and rural stations of the daily minimum and maximum temperature, an increase in the UHI intensity (UHII) of $0.53°C$ and $0.20°C$,

TABLE 4.1 Summary of the methods of analysis on the topic of urban heat islands.

References	Year	Period of analysis	Location	Type of study
Park et al. (2023)	2023	1997–2021	Seoul, South Korea	The investigation is based on 25 years of observations focused on the interactions between UHIs and temperature data in Seoul (South Korea)
Barrao et al. (2022)	2022	March 2015 to February 2021	Zaragoza, Spain	The study is based on a campaign of measurements performed from March 2015 to February 2021 with the aim of better understanding the increasingly frequent extreme thermal events within cities.
Sun et al. (2020)	2020	2013–18	Shanghai, China	The study is conducted through a machine learning method—the BRT—to understand the impact of urban geometry on the UHI phenomenon.
Sheng et al. (2022)	2022	2019–20	Shanghai, China	In this investigation, the prominent features of hybrid landscapes characterized by two-dimensional maps were selected to analyze their cooling effects.
Ramsay et al. (2023)	2023	Multispectral imagery 1993–2019 Surface temperature, in near-infrared, and red bands 1970–2020	Makassar, Indonesia	The study is based on 29 years of Landsat satellite-derived surface temperature, corroborated by in situ temperature measurements, to provide a detailed spatial and temporal assessment of UHIs.

| Mentaschi et al. (2022) | 2022 | 2003–20 | Tokyo, Delhi, Jakarta, Shanghai, Cairo, Sao Paulo, Mexico City, New York, Moscow, Buenos Aires, London, Lagos, Paris, Sydney | Several (i.e., 14) megacities spread across all continents are analyzed to create a long-term high-resolution global dataset of diurnal SUHI. An unprecedented global-scale overview of the spatiotemporal variability of urban-rural temperature differences is proposed. |
| Santamouris (2020) | 2020 | Various | Sydney, Australia | A literature review of the most recent advances and knowledge on the specific impact of current and projected urban warming on a range of issues is provided. New findings related to the characteristics and magnitude of urban warming are proposed, as well as the analysis of recent knowledge on the synergies between UHIs and heat waves. |

respectively, during heat waves was evidenced. This analysis was conducted based on synoptic patterns on heat wave days with strong and weak intensity UHI, based on the comparison of UHII under heat wave and nonheat wave conditions. The results show that UHII increases under heat wave conditions, but both negative and positive interactions are observed, depending on weather conditions and synoptic patterns. The first ones mainly depend on relative humidity and cloud fraction and the second ones on relatively warm, dry, calm, and clear weather conditions. It was noted that the latter have been more frequent in recent years. In addition, the intensity of UHIs tends to increase during more intense heat waves. If favorable conditions, as indicated by the study, continue to occur in the coming years, in addition to predicting worsening heat waves in the city of Seoul, it will be possible to predict, based on the change in a synoptic model, thermal changes in urban environments in the future. Barrao et al. (2022) analyzed the UHI phenomenon in the context of climate change, with the aim of better understanding the increasingly frequent extreme thermal events within cities. Spatial diversity plays a key role in the careful analysis of the phenomenon, and, in this study, using a thermos-hygrometric network of hourly observatories from 21 sensors, the UHI of the Zaragoza urban area in Spain was investigated. The campaign of measurements lasted from March 2015 to February 2021, and based on the amount of data collected, it was necessary to carry out a synoptic analysis capable of providing accurate data to elaborate the necessary information.

The study revealed the differences in the UHII between the city center and suburban areas, showing a temperature increase, mainly during the nighttime hours, of around 2°C for urban areas. In the innermost areas of the city center, sensors recorded differences of even more than 4°C between the center and the suburbs. Little or no variability in the intensity of UHIs was recorded for spaces further from the city center, such as peri-urban neighborhoods, or in those where there is a high presence of green areas and urban parks. From this analysis, it was possible to develop future urban plans for the city of Zaragoza or other Mediterranean cities with similar urban structures.

In the study of Sun et al. (2020), urban architecture and its impact on the phenomenon of UHI are analyzed, specifically in Shanghai, China, one of the aforementioned megacities. A machine learning method, the boosted regression tree (BRT), is used to analyze the impact of urban architecture and its effects on land surface temperature (LST), considering intensity, roughness, fragmentation, complexity, and diversity. The study shows the most influential factors on the thermal characteristics of the environment. Ten different architectural types are considered for the definition of the urban space and are modeled based on the typical characteristics of the Shanghai metropolitan area. According to results obtained in the summer period, the most influential factors on the LST are the building coverage ratio, with a contribution of 39.3%, the mean architecture height, with a contribution of 16.5%, the mean architecture height standard deviation affecting 12.3%, and the mean architecture projection area with 10.4%. Other important parameters are the intensity and roughness of urban center development, in contrast to the parameters of intensity, roughness, and complexity, which are not particularly affected. The building coverage ratio is the parameter that most influences urban surface thermal characteristics, contributing to a variation in the temperature level of $+2.7°C$; surface temperature rises as this parameter increases. Thereafter, the height of buildings is the second most significant parameter with a temperature growth of $0.9°C$; as this parameter increases, the urban surface temperature decreases, but this effect reduces markedly from the height of 26 m. According to this study, decreasing development intensity and architectural base area and increasing building height and roughness can improve thermal conditions within urban centers at the neighborhood scale. Such a study, through the BRT method, provides important data for rational architectural planning and management by understanding the relationship between 3D urban architectural patterns and the effect on UHIs.

In the same city, Sheng et al. (2022) analyzed the cooling effects of green spaces and water bodies on UHIs and on the physical characteristics, layout, and surrounding environments. This study aims to develop a flexible technical framework, which can be applied to different cities to understand and analyze depending on the characteristics of different built environments, the cooling intensity, and the efficiency of hybrid landscapes. The study was conducted by dividing the research samples into five groups according to the different construction characteristics of buildings and their proportions and shadows, and the marginal effects of hybrid landscape elements on relative temperatures and surface temperatures were calculated. Several conclusions emerged from this investigation, namely, in high-density neighborhoods, the warming or cooling effect of hybrid landscapes is not significant, and in densely populated neighborhoods, green areas and water bodies have no cooling effect; on the contrary, in urban forests, the effect is visible and does not depend on the built environment. Finally, it was noted that water bodies have positive effects on reducing surface temperature, but the effect on thermal environments is uncertain and can sometimes cause the opposite effect, generating temperature increments. The study aims to provide an analysis model that can be applied to cities in similar climatic conditions to mitigate the UHI phenomenon.

In tropical climates, Ramsay et al. (2023) analyzed the scale and shape of UHIs, as heat, in such areas, has a great influence in terms of morbidity and mortality, particularly in informal urban settlements. From the analysis, it is evidenced that in the parts of the city that have been urbanized for a long time, the recorded surface heat islands (SUHI) were up to $9.2°C$, while in the informal settlements, they were above $6.3°C$. In the newly urbanized areas, the increase in UHI occurred in greater weight before these became urbanized for 50%. In the areas with the longest urbanization, the UHI set remained stable during urban expansion. The contribution of green and blue spaces, that have protected some informal settlements from the increases in temperatures recorded in urban centers, has been important, and the presence of vegetation is believed to be crucial in mitigating the increasing heat loads due to anthropogenic climate change and urban expansion. A remarkable aspect was also found by evaluating the distance of urban areas from the coast. For areas more than 4 km away from coastlines, and with a normalized difference in vegetation index of less than 0.2, surface temperatures were higher. Based on this study, it was concluded that green spaces are essential for urban design, redevelopment, and the improvement of informal settlements, in the context of tropical cities.

The effects of SUHIs have a greater intensity variation than the UHIs, as confirmed by Mentaschi et al. (2022). The authors developed a long-term high-resolution global dataset concerning diurnal SUHI, providing an overview of the spatiotemporal variability of temperature differences between urban centers and rural areas on a global scale. Based on data for the period from 2003 to 2020, the study presents values for urban areas around the world and shows how three extreme SUHI days are more than twice the number of days when high temperatures of the average SUHI are recorded in the warm season. The value recorded for extreme SUHI days locally exceeds the average warm season values by 10 to 15K, values that within cities can vary by an order of magnitude from cooler places. Currently, SUHI values have increased on average for the warm season worldwide and are higher in percentages by +31% compared to 2003 or in terms of average temperature, higher by 1.04K also referring to the same year. This phenomenon is related to the increase in population within cities and thus increasing urbanization worldwide, in conjunction with rising temperatures and the frequency and intensity of hot seasons due to global warming that will lead to increased health risks in the most vulnerable neighborhoods in the coming years. This study presents a current scenario, in which it is deemed possible to generate fine-grained UHI models to support informed and targeted urban policies toward proper mitigation strategies. As underlined by this study, the possibility of using high-resolution models allows for their application to urban environments worldwide to fill the gaps in this field. In addition, through a cloud-based geospatial analysis platform in the Google Earth Engine, it is possible to consult the dataset produced to make more and more data available to anyone interested in improving knowledge of the heat island phenomenon within cities.

Recent advances and modern knowledge on the impact of current and projected urban warming were analyzed by Santamouris (2020), considering it in terms of energy, peak electricity, morbidity, mortality, air quality, and urban vulnerability. New findings on the characteristics and magnitude of urban warming are also discussed, and recent knowledge on heat waves and

UHI interactions is examined. Particular attention is focused on how current studies provide different information regarding the UHI phenomenon, but also how these are rich in quantified data, but without a linkage between different scientific disciplines, which does not allow for a comprehensive and holistic consideration of the issue.

As highlighted in the various studies, the complexity of the UHI phenomenon also translates into the difficult implementation of solutions on a large scale for the mitigation of its effects. Climate change influences this phenomenon in terms of the duration or frequency and intensity of heat waves, which have continuously increased in the last few years. Often the lack of quantitative and qualitative data does not allow for the comprehension of the specific factors that contribute to the formation of UHIs, and this leads to a continuous search for data despite monitoring aimed at providing increasingly accurate and detailed information about UHIs.

The factors influencing the phenomenon are several, and the periods in which they are verified are different. In addition, the geometric variability and urban structure of cities, together with population increasing density, make viable mitigation strategies valid only in certain contexts. The most globally efficient ones, however, point to albedo management for urban surfaces, leading to increased reflection of sunlight, and reducing its absorption. Others, as detailed in Section 4.5, involve increasing the amount of green cover in city areas on a large scale, both at street level and on the roofs or walls of buildings. Today, as mentioned, important progress has been made, by satellite measurements and high-resolution equipment, that allows for registering more accurate and real-time mapping of UHIs than in the past years, when the common practice was analyzing the phenomenon by measuring the temperature differences between downtown and rural areas. The common goal must even be the reduction of heat caused by inhabitants in urban centers, and, given the increasing population expected in the next years in these areas, further studies and analysis of the phenomenon are needed to enable designers to make increasingly informed choices.

4.5   Mitigation strategies to face urban heat islands

The UHI phenomenon affects our cities in an increasing and even more serious manner, causing irreversible effects on our lives. The uncontrolled increase of urban density without a defined plan has pushed the conditions of many cities to the limit of poor livability, especially during the summer, because the global rise in temperatures is added to urban overheating. The growing decrease of green spaces in urban contexts and the increasing need for air conditioning in indoor spaces affect the external microclimate, making it less and less favorable. This has repercussions on the inhabitants, their health, and their comfort conditions. The UHI phenomenon and climate change are summed to the growth of the population that, predominately, will live in the city centers, causing the increase in housing density. On the one hand, the building construction is on the rise; on the other hand, the green areas in the city centers are decreasing, contributing to a worsening outdoor microclimate because these are often replaced by cemented/asphalted built environments. As confirmed by Zhang et al. (2018), urban vegetation has a regulatory role in cities, both during the summer and winter, with positive microcirculatory effects. In addition, it has beneficial repercussions on outdoor thermal comfort and human psychology (Shooshtarian & Ridley, 2017).

In view of reducing urban overheating, several researchers have evaluated the reforestation of cities as a strategy to mitigate the outdoor microclimate. Yilmaz et al. (2018), for example, have analyzed different scenarios of refor-estation in industrial, rural, and urban areas, comparing several vegetative species. Their outcomes show the advantageous effects of afforestation on outdoor thermal comfort in regions with cold climates, both during the summer and winter. Some other studies are focused on the relationship between the thermal comfort level and the addition of different plats/tree types in cit-ies. Toy and Yilmaz (2010) assessed that for continental climate, landscape design should be oriented to the plantation of deciduous trees, rather than coniferous, to prevent the recreational areas from cold stress in spring and autumn, and heat stress during the summer. Tan et al. (2016) made a relation between the sky view factor and the cooling effect of trees, and concluded that the cooling effect and reduction of sensible heat are doubled when the vegetation is arranged in wind corridors and not in leeward areas. On a general evaluation, green infrastructures in European cities can reduce urban temperatures from $1°C$ to around $3°C$, assuring a tree coverage of at least 16% to achieve a $1°C$ drop (Marando et al., 2022). As mentioned above, the rise of latent heat due to evapotranspiration must be considered in suitable indicators of outdoor comfort. According to Iungman et al. (2023), each European city has a different cooling response to the urban greenery addition, as shown in Fig. 4.5, by increasing the tree coverage by 30%. The potential cooling capacity in European cities can vary between $-0.02°C$ and $0.86°C$. Italy, Romania, Portugal, and Spain are the countries with the high-est potential in reducing the UHI phenomenon through the increase in tree coverage (which led to the great benefits of increased reflection of solar radi-ation and shadows at the pedestrian level).

Implementing mitigation strategies to reduce overheating also means implementing building-scale measures, which have an effect not only on the energy performance of the building itself but also on the reduction of the UHI phenomenon. Greenery on the building envelope, both vertical and hori-zontal, can have benefits on urban environments, due to the evapotranspira-tion effect of plants, which causes a temperature reduction of outdoor air and better air quality. These benefits provide mainly a reduced use of cooling systems in buildings and thus less rejection heat from air cooling systems. Indeed, and this is important to be underlined, heat pumps fueled by

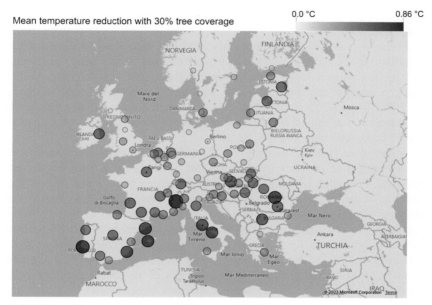

FIGURE 4.5 The cooling capacity of EU cities according to tree coverage cooling capacity of EU cities with 30% tree coverage. *Data from Iungman, T., Cirach, M., Marando, F., Pereira Barboza, E., Khomenko, S., Masselot, P., Quijal-Zamorano, M., Mueller, N., Gasparrini, A., Urquiza, J., Heris, M., & Thondoo, M. (2023). Nieuwenhuijsen, cooling cities through urban green infrastructure: A health impact assessment of European cities.* The Lancet 401(10376), 577–589.

renewables probably are a free of costs solution but not a free-of-emitted heat technology (Ascione, 2017).

At the same time, the addition of a terrain substrate and a vegetative layer, with a thermal insulation layer, as typical in green roofs and living walls, causes a decrease in the building surface' temperature, contributing to lessening the heat transfer between the outside and the inside. The benefits of green roofs as a passive strategy for the building envelope are confirmed by Mazzeo et al. (2023) who, for a Mediterranean area, evaluated the green roof impact on the UHI mitigation and on the building energy demand and thermal comfort. By considering different configurations of green roofs, it was predicted, through an artificial neural network (ANN), the building surface temperature of the inner and outer layers, and the indoor air temperature, confirming the positive effect of such a strategy. With a similar approach, Asadi et al. (2020) investigated the potentiality of green roofs in mitigating the UHI in Austin, Texas. It was pointed out that the green roofs applied to buildings having heights between 15 and 25 m, with the highest value of solar radiation and lowest distance from water bodies, have the highest cooling effect on LSTs. In general, in hot, humid, and temperate climate cities, green roofs can produce a reduction of the median surface temperature of 30°C and 28°C, respectively. As green roofs, even green walls can decrease the building energy need and, if applied to a large scale, contribute to the reduction of the UHI phenomenon. Both during the heating and cooling periods in different climatic regions, green walls could cut the building energy demand by 16.5% and 51%, respectively, and mitigate the UHI of up to 5°C (Susca et al., 2022). Anyway, where the rainfalls in the summertime are low, the economic and environmental costs of irrigation should be considered too (Ascione, 2017).

Some other passive strategies for the building envelope are investigated for the mitigation of urban overheating and reduction of the building energy demand. A decrease in the ambient temperature can be obtained through the addition of radiative materials on building surfaces, and a lowering of the building energy consumption coupled with an increase in surface solar reflection can be gained, thanks to thermochromic coatings. High-reflectance materials produce a lowering of surface peak temperatures up to 15°C (Manni et al., 2021). The use of cool coatings at the urban level is a common research topic in many studies. Lopez-Cabeza et al. (2022) confirmed the importance of using the right materials in urban environments by evaluating their albedo, as this has a significant impact on the well-being of citizens, the quality of the outdoor environment, and energy demand. The lower heat accumulation of surfaces due to the use of high albedo coatings allows for a reduction in surface temperatures. These materials can, at the same time, have a negative effect on comfort, and in particular on the physiological equivalent temperature, and to face this disadvantage, the authors suggest the use of materials with albedo values of about 0.4 for walls and higher than 0.7 for pavements.

However, the cooling potential of each mitigation strategy could be controversial according to the climatic region. Chen et al. (2023) evaluated the effect of rooftop mitigation strategies in subtropical climates. Cool roofs, green roofs (GR), rooftop photovoltaic panels (RPVP), and PV panels plus green roofs (PVP + GR) were numerically tested in Guangzhou (China). The results showed that RPVP and PVP + GR are viable strategies for reducing temperatures by mitigating the effect of UHI during the indoor daytime and particularly from 12 to 5 p.m. according to local sidereal time, compared to RPVP and GRs that have lower potential but can additionally lead to an increase in specific humidity. Even Catalbas et al. (2021) analyzed the impact of PV addition on roofs in the urban microclimate, and they focused their attention on photovoltaic green roof systems in an industrial zone in Turkey. It results in an innovative strategy for reducing CO_2 emissions during electricity generation. The aim was to evaluate the long-term cost-effectiveness and carbon emission benefits of green roof photovoltaic applications. This combination can become a model of green technology and provide several benefits, such as reducing energy consumption, improving sound insulation with noise reduction, and reducing air pollution, in addition to mitigating the UHI effect. The results obtained have made it clear that this application is not only suitable for individual homes but also for industrial areas, as they provide both public and individual benefits, although the initial investment costs to be incurred are slightly high at present.

Some other innovative technologies are recently attracting the interest of researchers for their high mitigation potential, nanofluorescent, quantum dots and thermochromic materials, and photonic and plasmonic structures, as provided by Santamouris and Yun (2020). Nanofluorescent materials have the characteristic of emitting extra nonthermal electromagnetic radiation determining a lower surface temperature if compared to the traditional cool materials. Nanofluorescent materials together with quantum dots are among the more promising materials for UHI mitigation. Thermochromic materials have positive effects on the reduction of peak ambient and surface temperatures. Photonic materials have a radiative cooling effect during the daytime and exhibit surface temperatures lower than ambient temperatures even if under the sun. These materials are suitable in dry climatic zones and lightly polluted zones and, if applied to the city scale, can alter the vertical stratification of temperatures in the lower parts of the atmosphere. Even if potentially effective and promising, all these solutions are still objects of several research to translate them into building products. Table 4.2 shows the investigated studies about mitigation strategies to reduce UHIs, evidencing the year, the location, the study's approach, and the main contents.

4.6   Conclusive remarks

The phenomenon of UHI largely affects our cities, and, by considering the increasing urban density and population, it can be seen that the consequences of urban overheating on the citizens' health and on the urban microclimate are destined to get worse. This chapter pointed out the main criticalities that cause the UHI in cities, comparing them with the most peripheral and rural areas, mainly with reference to the impacts that increased temperatures may have on the health of low-income populations. Building materials, urban configuration, anthropogenic heat gains, and natural morphology can contribute to heating cities, increasing their temperatures and further increasing the socioeconomic gap between the different segments of the citizens. Energy poverty, and therefore, the impossibility of accessing energy resources and equipment will be even more accentuated for low incomes, with harmful effects on the health of the weaker or the elderly. What actions should be taken to combat overheating in cities? How can this phenomenon be slowed down? The first way is the deep analysis and knowledge of the phenomenon to understand the mechanisms of evolution. The second is the adoption and promotion of mitigation strategies both on a building and urban scale. Intervening in urban design by bringing greenery and water ponds back to

TABLE 4.2 Significant studies about mitigation strategies for the urban heat island phenomenon.

References	Year	Mitigation strategy	Type of study	Location	Main content and conclusions
Yilmaz et al. (2018)	2018	Green infrastructure	Numerical and experimental	Turkey	Different vegetation characteristics were examined in rural areas, city centers, and industrial areas. It was found that afforestation in cold climatic regions has a positive effect both during summer and winter, in the outdoor thermal comfort.
Toy and Yilmaz (2010)	2010	Green infrastructures	Experimental	Turkey	In continental climate, the deciduous plants can prevent climatic stress both in the mid seasons and in the summer seasons
Tan et al. (2016)	2016	Green infrastructures	Numerical and experimental	Hong Kong (China)	Two different designs for urban greenery are evaluated in two climate-sensitive areas. It was pointed out that the cooling effect of trees is largely affected by the sky view factor.
Marando et al. (2022)	2022	Green infrastructures	Numerical	Europe	Green infrastructures in urban spaces can cool EU cities by 1°C to 3°C.
Mazzeo et al. (2023)	2023	Green roofs	Numerical approach	Palermo (Italy)	An ANN was developed to predict the surface temperature reduction and the air temperature reduction after the addition of green roofs.
Asadi et al. (2020)	2020	Green roofs	Numerical approach	Austin (Texas)	An ANN was developed to understand the relationship between green roofs and LST reduction.
Jamei et al. (2021)	2021	Green roofs	Review study	Hot−humid, temperate climate, and hot−dry cities	Green roofs can cause a reduction of 30°C and 28°C in median surface temperature in cities with hot−humid and temperate climates, respectively.

Reference	Topic	Approach	Location	Findings
Susca et al. (2022)	Green walls	Review study	Several climatic zones	Green walls can reduce UHI of 5°C and decrease the cooling energy demand of the building by 51% and the heating one by 16.5%.
Manni et al. (2021)	Cool coatings for facades, thermochromic and radiative cooling technologies	Experimental approach		Cool coating for facades can cause a lowering of the surface peak temperatures of up to 15°C; thermochromic increases the solar reflection of surfaces by 23% and reduce the building energy demand by 40%; radiative coatings cause a reduction of the ambient temperature of 5°C.
Lopez-Cabeza et al. (2022)	Cool coatings	Numerical and experimental approach	Seville (Spain)	High-reflectance coatings can have negative effects on users' comfort but can reduce surface temperatures of up to 25°C.
Chen et al. (2023)	Green roofs, cool roofs, rooftop PV panels, PV panels combined with green roofs	Numerical approach	Subtropical climate	The rooftop mitigation strategies are evaluated in subtropical cities. The outcomes show that PV panels combined with green roofs have a cooling effect on the city scale, cool roofs in urban and downwind areas, and green roofs in urban areas.
Catalbas et al. (2021)	Photovoltaic green roofs	Numerical approach	Turkey	Photovoltaic green roofs can have benefits even in industrial areas in addition to residential areas.
Santamouris and Yun (2020)	Innovative materials	Experimental approach	–	Photonic and plasmonic materials, quantum dots, and thermochromic, nanofluorescent materials are innovative solutions with high mitigation potential.

the city and adopting the right building materials are some of the actions that could lead to more livable cities. A future target should be a wider horizon than single buildings that look to the whole community to share energy, needs, capabilities, and scopes, with a view to targets of energy affordability to contrast the energy poverty and to give equal possibilities for a good and right life. Today, some international projects are being done in this direction. We must transform "living space" into "livable spaces," with a high standard of comfort, but compatible with the right and responsible use of energy, emissions, and environmental impacts. Fighting urban overheating, climate change, and energy poverty is our main aim, to have a decent and pleasant life for our generation and a sustainable future for the next generations. In this regard, it will be very important at the beginning of the work to identify actors, indicators, and parameters and map/monitor available materials, standards, and codes. The challenges are many, and we are close to the "too late" line. Now is the moment, considering the rethinking of living that the COVID pandemic has imposed. City centers and suburbs, novel architecture and existing buildings, heating- or cooling-dominated climates, reduction of energy demands or improvement of comfort, and the extension of life comfort or reduction of energy needs are dichotomic words that should be composed. We must recompose contrasting targets and unify the view. We should renew (according to the requirement of the new millennium, in terms of structural security, safety, sustainability, and functionality) and protect the existing beauty of historical centers, and, at the same time, we shall export amenities to peripheral suburbs, characterized by criticalities, sick syndrome building, insufficient heating, and impracticable conditions in summer. We must transform the massive housing stock, the suburbs, city peripheries, and margins into green towns, sustainable and beautiful. This is mandatory for our common future.

References

Asian Development Bank. (2011). *Asia 2050: Realizing the Asian Century: Executive summary*.

Adilkhanova, I., Santamouris, M., & Yun, G. Y. (2023). Coupling urban climate modeling and city-scale building energy simulations with the statistical analysis: Climate and energy implications of high albedo materials in Seoul. *Energy and Buildings*, *290*. Available from https://doi.org/10.1016/j.enbuild.2023.113092, https://www.journals.elsevier.com/energy-and-buildings.

Aghamohammadi, N., Fong, C. S., Farid, N. D. N., Ramakreshnan, L., & Mohammadi, P. A. (2022). Heat and mental health in cities. *Springer Science and Business Media LLC*, 81−107. Available from https://doi.org/10.1007/978-981-19-4707-0_4.

Asadi, A., Arefi, H., & Fathipoor, H. (2020). Simulation of green roofs and their potential mitigating effects on the urban heat island using an artificial neural network: A case study in Austin, Texas. *Advances in Space Research*, *66*(8), 1846−1862. Available from https://doi.org/10.1016/j.asr.2020.06.039, http://www.journals.elsevier.com/advances-in-space-research/.

Ascione, F. (2017). Energy conservation and renewable technologies for buildings to face the impact of the climate change and minimize the use of cooling. *Solar Energy*, *154*, 34−100.

Available from https://doi.org/10.1016/j.solener.2017.01.022, http://www.elsevier.com/inca/ publications/store/3/2/9/index.htt.

Ascione, F., De Masi, R. F., Mastellone, M., Santamouris, M., Tariello, F., & Vanoli, G. P. (2022). The trend of heat-related mortality in European cities. *Springer Science and Business Media LLC*, 293−320. Available from https://doi.org/10.1007/978-981-19-4707-0_15.

Awaworyi Churchill, S., Smyth, R., & Trinh, T. A. (2022). Energy poverty, temperature and climate change. *Energy Economics*, 114. Available from https://doi.org/10.1016/j.eneco.2022.106306, http://www.elsevier.com/inca/publications/store/3/0/4/1/3/.

Barrao, S., Serrano-Notivoli, R., Cuadrat, J. M., Tejedor, E., & Saz Sánchez, M. A. (2022). Characterization of the UHI in Zaragoza (Spain) using a quality-controlled hourly sensor-based urban climate network. *Urban Climate*, 44. Available from https://doi.org/10.1016/j. uclim.2022.101207, http://www.journals.elsevier.com/urban-climate/.

Bienvenido-Huertas, D., Sánchez-García, D., & Rubio-Bellido, C. (2020). Analysing natural ventilation to reduce the cooling energy consumption and the fuel poverty of social dwellings in coastal zones. *Applied Energy*, 279. Available from https://doi.org/10.1016/j.apenergy. 2020.115845, https://www.journals.elsevier.com/applied-energy.

Boardman, B., & Poverty, F. (1991). *From cold homes to affordable warmth*. London: Belhaven Press.

Carfora, A., Scandurra, G., & Thomas, A. (2022). Forecasting the COVID-19 effects on energy poverty across EU member states. *Energy Policy*, 161. Available from https://doi.org/ 10.1016/j.enpol.2021.112597, http://www.journals.elsevier.com/energy-policy/.

Castaño-Rosa, R., Solís-Guzmán, J., Rubio-Bellido, C., & Marrero, M. (2019). Towards a multiple-indicator approach to energy poverty in the European Union: A review. *Energy and Buildings*, 193, 36−48. Available from https://doi.org/10.1016/j.enbuild.2019.03.039.

Catalbas, M. C., Kocak, B., & Yenipınar, B. (2021). Analysis of photovoltaic-green roofs in OSTIM industrial zone. *International Journal of Hydrogen Energy*, 46(27), 14844−14856. Available from https://doi.org/10.1016/j.ijhydene.2021.01.205, http://www.journals.elsevier. com/international-journal-of-hydrogen-energy/.

Chen, B., Wang, W., You, Y., Zhu, W., Dong, Y., Xu, Y., Chang, M., & Wang, X. (2023). Influence of rooftop mitigation strategies on the thermal environment in a subtropical city. *Urban Climate*, 49. Available from https://doi.org/10.1016/j.uclim.2023.101450, http://www. journals.elsevier.com/urban-climate/.

Demographia. (2022). *Demographia World Urban Areas 18th Annual*. Available from http:// www.demographia.com/db-worldua.pdf.

ECMWF. (2023). *Unpublished content Urban heat islands and heat mortality demonstrating heat stress*. Available from https://stories.ecmwf.int/urban-heat-islands-and-heat-mortality/index.html.

European Commission. (2012). Eu-Silc module on housing conditions assessment of the implementation. Available from https://ec.europa.eu/eurostat/documents/1012329/1012401/2012 + Module + assessment.pdf

European Environment Agency. (2016). Trend in heating and cooling degree days (1981−2014). Available from https://www.eea.europa.eu/data-and-maps/figures/trend-in-heating-and-cooling.

Falasca, S., Zinzi, M., Ding, L., Curci, G., & Santamouris, M. (2022). On the mitigation potential of higher urban albedo in a temperate oceanic metropolis. *Sustainable Cities and Society*, 81. Available from https://doi.org/10.1016/j.scs.2022.103850, http://www.elsevier. com/wps/find/journaldescription.cws_home/724360/description#description.

Feeny, S., Trinh, T. A., & Zhu, A. (2021). Temperature shocks and energy poverty: Findings from Vietnam. *Energy Economics*, 99. Available from https://doi.org/10.1016/j.eneco.2021. 105310, http://www.elsevier.com/inca/publications/store/3/0/4/1/3/.

Ferrara, A. (2018). Jakarta in testa. Città gigantesche e tristi record di criminalità, inquinamento, povertà Ansa.it. *Megalopoli, Ecco Le Sei Città Più Grandi Del Mondo Entro Il 2030.* Available from https://www.ansa.it/canale_lifestyle/notizie/societa_diritti/2018/10/16/megalopoli-ecco-le-sei-citta-piu-grandi-del-mondo-entro-il-2030_2bd8b2cb-3261-4dd0-a512-118148d87b20.html.

Garshasbi, S., Feng, J., Paolini, R., Jonathan Duverge, J., Bartesaghi-Koc, C., Arasteh, S., Khan, A., & Santamouris, M. (2023). On the energy impact of cool roofs in Australia. *Energy and Buildings*, *278*, 112577. Available from https://doi.org/10.1016/j.enbuild.2022.112577.

Haddad, S., Paolini, R., Synnefa, A., De Torres, L., Prasad, D., & Santamouris, M. (2022). Integrated assessment of the extreme climatic conditions, thermal performance, vulnerability, and well-being in low-income housing in the subtropical climate of Australia. *Energy and Buildings*, *272*. Available from https://doi.org/10.1016/j.enbuild.2022.112349, https://www.journals.elsevier.com/energy-and-buildings.

Halkos, G. E., & Gkampoura, E. C. (2021). Evaluating the effect of economic crisis on energy poverty in Europe. *Renewable and Sustainable Energy Reviews*, *144*. Available from https://doi.org/10.1016/j.rser.2021.110981, https://www.journals.elsevier.com/renewable-and-sustainable-energy-reviews.

Hussain, S. A., Razi, F., Hewage, K., & Sadiq, R. (2023). The perspective of energy poverty and 1st energy crisis of green transition. *Energy*, *275*. Available from https://doi.org/10.1016/j.energy.2023.127487, https://www.journals.elsevier.com/energy.

Imhoff, M. L., Zhang, P., Wolfe, R. E., & Bounoua, L. (2010). Remote sensing of the urban heat island effect across biomes in the continental USA. *Remote Sensing of Environment*, *114*(3), 504−513. Available from https://doi.org/10.1016/j.rse.2009.10.008.

Iungman, T., Cirach, M., Marando, F., Pereira Barboza, E., Khomenko, S., Masselot, P., Quijal-Zamorano, M., Mueller, N., Gasparrini, A., Urquiza, J., Heris, M., Thondoo, M., & Nieuwenhuijsen, M. (2023). Cooling cities through urban green infrastructure: A health impact assessment of European cities. *The Lancet*, *401*(10376), 577−589. Available from https://doi.org/10.1016/S0140-6736(22)02585-5, http://www.journals.elsevier.com/the-lancet/.

Jamei, E., Chau, H. W., Seyedmahmoudian, M., & Stojcevski, A. (2021). Review on the cooling potential of green roofs in different climates. *Science of the Total Environment*, *791*. Available from https://doi.org/10.1016/j.scitotenv.2021.148407, http://www.elsevier.com/locate/scitotenv.

Karagianis, L. (2014). *The future is cities*. MIT Spectrum. Available from https://spectrum.mit.edu/winter-2014/the-future-is-cities/.

Kumareswaran, K., Rajapaksha, I., & Jayasinghe, G. Y. (2021). Energy poverty, occupant comfort, and wellbeing in internally displaced people's residences in Sri Lanka. *Energy and Buildings*, *236*. Available from https://doi.org/10.1016/j.enbuild.2021.110760, https://www.journals.elsevier.com/energy-and-buildings.

Li, X., Smyth, R., Xin, G., & Yao, Y. (2023). Warmer temperatures and energy poverty: Evidence from Chinese households. *Energy Economics*, *120*. Available from https://doi.org/10.1016/j.eneco.2023.106575, http://www.elsevier.com/inca/publications/store/3/0/4/1/3/.

Lopez-Cabeza, V. P., Alzate-Gaviria, S., Diz-Mellado, E., Rivera-Gomez, C., & Galan-Marin, C. (2022). Albedo influence on the microclimate and thermal comfort of courtyards under Mediterranean hot summer climate conditions. *Sustainable Cities and Society*, *81*. Available from https://doi.org/10.1016/j.scs.2022.103872, http://www.elsevier.com/wps/find/journaldescription.cws_home/724360/description#description.

Manni, M., Kousis, I., Lobaccaro, G., Fiorito, F., Cannavale, A., & Santamouris, M. (2021). *Urban overheating mitigation through facades: The role of new and innovative cool coatings.*

Rethinking building skins: Transformative technologies and research trajectories (pp. 61−87). Italy: Elsevier. Available from https://www.sciencedirect.com/book/9780128224779, 10.1016/B978-0-12-822477-9.00013-9.

Marando, F., Heris, M. P., Zulian, G., Udías, A., Mentaschi, L., Chrysoulakis, N., Parastatidis, D., & Maes, J. (2022). Urban heat island mitigation by green infrastructure in European Functional Urban Areas. *Sustainable Cities and Society, 77*, 103564. Available from https://doi.org/10.1016/j.scs.2021.103564.

Mazzeo, D., Matera, N., Peri, G., & Scaccianoce, G. (2023). Forecasting green roofs' potential in improving building thermal performance and mitigating urban heat island in the Mediterranean area: An artificial intelligence-based approach. *Applied Thermal Engineering, 222*. Available from https://doi.org/10.1016/j.applthermaleng.2022.119879, http://www.journals.elsevier.com/applied-thermal-engineering/.

Mentaschi, L., Duveiller, G., Zulian, G., Corbane, C., Pesaresi, M., Maes, J., Stocchino, A., & Feyen, L. (2022). Global long-term mapping of surface temperature shows intensified intra-city urban heat island extremes. *Global Environmental Change, 72*. Available from https://doi.org/10.1016/j.gloenvcha.2021.102441, http://www.elsevier.com/inca/publications/store/3/0/4/2/5.

Oliveras, L., Peralta, A., Palència, L., Gotsens, M., López, M. J., Artazcoz, L., Borrell, C., & Marí-Dell'Olmo, M. (2021). Energy poverty and health: Trends in the European Union before and during the economic crisis, 2007−2016. *Health and Place, 67*. Available from https://doi.org/10.1016/j.healthplace.2020.102294, http://www.elsevier.com/locate/healthplace.

Page, L. A., Hajat, S., Sari Kovats, R., & Howard, L. M. (2012). Temperature-related deaths in people with psychosis, dementia and substance misuse. *British Journal of Psychiatry, 200* (6), 485−490. Available from https://doi.org/10.1192/bjp.bp.111.100404United, http://bjp.rcpsych.org/content/200/6/485.full.pdf + html, Kingdom.

Pan, L., Biru, A., & Lettu, S. (2021). Energy poverty and public health: Global evidence. *Energy Economics, 101*. Available from https://doi.org/10.1016/j.eneco.2021.105423, http://www.elsevier.com/inca/publications/store/3/0/4/1/3/.

Park, K., Jin, H. G., & Baik, J. J. (2023). Contrasting interactions between urban heat islands and heat waves in Seoul, South Korea, and their associations with synoptic patterns. *Urban Climate, 49*. Available from https://doi.org/10.1016/j.uclim.2023.101524, http://www.journals.elsevier.com/urban-climate/.

Ramsay, E. E., Duffy, G. A., Burge, K., Taruc, R. R., Fleming, G. M., Faber, P. A., & Chown, S. L. (2023). Spatio-temporal development of the urban heat island in a socioeconomically diverse tropical city. *Environmental Pollution, 316*. Available from https://doi.org/10.1016/j.envpol.2022.120443, https://www.journals.elsevier.com/environmental-pollution.

Sanchez-Guevara, C., Núñez Peiró, M., Taylor, J., Mavrogianni, A., & Neila González, J. (2019). Assessing population vulnerability towards summer energy poverty: Case studies of Madrid and London. *Energy and Buildings, 190*, 132−143. Available from https://doi.org/10.1016/j.enbuild.2019.02.024.

Santamouris, M. (2020). Recent progress on urban overheating and heat island research. Integrated assessment of the energy, environmental, vulnerability and health impact. Synergies with the global climate change. *Energy and Buildings, 207*. Available from https://doi.org/10.1016/j.enbuild.2019.109482, https://www.journals.elsevier.com/energy-and-buildings.

Santamouris, M. (2022). The impact and influence of mitigation technologies on heat-related mortality in overheated cities. *Springer Science and Business Media LLC*, 155−169. Available from https://doi.org/10.1007/978-981-19-4707-0_7.

Santamouris, M., & Yun, G. Y. (2020). Recent development and research priorities on cool and super cool materials to mitigate urban heat island. *Renewable Energy, 161*, 792−807.

Available from https://doi.org/10.1016/j.renene.2020.07.109, http://www.journals.elsevier. com/renewable-and-sustainable-energy-reviews/.

Sheng, S., Xiao, H., & Wang, Y. (2022). The cooling effects of hybrid landscapes at the district scale in mega-cities: A case study of Shanghai. *Journal of Cleaner Production, 366*. Available from https://doi.org/10.1016/j.jclepro.2022.132942, https://www.journals.elsevier. com/journal-of-cleaner-production.

Shooshtarian, S., & Ridley, I. (2017). The effect of physical and psychological environments on the users thermal perceptions of educational urban precincts. *Building and Environment, 115*, 182−198. Available from https://doi.org/10.1016/j.buildenv.2016.12.022, http://www. elsevier.com/inca/publications/store/2/9/6/index.htt.

Simshauser, P. (2023). The 2022 energy crisis: Fuel poverty and the impact of policy interventions in Australia's National Electricity Market. *Energy Economics, 121*, 106660. Available from https://doi.org/10.1016/j.eneco.2023.106660.

Sun, F., Liu, M., Wang, Y., Wang, H., & Che, Y. (2020). The effects of 3D architectural patterns on the urban surface temperature at a neighborhood scale: Relative contributions and marginal effects. *Journal of Cleaner Production, 258*. Available from https://doi.org/10.1016/j. jclepro.2020.120706, https://www.journals.elsevier.com/journal-of-cleaner-production.

Susca, T., Zanghirella, F., Colasuonno, L., & Del Fatto, V. (2022). Effect of green wall installation on urban heat island and building energy use: A climate-informed systematic literature review. *Renewable and Sustainable Energy Reviews, 159*, 112100. Available from https:// doi.org/10.1016/j.rser.2022.112100.

Sánchez-Guevara Sánchez, C., Mavrogianni, A., & Neila González, F. J. (2017). On the minimal thermal habitability conditions in low income dwellings in Spain for a new definition of fuel poverty. *Building and Environment, 114*, 344−356. Available from https://doi.org/10.1016/j. buildenv.2016.12.029, http://www.elsevier.com/inca/publications/store/2/9/6/index.htt.

Sánchez-Guevara Sánchez, C., Sanz Fernández, A., Núñez Peiró, M., & Gómez Muñoz, G. (2020). Feminisation of energy poverty in the city of Madrid. *Energy and Buildings, 223*. Available from https://doi.org/10.1016/j.enbuild.2020.110157, https://www.journals.elsevier. com/energy-and-buildings.

The World Bank. (2023a). *Data Bank world development indicators*. Available from https://data-topics.worldbank.org/world-development-indicators/.

The World Bank. (2023b). *World Development Indicators*. Available from https://databank. worldbank.org/source/world-development-indicators.

Tabata, T., & Tsai, P. (2020). Fuel poverty in summer: An empirical analysis using microdata for Japan. *Science of the Total Environment, 703*. Available from https://doi.org/10.1016/j. scitotenv.2019.135038, http://www.elsevier.com/locate/scitotenv.

Tan, Z., Lau, K. K. L., & Ng, E. (2016). Urban tree design approaches for mitigating daytime urban heat island effects in a high-density urban environment. *Energy and Buildings, 114*, 265−274. Available from https://doi.org/10.1016/j.enbuild.2015.06.031.

Thomson, H., Simcock, N., Bouzarovski, S., & Petrova, S. (2019). Energy poverty and indoor cooling: An overlooked issue in Europe. *Energy and Buildings, 196*, 21−29. Available from https://doi.org/10.1016/j.enbuild.2019.05.014.

Toy, S., & Yilmaz, S. (2010). Thermal sensation of people performing recreational activities in shadowy environment: A case study from Turkey. *Theoretical and Applied Climatology, 101*(3), 329−343. Available from https://doi.org/10.1007/s00704-009-0220-z, http://link. springer.de/link/service/journals/00704/.

Tsilini, V., Papantoniou, S., Kolokotsa, D. D., & Maria, E. A. (2015). Urban gardens as a solution to energy poverty and urban heat island. *Sustainable Cities and Society, 14*(1), 323−333.

Available from https://doi.org/10.1016/j.scs.2014.08.006, http://www.elsevier.com/wps/find/journaldescription.cws_home/724360/description#description.

Ugembe, M. A., Brito, M. C., & Inglesi-Lotz, R. (2022). Measuring energy poverty in Mozambique: Is energy poverty a purely rural phenomenon. *Energy Nexus, 5*. Available from https://doi.org/10.1016/j.nexus.2022.100039, https://www.journals.elsevier.com/energy-nexus.

Wang, J., Meng, Q., Tan, K., & Santamouris, M. (2022). Evaporative cooling performance estimation of pervious pavement based on evaporation resistance. *Building and Environment, 217*. Available from https://doi.org/10.1016/j.buildenv.2022.109083, http://www.elsevier.com/inca/publications/store/2/9/6/index.htt.

Wang, W., & He, B. J. (2023). Co-occurrence of urban heat and the COVID-19: Impacts, drivers, methods, and implications for the post-pandemic era. *Sustainable Cities and Society, 90*. Available from https://doi.org/10.1016/j.scs.2022.104387, http://www.elsevier.com/wps/find/journaldescription.cws_home/724360/description#description.

Yilmaz, S., Mutlu, E., & Yilmaz, H. (2018). Alternative scenarios for ecological urbanizations using ENVI-met model. *Environmental Science and Pollution Research, 25*(26), 26307–26321. Available from https://doi.org/10.1007/s11356-018-2590-1, http://www.springerlink.com/content/0944-1344.

Zhang, L., Zhan, Q., & Lan, Y. (2018). Effects of the tree distribution and species on outdoor environment conditions in a hot summer and cold winter zone: A case study in Wuhan residential quarters. *Building and Environment, 130*, 27–39. Available from https://doi.org/10.1016/j.buildenv.2017.12.014, http://www.elsevier.com/inca/publications/store/2/9/6/index.htt.

Chapter 5

The impact of heat mitigation and adaptation technologies on urban health

Nasrin Aghamohammadi[1,2] **and Logaraj Ramakreshnan**[3]
[1]*School of Design and the Built Environment, Curtin University Sustainability Policy Institute, Bentley, WA, Australia,* [2]*Centre for Epidemiology and Evidence-Based Practice, Department of Social and Preventive Medicine, Faculty of Medicine, University of Malaya, Kuala Lumpur, Malaysia,* [3]*Institute for Advanced Studies, University of Malaya, Kuala Lumpur, Malaysia*

5.1 Introduction

Cities are the engines of change and transformation that drive economic development due to greater opportunities for education, employment, and prosperity (Filho et al., 2021). However, the emergence of major cities has enormous environmental consequences related to traffic congestion, poor sanitation, waste management, urban sprawl, pollution levels, and overexploitation of natural resources. The urbanization-related emissions from merely 100 cities are reported to account for 20% of the global carbon footprint that contributes to global warming and climate change (Moran et al., 2018). Even though net zero pledges have received a lot of attention recently to counterbalance carbon emissions and temperature reductions, scientists and nongovernmental organizations are increasingly warning that these assertions frequently amount to dangerous greenwashing (Newman, 2023). They contend that some businesses and governments are using net zero pledges as a means of postponing the immediate action needed to combat global warming and climate change.

The overheating of the cities is a critical ramification of urbanization in this 21st century expressed in terms of heightened average city temperatures than those in nearby rural areas. It is known as the urban heat island (UHI) effect and is directly caused by the anthropogenic heat sources and land cover and land use changes that have replaced green and blue spaces and vacant land with impervious urban infrastructures of various thermal properties

Mitigation and Adaptation of Urban Overheating. DOI: https://doi.org/10.1016/B978-0-443-13502-6.00005-1

(Koch et al., 2020). UHI is also intensified by global climate changes and manifested into more severe and frequent heat waves (Leal Filho et al., 2018). With global temperatures on the rise, cities are experiencing more frequent and intense heat waves, which can have serious consequences for public health. Owing to this, the impact of heat mitigation on urban health is a growing area of research and concern worldwide. Even though the UHI effect is documented in more than 400 cities in the world, it is anticipated to occur in every human settlement (Antoszewski et al., 2020). Particularly, the effect is more pronounced at night when man-made structures dissipate heat much more slowly than the natural elements (Morakinyo et al., 2020).

The persistence of heat in megacities, exclusively in the city centers, creates conditions that jeopardize human health and well-being. Heat waves are associated with a range of health problems, including dehydration, heat exhaustion, and heat stroke. These conditions can be dangerous for vulnerable populations, such as older adults, children, and people with preexisting medical conditions. The excess heat in the cities affects human health in terms of compromised human thermal comfort levels (Yang et al., 2019), heat-related morbidity and mortality (Green et al., 2019), degraded air quality (Swamy et al., 2020), additional cooling energy consumption (Frayssinet et al., 2018), and collateral economic and social costs (Guardaro et al., 2022). These issues can threaten the ecological environment, biodiversity, and quality of life in the cities, while further making the cities drivers of climate change and epicenters of noncommunicable disease epidemics (Nieuwenhuijsen, 2021). In this sense, the viability of cities can only be ensured if they become resilient to the elevating heat levels. Heat mitigation strategies aim to reduce the impact of heat on urban populations. These strategies can include the installation of green roofs and walls, increasing urban vegetation covers as well as the creation or proclamation of urban parks and wetlands as cooling centers (Almaaitah et al., 2021). Scientific research has shown that these strategies can have a significant impact on urban health. A study conducted in the United Kingdom found that the creation of a network of green spaces reduced heat stress and improved the overall health outcomes in the city (Lafortezza et al., 2009). Overall, heat mitigation strategies are an important tool for protecting urban populations from the health impacts of extreme heat in the cities. With regard to Sustainable Development Goal 11 (Sustainable Cities and Communities), as temperatures continue to rise, these strategies will likely become increasingly important for ensuring the health and well-being of urban communities (Meftahi et al., 2022).

5.2 Urban heat: energy, environmental, and urban health implications

Cities trap, accumulate, and retain heat due to the heat-absorbing properties of urban infrastructure such as roads, buildings, and other paved surfaces.

Urban heat imposes severe impacts on urban energy demand, environment, as well as public health and therefore is a serious menace to urban sustainability. Urban heat can have significant energy implications. The demand for air conditioning elevates with temperature, which can put a strain on the energy grid and increase energy costs. The heat stress in overheated cities can lead to overconsumption of energy due to the multiplication of active cooling devices to create a thermally comfortable environment for the building occupants (Frayssinet et al., 2018). The UHI effect can also reduce the efficiency of energy infrastructure, such as solar panels, which can further exacerbate energy demand (Ma et al., 2022). Several strategies can help to reduce the energy implications of the UHI effect. For example, urban planning and design can play a significant role in reducing heat retention in cities. Increasing the number of green spaces, promoting the use of reflective surfaces, and designing buildings with energy-efficient materials and technologies can reduce the UHI effect and lower the energy consumption (Lehmann, 2014). Another strategy is to promote the use of renewable energy sources such as solar and wind power in urban areas. By generating energy locally, cities can reduce their reliance on the energy grid and lower their carbon footprint (Wang et al., 2020). Energy-efficient buildings and renewable energy sources can help mitigate these impacts by reducing energy consumption and reliance on the energy grid (Newman, 2023).

In a comprehensive review of contemporary case studies, Li et al. (2019) discovered that UHI could lead to a median of 19% increase in building cooling energy consumption and a median of 18.7% decrease in building heating energy consumption. The energy demands for active cooling also vary across regions. In the Mediterranean region, the energy demands for cooling the buildings elevated by 46% in the city centers and by 12% in the peripheral neighborhood (Zinzi & Carnielo, 2017). In Europe, electricity consumption is reported to increase by 1.66% in hot climatic countries and by 0.542% in mild climatic countries when the temperature increases by 1°C in summer (Santamouris, 2013). The heightened energy demands and operation of active cooling devices, in turn, release more heat into the urban areas that exacerbate the UHI effect. These associations are more pronounced in the urban centers of the cities compared to the urban peripheries and showed a decreasing urban—rural trend, consistent with the intracity pattern of the UHI intensity. In addition, demand-side management strategies, such as encouraging energy-efficient behavior and promoting energy conservation, can also help reduce energy consumption in urban areas (Solà et al., 2021). Despite the significant energy implications of UHI, there are several strategies to reduce these impacts. By promoting sustainable urban design, renewable energy sources, and energy-efficient behavior, cities can work to mitigate the energy implications of the UHI effect and create more sustainable and liveable communities.

The UHI effect can have significant environmental implications for urban areas. UHI can lead to increased water consumption, which can strain local water resources and lead to the loss of natural water storage areas such as wetlands and blue bodies (Wagner et al., 2022). Urban development can lead to the loss of green spaces and natural habitats, which can reduce biodiversity and disrupt local ecosystems. The impact can be amplified by the UHI effect by creating conditions that are less conducive to certain plants, insects, birds, and animals (Čeplová et al., 2017). Notwithstanding this, another highly investigated effect of UHI is its impact and synergy with air pollution. High urban heat influences certain atmospheric chemistry cycles that lead to enhanced ground-level ozone production, higher emission of biogenic volatile organic compounds (VOCs), and higher evaporation of synthetic VOCs from vehicle engines (Ulpiani, 2021). Besides, the long activation of air conditioning units during intense heat islands accelerates the emission of carbon dioxide (CO_2), ozone-precursors, and exhaust heat into the urban air, which leads to UHI amplification (Ulpiani, 2021). The UHI-caused warmer air in the atmosphere also enhances turbulent mixing that promotes primary pollutant dispersion to higher atmospheric layers (Li et al., 2018). The increased use of air conditioning and other cooling technologies also contributes to air pollution by increasing energy consumption and emissions. The combined effect of UHI and air pollution aggravated human health, putting vulnerable groups such as children, older adults, pregnant women, and patients at high risk (He et al., 2022). A number of studies, including cross-country investigations, reported a compelling association between UHI, air pollution, and respiratory illness (Grigorieva & Lukyanets, 2021; Solomon et al., 2022). In addition, high temperatures can exacerbate the preexisting medical conditions such as cardiovascular diseases and respiratory illnesses, mainly asthma and chronic obstructive pulmonary disease (Zafirah et al., 2021). The increasing mortality rate due to cardiovascular and respiratory ailments caused by air pollution and heat is another leading health impact of UHI (Aghamohammadi et al., 2022; Shirinde & Wichmann, 2022). Another comprehensive review encompassing case studies from 19 countries suggested that heat stress and consistent exposure to high temperatures are associated with mental health issues such as depression, anxiety, insomnia, and even suicide risks (Aghamohammadi et al., 2022). High temperatures can lead to heat exhaustion and heat stroke, particularly in vulnerable populations and those with chronic health conditions (Mohammad Harmay & Choi, 2023). On the other hand, the UHI effect can also create conditions that are conducive to the breeding of disease-carrying insects, which can increase the risk of vector-borne diseases such as dengue fever, chikungunya virus, and West Nile virus infections (Medlock & Leach, 2015).

Blue-collar laborers undertaking heavy work outside the buildings are prone to heat stresses that increase the probability of heat-related morbidity and mortality (Dunne et al., 2013). Such a situation limits the workers' operating hours to avoid heat stresses, which in turn affects labor productivity. Indeed, estimations revealed that heat-induced global productivity loss may reach up to 20% by 2050 (Zander et al., 2015). The households in cities with severe UHI effects may migrate to more liveable cities, which could affect the regional economic and social sustainability. Hence, pragmatic actions are necessary to overcome the UHI issue to eradicate heat-related impacts on human health as well as to ensure the sustainability of the economy and society. Climate-sensitive urban designs that integrate natural elements such as urban green and water bodies offer viable remedies to UHI. In particular, urban greenery is one of the most effective countermeasures due to cooling by shading, promoting airflows, and intercepting precipitation and evapotranspiration (Liu et al., 2021). Meanwhile, the promotion of low-emission transportation options such as walking, cycling, and public transportation can reduce emissions and air pollution (Ramakreshnan et al., 2021). Encouraging energy-efficient behavior and promoting sustainable consumption can also help reduce the environmental impact of urban areas (Solà et al., 2021). Besides, public health interventions such as public educational campaigns on heat safety and targeted outreach to vulnerable populations can help in reducing the risk of heat-related illnesses (Tomlinson et al., 2011). In conclusion, the UHI effects can have significant energy, environmental, and health implications for urban populations. However, by promoting sustainable urban design, public health interventions, and targeted outreach to vulnerable populations, cities can work to reduce the health risks associated with the UHI effect, reduce their environmental footprint, and create more liveable and sustainable communities.

5.3   Urban heat mitigation via urban greeneries

The idea of using urban greeneries, other natural elements, and ecosystem services to address environmental and societal challenges is not a new concept in the global arena. Many scholarly investigations are devoted to the roles of greeneries and ecosystem services in addressing environmental and urban health issues using different metaphors such as green infrastructure, urban forestry, urban parks and green spaces, ecosystem services, biophilic urbanism, and nature-based solutions (Ferreira et al., 2020). In the context of global warming, integration of green elements is anticipated to provide around 30% of the cost-effective mitigation by 2030 to stabilize warming to below 2°C, while bringing additional environmental and social benefits (Seddon et al., 2019). With regard to this, more than two-thirds of Paris Agreement signatories reported having included green or other natural elements in some form in their nationally determined contributions (Seddon et al., 2019). Focusing on UHI mitigation, several strategies have been proposed, developed, and implemented, such as urban parks, street trees, green roofs, and green walls that provide both active and passive cooling effects to the buildings and cities (Wong et al., 2021). The presence of urban greenery tackles the UHI problem by enhancing the microclimate based on three

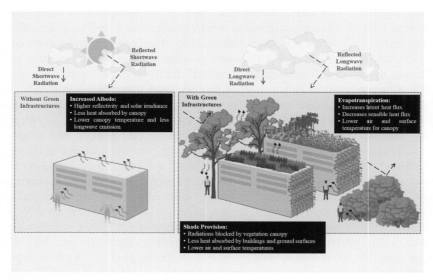

FIGURE 5.1 The cooling services provided by the green infrastructure based on shading, increased albedo, and evapotranspiration, which collectively decrease the sensible heat gain and surface temperatures. Red and blue boxes indicate the warming and cooling mechanisms with and without green infrastructures, respectively (Wong et al., 2021). *Own illustrations adapted from Wong et al. (2021).*

fundamental mechanisms (Fig. 5.1) such as (1) solar shading, (2) increasing albedo, and (3) increasing the evapotranspiration. The integration of vegetation into buildings and urban areas leads to the improvement of building energy balance and occupant thermal comfort levels as well as improvement of the outdoor environment and pedestrian thermal comfort levels (Hayes et al., 2022). The integration of urban greenery at the individual building level, street level, neighborhood level and microscale level can provide various cooling effects and other ecosystem services such as air pollutant filtration, carbon sequestration, acoustic improvement, increased biodiversity, increased ecosystem resilience, water purification, as well as and social benefits that also able to enhance urban livability levels (Koch et al., 2020; Priya & Senthil, 2021).

5.3.1 Urban parks

Urban forests are typical ecosystems within a city, which provide ground cooling by facilitating the flow of cool air (Lee & Park, 2020). The three-dimensional spatial arrangement of the urban forest includes tree size, height, crown closure, leave area index, and species composition, which promotes the energy exchanges between the urban park and the atmosphere

(Ren et al., 2018). A comprehensive metaanalysis covering 24 case studies from tropical and temperate climates demonstrated that urban parks provide air temperature cooling of 0.94°C (Bowler et al., 2010). Another review of 89 studies conducted in a range of geographic and climatic regions outlined that urban parks can provide cooling of 1.5°C−3.5°C, with no apparent correlation to the climatic region (Saaroni et al., 2018). Nevertheless, studies have shown that the cooling potential of urban parks is amplified when assessing surface temperature rather than air temperature due to the better thermal conductivity of solid surfaces compared with air (Wong et al., 2021). Based on the review conducted by Ramakreshnan and Aghamohammadi (2024) for Asian cities, the cooling effects of urban parks are highly characterized by the forest types, size, shape and their arrangements, vegetation architecture, threshold distances, geographical attributes, and the energy flux between vegetation and other structures. Besides their contribution to heat mitigation, urban parks offer a multitude of social benefits such as social activities, leisure, and cultural education (Wang et al., 2019). Ren et al. (2022) found that the street with the highest tree cover registered significantly lower physiological equivalent temperature, systolic blood pressure, diastolic blood pressure, and pulse rate, suggesting their contribution to thermal comfort and other physiological parameters.

5.3.2 Green walls

A green wall is a form of a vertical greening system and is suitable to be applied on buildings in congested cities when there is limited availability of space for urban parks or street trees (Koch et al., 2020). Green walls can be further categorized as green facades and living walls based on their systems and construction characteristics (Manso & Castro-Gomes, 2015). As illustrated in Fig. 5.2, direct (A), indirect (B), and indirect with air channel (C) types of green façades have epiphytes that climb up the vertical face of the host wall with or without the help of support, which can remain in contact with the wall or be separated from it to form an air channel (Ávila-Hernández et al., 2023). The living wall, either continuous (D) or modular (E), is another complex structure involved in the recent innovation of wall cladding composed of prevegetated panels fixed to a structural wall of a free-standing frame (Al-Kayiem et al., 2020). Continuous systems do not require the substrate due to a geotextile membrane, whereas modular living walls require a substrate and an irrigation system with the necessary nutrients, thus bearing a high maintenance cost.

A metaanalysis of green wall literature by Koch et al. (2020) has underscored their capability to reduce temperature extremes, both in warm and cold situations. However, for summer situations, the cooling effect could be overestimated by about 5°C. Another review by Wong et al. (2021)

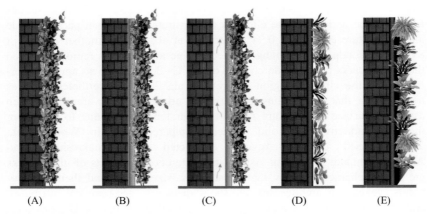

(A)   (B)   (C)   (D)   (E)

FIGURE 5.2   The types of green facades and living walls: (A) direct, (B) indirect, and (C) indirect with air channel types of green facades as well as (D) continuous and (E) modular living walls. *From Ávila-Hernández et al., 2022.*

suggested that the green walls are able to reduce peak air temperature by an average of 3°C (range: 2°C–4°C) and peak surface temperature by an average of 16°C (range: 10.7°C–18.8°C). In Mediterranean climatic conditions, green walls are reported to present an annual energy saving of 28.5% compared to a wall without vegetation (Vox et al., 2022). In another study conducted in four cities in China, it was identified that green walls decrease the cooling load of a room between 11.7% and 18.4% (Zhang et al., 2022). In the review of Ramakreshnan & Aghamohammadi (2024), it was found that the majority of studies performed on green wall intervention are based on parametric simulations rather than real-scale, in-situ experimentations. This is mainly due to the difficulty of studying green wall performance using large-scale projects with a similar control case. In another comprehensive review by Balany et al. (2020), it is reported that a maximum temperature reduction of 8.4°C can be expected from green walls in the urban canyons of the humid climate of Hong Kong. Nonetheless, in-situ experimentations are rather few; it is then not sufficient to draw a concrete conclusion regarding the cooling ability of green walls.

5.3.3   Green roofs

Green roofs are the roofs of buildings partially or completely covered with greeneries over a waterproofing membrane. Green roofs are classified into three main types such as extensive, semiintensive, and intensive green roofs with respect to their level of complexity in terms of weight, system build-up height, substrate layer, maintenance, cost, plant types, and irrigation (Besir & Cuce, 2018; Ismail et al., 2018; Shahmohammad et al., 2022). Installation of green roofs into a rooftop brings three changes such as amplifying the thermal mass, adding porous substances with moisture-holding capacity, and increasing the thermal conductivity. Nonetheless, these properties are dependent on the plant communities, biomass structure, foliage cover, transpiration rate, substrate composition, thickness, and moisture content (Jim, 2015).

On average, green roofs are estimated to be able to decrease the peak air temperature by an average of 3°C (range: 1.5°C−4.1°C) and peak surface temperature by an average of 17°C (range: 11°C−22.4°C) (Wong et al., 2021). However, its cooling potential is highly influenced by the climate settings of the cities. For instance, peak surface temperature reduction by 20°C−30°C was recorded in the summer of Mediterranean climate cities, whereas a much lesser effect is recorded during the winter (10°C−13°C) (Bevilacqua et al., 2016). Similarly, the potential of green roofs for urban temperature reduction is observed to be greater for hot and dry climates (Alexandri & Jones, 2008). Green roofs can also reduce energy consumption in buildings with poor thermal insulation values both in summer and winter. For example, Cascone et al. (2018) simulated a green roof scenario for Catania city and identified that summer cooling energy consumption and winter heating energy consumption can be reduced by up to 35% and 10%, respectively. In terms of energy consumption, Chagolla-Aranda et al. (2017) suggested that green roofs are able to reduce electrical consumption by air conditioning by up to 10.3%, as compared with conventional roofs in a semi-warm climate of Mexico. Moreover, the energy saving of green roofs is dependent on the type of green roofs, depth, and type of substrate, climate, plants, irrigation conditions, and the building characteristics (Reyes et al., 2016).

5.4 Urban health implications of urban heat mitigation via greeneries

The integration of urban greeneries into urban heat mitigation approaches in terms of urban parks, green walls, and green roofs also render many implications for urban health. Studies have revealed that urban greeneries along with other nature-based solutions and ecosystem services have a positive impact on various dimensions of human health and well-being, which can be expressed in terms of physical health, mental well-being, and social and cultural implications (Fonseca et al., 2023).

Urban greeneries play an eminent role in filtering air pollutants and absorbing gases and retaining particles that can be harmful to humans. The leaves of vegetation incorporated into urban greenery systems are able to perform a sink function for significant quantities of health-damaging fine particles less than 10 μm diameter, thus reducing the probability of respiratory ailments (Ottelé et al., 2011). Similar findings were reported by Srbinovska et al. (2021), who found that green walls are able to reduce the concentration of $PM_{2.5}$ and PM_{10} on average by 25% and 37% when compared to nongreen areas. Meantime, others studies have evaluated the capability of green walls and roofs to filter even dangerous air pollutants such as ozone (O_3), $PM_{2.5}$, and nitrogen dioxide (NO_2), from wildfire emissions (Pettit et al., 2020). Besides, the green walls or greeneries incorporated into the buildings were reported to enhance indoor air quality. Since most urbanites spend 80% of their time in confined environments, the health risks of indoor air pollution can be greater than those of outdoor air pollution (Kazemi et al., 2020). Green walls on buildings are considered a viable phytoremediation solution to remove air pollutants such as VOCs and PMs without consuming much space. For instance, Abedi et al. (2022) reported that plants have a single-pass removal efficiency of 13.5%−28.1% of formaldehyde and a clean air delivery rate of 0.39−5.46 m^3/h. On the other hand, urban greeneries are also able to improve human thermal comfort levels and reduce the energy and the costs associated with cooling. For example, a square meter of Azolla per person on a green wall is anticipated to reduce fresh air ventilation demand by almost 30% (Parhizkar et al., 2020). In subtropical China, the combination of green walls with other sunshades and water surfaces elevated thermal comfort levels in a sunken courtyard (He et al., 2022). Urban greeneries have also been reported to act as a noise barrier from traffic. For example, studies have shown that the green walls accommodated in inner city buildings are able to attenuate traffic noise from 2.6 to 5.1 dBA (Van Renterghem et al., 2013). In addition, the availability of urban parks in local neighborhoods creates a platform for urbanites to engage in recreational and sporting activities, thus enhancing physical fitness and health (Petrunoff et al., 2022).

Many studies suggested that exposure to greenery and the use of green space is essential for promoting mental health by facilitating visual exposure to greenery and stress reduction. Even recent studies carried out during the COVID-19 pandemic suggested that more public green areas were associated with less severe COVID-19 clinical outcomes in terms of contagion, hospitalizations, and deaths (Falco et al., 2023). Based on a systematic review of 223 eligible studies, Mygind et al. (2021) found a positive association between the availability of and spending time in green space and children's intra- and interpersonal socioemotional function and development. In another study by Siah et al. (2022), the findings demonstrated that the participants who engaged in the green environment and forest in urban areas reported significantly higher positive mental effects. According to Sousa et al. (2022/2022), contact with nature has shown a decrease in cortisol levels, which is the main hormone that causes stress, thus improving the mental health of the urbanites. From another perspective, the general public always has positive attitudes toward green elements in their neighborhood. For example, the psychological well-being of 82% of the participants was found to be improved with the installation of a green wall in a hospital in Seville, Spain (Pérez-Urrestarazu et al., 2017). Urban greeneries also foster social interactions and provide recreational opportunities for the urbanites, which help in improving their well-being. On a different note, some studies have indicated that the selection of certain types of urban greeneries could have less social acceptance. For instance, most of the participants in a study conducted in Dhaka, Bangladesh, preferred a rooftop garden or agriculture plot as the most implementable adaptation strategy, as they found that green walls were not culturally appreciated in the city (Zinia & McShane, 2018). Similarly, another study by Liberalesso et al. (2020) in Lisbon revealed that even though the installation of green roofs and walls in hostels is supported by the public, the male occupants were not willing to pay higher rates to stay in these hostels, as they consider it least important and less necessary.

5.5 The exposure and vulnerability of the urban populations to urban heat

The exposure and vulnerability of urban populations to urban heat is a critical issue that must be addressed by policymakers and urban planners. Urban areas are characterized by high densities of the human population, buildings, and man-made structures that trap and retain heat. Moreover, intraurban surface geometry and urban factors such as sky-view factor, aspect ratio, and surface albedo have a stronger relationship with air temperature, thermal comfort, and health (Thorsson et al., 2011; Unger, 2009). Therefore, to address the exposure and vulnerability of urban populations to urban heat, it is important to adopt a holistic approach that includes urban planning, public health interventions, and community engagement. This may involve strategies such as improving access to green spaces and cooling technologies, increasing community awareness of the risks of urban heat, and addressing social and environmental inequities that contribute to heat vulnerability.

Improving city design and housing quality by implementing building codes that require adequate insulation, ventilation, and cooling systems, as well as retrofitting the existing buildings with these features could be viable solutions to mitigate UHI. Poorly insulated buildings and buildings with large windows or poor orientation are increasing the likelihood of heat exposure. Vulnerable groups, such as older adults, children, and those with preexisting health conditions are particularly vulnerable to the effects of such heat. Accessibility to cooling technologies as well as mapping UHI can help in identifying city areas that are particularly vulnerable to heat and targeted interventions and resources can be allocated accordingly. As discussed earlier, increasing the amount of green infrastructure such as parks and green walls or roofs can help reduce the exposure to heat via the provision of shade and cooling effects. The use of cool roofs and pavements can also help reduce exposure to heat by reflecting heat and reducing surface temperatures (Yang et al., 2018). Policies that promote the availability and affordability of cooling technologies, particularly for vulnerable populations, can help reduce exposure to heat and its associated health risks.

Urban design that prioritizes walkability, bikeability, and public transportation can also help reduce exposure to heat by reducing the use of private vehicles and associated emissions.

5.6 Limitations of heat mitigation technologies on urban health

While heat mitigation technologies can have positive contributions to urban health, they also have some limitations. Heat mitigation technologies such as green roofs, cool roofs, reflective materials, and green spaces can be very expensive to implement and maintain, which may limit their availability and accessibility to all urban communities (William et al., 2016). The use of certain heat mitigation technologies, such as white roofs and pavements, can lead to increased glare and light pollution, which can negatively affect human health and the environment (Yang et al., 2015). In extreme heat events, heat mitigation technologies may not be effective enough to protect urban populations from heat, which can be particularly problematic for vulnerable populations. While heat mitigation technologies can improve specific aspects of urban health, such as reducing heat-related illnesses and improving air quality, they may have limited impact on other health issues such as access to healthcare, social determinants of health and health disparities. Heat mitigation technologies may compete with other urban priorities, such as transportation, housing, and economic development, which may limit their adoption and implementation. Overall, while heat mitigation technologies can have positive impacts on urban health, their limitations should be carefully considered when developing and implementing urban heat mitigation strategies.

5.7 Conclusion

In conclusion, UHI is a complex issue that has significant energy, environmental, and urban health implications. Being a product of urbanization and climate change, UHI and its adaptation and mitigation strategies have drawn considerable attention in recent years. Mitigation strategies, including energy-efficient buildings, sustainable urban design, and other UHI reduction efforts, can help reduce the impact of heat on cities and their populations. Ideally, climate-sensitive urban designs that integrate urban greeneries can be a promising countermeasure due to their cooling performances. Even though there are ample studies on the cooling performance of urban parks, street trees, green roofs, and green walls, more real-scale, in-situ experimentations are needed to explore their real performances across different cities, regions, climates, and topographical settings. Besides UHI mitigation, such studies are needed to explore and verify their impacts on various dimensions of human health and well-being. Since urban greeneries are currently receiving growing attention from urban planners and policymakers as an important strategy to overcome the rising temperatures in urban areas, continuous research in this area is crucial to expand their application and optimize their performance.

Acknowledgment

The authors would like to extend their gratitude to Universiti Malaya for providing the opportunities and space to conduct this study.

Declaration of competing interest

The authors declare that they have no known competing financial interests or personal relationships that could have appeared to influence the work reported in this paper.

References

Abedi, S., Yarahmadi, R., Farshad, A. A., Najjar, N., Ebrahimi, H., & Soleimani-Alyar, S. (2022). Evaluation of the critical parameters on the removal efficiency of a botanical biofilter system. *Building and Environment*, *212*. Available from https://doi.org/10.1016/j.buildenv. 2022.108811.

Aghamohammadi N., Ramakreshnan L., Fong C.S., Kumar P., (2022). A global synthesis of heat-related mortality in overheated cities. Springer Science and Business Media LLC, 21−38. Available from https://doi.org/10.1007/978-981-19-4707-0_2.

Alexandri, E., & Jones, P. (2008). Temperature decreases in an urban canyon due to green walls and green roofs in diverse climates. *Building and Environment*, *43*(4), 480−493. Available from https://doi.org/10.1016/j.buildenv.2006.10.055.

Al-Kayiem, H. H., Koh, K., Riyadi, T. W. B., & Effendy, M. (2020). A comparative review on greenery ecosystems and their impacts on sustainability of building environment. *Sustainability*, *12*(20). Available from https://doi.org/10.3390/su12208529.

Almaaitah, T., Appleby, M., Rosenblat, H., Drake, J., & Joksimovic, D. (2021). The potential of Blue-Green infrastructure as a climate change adaptation strategy: A systematic literature review. *Blue-Green Systems*, *3*(1), 223−248. Available from https://doi.org/10.2166/bgs.2021.016.

Antoszewski, P., Świerk, D., & Krzyżaniak, M. (2020). Statistical review of quality parameters of blue-green infrastructure elements important in mitigating the effect of the urban heat island in the temperate climate (C) zone. *International Journal of Environmental Research and Public Health*, *17*(19). Available from https://doi.org/10.3390/ijerph17197093.

Ávila-Hernández, A., Simá, E., & Ché-Pan, M. (2023). Research and development of green roofs and green walls in Mexico: A review. *Science of The Total Environment*, *856*. Available from https://doi.org/10.1016/j.scitotenv.2022.158978.

Balany, F., Ng, A. W. M., Muttil, N., Muthukumaran, S., & Sing Wong, M. (2020). Green infrastructure as an urban heat island mitigation strategy—A review. *Water*, *12*(12). Available from https://doi.org/10.3390/w12123577.

Besir, A. B., & Cuce, E. (2018). Green roofs and facades: A comprehensive review. *Renewable and Sustainable Energy Reviews*, *82*, 915−939. Available from https://doi.org/10.1016/j. rser.2017.09.106, https://www.journals.elsevier.com/renewable-and-sustainable-energy-reviews.

Bevilacqua, P., Mazzeo, D., Bruno, R., & Arcuri, N. (2016). Experimental investigation of the thermal performances of an extensive green roof in the Mediterranean area. *Energy and Buildings*, *122*, 63−79. Available from https://doi.org/10.1016/j.enbuild.2016.03.062.

Bowler, D. E., Buyung-Ali, L., Knight, T. M., & Pullin, A. S. (2010). Urban greening to cool towns and cities: A systematic review of the empirical evidence. *Landscape and Urban Planning*, *97*(3), 147−155. Available from https://doi.org/10.1016/j.landurbplan.2010.05.006, http://www.elsevier.com/inca/publications/store/5/0/3/3/4/7.

Cascone, S., Catania, F., Gagliano, A., & Sciuto, G. (2018). A comprehensive study on green roof performance for retrofitting existing buildings. *Building and Environment, 136*, 227−239. Available from https://doi.org/10.1016/j.buildenv.2018.03.052.

Čeplová, N., Kalusová, V., & Lososová, Z. (2017). Effects of settlement size, urban heat island and habitat type on urban plant biodiversity. *Landscape and Urban Planning, 159*, 15−22. Available from https://doi.org/10.1016/j.landurbplan.2016.11.004.

Chagolla-Aranda, M. A., Simá, E., Xamán, J., Álvarez, G., Hernández-Pérez, I., & Téllez-Velázquez, E. (2017). Effect of irrigation on the experimental thermal performance of a green roof in a semi-warm climate in Mexico. *Energy and Buildings*, *154*, 232−243. Available from https://doi.org/10.1016/j.enbuild.2017.08.082.

Dunne, J. P., Stouffer, R. J., & John, J. G. (2013). Reductions in labour capacity from heat stress under climate warming. *Nature Climate Change*, *3*(6), 563−566. Available from https://doi.org/10.1038/nclimate1827.

Falco, A., Piscitelli, P., Vito, D., Pacella, F., Franco, C., Pulimeno, M., Ambrosino, P., Arias, J., & Miani, A. (2023). COVID-19 epidemic spread and green areas Italy and Spain between 2020 and 2021: An observational multi-country retrospective study. *Environmental Research*, *216*. Available from https://doi.org/10.1016/j.envres.2022.114089.

Ferreira, V., Barreira, A., Loures, L., Antunes, D., & Panagopoulos, T. (2020). Stakeholders' engagement on nature-based solutions: A systematic literature review. *Sustainability*, *12*(2). Available from https://doi.org/10.3390/su12020640.

Filho, W. L., Wolf, F., Castro-Díaz, R., Li, C., Ojeh, V. N., Gutiérrez, N., Nagy, G. J., Savić, S., Natenzon, C. E., Al-Amin, A. Q., Maruna, M., & Bönecke, J. (2021). Addressing the urban heat islands effect: A cross-country assessment of the role of green infrastructure. *Sustainability*, *13*(2). Available from https://doi.org/10.3390/su13020753.

Fonseca, F., Paschoalino, M., & Silva, L. (2023). Health and well-being benefits of outdoor and indoor vertical greening systems: A review. *Sustainability*, *15*(5). Available from https://doi.org/10.3390/su15054107.

Frayssinet, L., Merlier, L., Kuznik, F., Hubert, J. L., Milliez, M., & Roux, J. J. (2018). Modeling the heating and cooling energy demand of urban buildings at city scale. *Renewable and Sustainable Energy Reviews*, *81*, 2318−2327. Available from https://doi.org/10.1016/j.rser.2017.06.040, https://www.journals.elsevier.com/renewable-and-sustainable-energy-reviews.

Green, H., Bailey, J., Schwarz, L., Vanos, J., Ebi, K., & Benmarhnia, T. (2019). Impact of heat on mortality and morbidity in low and middle income countries: A review of the epidemiological evidence and considerations for future research. *Environmental Research*, *171*, 80−91. Available from https://doi.org/10.1016/j.envres.2019.01.010.

Grigorieva, E., & Lukyanets, A. (2021). Combined effect of hot weather and outdoor air pollution on respiratory health: Literature review. *Atmosphere*, *12*(6). Available from https://doi.org/10.3390/atmos12060790.

Guardaro, M., Hondula, D. M., & Redman, C. L. (2022). Social capital: Improving community capacity to respond to urban heat. *Local Environment*, *27*(9), 1133−1150. Available from https://doi.org/10.1080/13549839.2022.2103654.

Harmay, N. S. M., & Choi, M. (2023). The urban heat island and thermal heat stress correlate with climate dynamics and energy budget variations in multiple urban environments. *Sustainable Cities and Society*, *91*. Available from https://doi.org/10.1016/j.scs.2023.104422.

Hayes, A. T., Jandaghian, Z., Lacasse, M. A., Gaur, A., Lu, H., Laouadi, A., Ge, H., & Wang, L. (2022). Nature-based solutions (NBSs) to mitigate urban heat island (UHI) effects in canadian cities. *Buildings*, *12*(7). Available from https://doi.org/10.3390/buildings12070925, https://www.mdpi.com/2075-5309/12/7/925/pdf?version = 1656642789.

He, B.-J., Zhao, D., Dong, X., Xiong, K., Feng, C., Qi, Q., Darko, A., Sharifi, A., & Pathak, M. (2022). Perception, physiological and psychological impacts, adaptive awareness and knowledge, and climate justice under urban heat: A study in extremely hot-humid Chongqing, China. *Sustainable Cities and Society*, *79*. Available from https://doi.org/10.1016/j.scs.2022.103685.

Ismail, W.Z. W., Abdullah, M.N., Hashim, H., & Rani, W.S. W. (2018). 9 26 2018/09/26 AIP conference proceedings 10.1063/1.5055460 15517616 American Institute of Physics Inc. Malaysia An overview of green roof development in Malaysia and a way forward. http://scitation.aip.org/content/aip/proceeding/aipcp.

Jim, C. Y. (2015). Diurnal and partitioned heat-flux patterns of coupled green-building roof systems. *Renewable Energy*, *81*, 262−274. Available from https://doi.org/10.1016/j.renene.2015.03.044.

Kazemi, F., Rabbani, M., & Jozay, M. (2020). Investigating the plant and air-quality performances of an internal green wall system under hydroponic conditions. *Journal of Environmental Management*, *275*. Available from https://doi.org/10.1016/j.jenvman.2020.111230.

Koch, K., Ysebaert, T., Denys, S., & Samson, R. (2020). Urban heat stress mitigation potential of green walls: A review. *Urban Forestry & Urban Greening*, *55*. Available from https://doi.org/10.1016/j.ufug.2020.126843.

Lafortezza, R., Carrus, G., Sanesi, G., & Davies, C. (2009). Benefits and well-being perceived by people visiting green spaces in periods of heat stress. *Urban Forestry & Urban Greening*, *8*(2), 97−108. Available from https://doi.org/10.1016/j.ufug.2009.02.003.

Leal Filho, W., Echevarria Icaza, L., Neht, A., Klavins, M., & Morgan, E. A. (2018). Coping with the impacts of urban heat islands. A literature based study on understanding urban heat vulnerability and the need for resilience in cities in a global climate change context. *Journal of Cleaner Production*, *171*, 1140−1149. Available from https://doi.org/10.1016/j.jclepro.2017.10.086.

Lee, P. S. H., & Park, J. (2020). An effect of urban forest on urban thermal environment in Seoul, South Korea, based on landsat imagery analysis. *Forests*, *11*(6). Available from https://doi.org/10.3390/F11060630, https://www.mdpi.com/1999-4907/11/6/630.

Lehmann, S. (2014). Low carbon districts: Mitigating the urban heat island with green roof infrastructure. *City, Culture and Society*, *5*(1), 1−8. Available from https://doi.org/10.1016/j.ccs.2014.02.002.

Li, H., Meier, F., Lee, X., Chakraborty, T., Liu, J., Schaap, M., & Sodoudi, S. (2018). Interaction between urban heat island and urban pollution island during summer in Berlin. *Science of the Total Environment*, *636*, 818−828. Available from https://doi.org/10.1016/j.scitotenv.2018.04.254.

Li, X., Zhou, Y., Yu, S., Jia, G., Li, H., & Li, W. (2019). Urban heat island impacts on building energy consumption: A review of approaches and findings. *Energy*, *174*, 407−419. Available from https://doi.org/10.1016/j.energy.2019.02.183.

Liberalesso, T., Júnior, R. M., Cruz, C. O., Silva, C. M., & Manso, M. (2020). Users' perceptions of green roofs and green walls: An analysis of youth hostels in Lisbon, Portugal. *Sustainability*, *12*(23). Available from https://doi.org/10.3390/su122310136.

Liu, Z., Cheng, W., Jim, C. Y., Morakinyo, T. E., Shi, Y., & Ng, E. (2021). Heat mitigation benefits of urban green and blue infrastructures: A systematic review of modeling techniques, validation and scenario simulation in ENVI-met V4. *Building and Environment, 200*, 107939. Available from https://doi.org/10.1016/j.buildenv.2021.107939.

Pérez-Urrestarazu, L., Blasco-Romero, A., & Rafael, F.-C. (2017). Media and social impact valuation of a living wall: The case study of the Sagrado Corazon hospital in Seville (Spain). *Urban Forestry & Urban Greening, 24*, 141−148. Available from https://doi.org/10.1016/j.ufug.2017.04.002.

Ma, T., Li, S., Gu, W., Weng, S., Peng, J., & Xiao, G. (2022). Solar energy harvesting pavements on the road: Comparative study and performance assessment. *Sustainable Cities and Society, 81*. Available from https://doi.org/10.1016/j.scs.2022.103868.

Manso, M., & Castro-Gomes, J. (2015). Green wall systems: A review of their characteristics. *Renewable and Sustainable Energy Reviews, 41*, 863−871. Available from https://doi.org/10.1016/j.rser.2014.07.203.

Medlock, J. M., & Leach, S. A. (2015). Effect of climate change on vector-borne disease risk in the UK. *The Lancet Infectious Diseases, 15*(6), 721−730. Available from https://doi.org/10.1016/S1473-3099(15)70091-5, http://www.journals.elsevier.com/the-lancet-infectious-diseases.

Meftahi, M., Monavari, M., Kheirkhah Zarkesh, M., Vafaeinejad, A., & Jozi, A. (2022). Achieving sustainable development goals through the study of urban heat island changes and its effective factors using spatio-temporal techniques: The case study (Tehran city). *Natural Resources Forum, 46*(1), 88−115. Available from https://doi.org/10.1111/1477-8947.12245, http://onlinelibrary.wiley.com/journal/10.1111/(ISSN)1477-8947.

Morakinyo, T. E., Ouyang, W., Lau, K. K. L., Ren, C., & Ng, E. (2020). Right tree, right place (urban canyon): Tree species selection approach for optimum urban heat mitigation - development and evaluation. *Science of the Total Environment, 719*. Available from https://doi.org/10.1016/j.scitotenv.2020.137461, http://www.elsevier.com/locate/scitotenv.

Moran, D., Kanemoto, K., Jiborn, M., Wood, R., Többen, J., & Seto, K. C. (2018). Carbon footprints of 13 000 cities. *Environmental Research Letters, 13*(6). Available from https://doi.org/10.1088/1748-9326/aac72a.

Mygind, L., Kurtzhals, M., Nowell, C., Melby, P. S., Stevenson, M. P., Nieuwenhuijsen, M., Lum, J. A. G., Flensborg-Madsen, T., Bentsen, P., & Enticott, P. G. (2021). Landscapes of becoming social: A systematic review of evidence for associations and pathways between interactions with nature and socioemotional development in children. *Environment International, 146*. Available from https://doi.org/10.1016/j.envint.2020.106238, http://www.elsevier.com/locate/envint.

Newman, P. W. G. (2023). Net zero in the maelstrom: Professional practice for net zero in a time of turbulent change. *Sustainability, 15*(6). Available from https://doi.org/10.3390/su15064810.

Nieuwenhuijsen, M. J. (2021). New urban models for more sustainable, liveable and healthier cities post covid19; reducing air pollution, noise and heat island effects and increasing green space and physical activity. *Environment International, 157*. Available from https://doi.org/10.1016/j.envint.2021.106850.

Ottelé, M., Ursem, W. J. N., Fraaij, A. L. A., & van Bohemen, H. D. (2011). The development of an ESEM based counting method for fine dust particles and a philosophy behind the background of particle adsorption on leaves. *WIT Transactions on Ecology and the Environment, 147*, 219−230. Available from https://doi.org/10.2495/AIR110201, http://library.witpress.com/pages/listBooks.asp?tID = 4.

Parhizkar, H., Khoraskani, R. A., & Tahbaz, M. (2020). Double skin façade with Azolla; ventilation, indoor air quality and thermal performance assessment. *Journal of Cleaner Production, 249*.

Available from https://doi.org/10.1016/j.jclepro.2019.119313, https://www.journals.elsevier.com/journal-of-cleaner-production.

Petrunoff, N. A., Edney, S., Yi, N. X., Dickens, B. L., Joel, K. R., Xin, W. N., Sia, A., Leong, D., van Dam, R. M., Cook, A. R., Sallis, J. F., Chandrabose, M., Owen, N., & Müller-Riemenschneider, F. (2022). Associations of park features with park use and park-based physical activity in an urban environment in Asia: A cross-sectional study. *Health and Place*, *75*. Available from https://doi.org/10.1016/j.healthplace.2022.102790, http://www.elsevier.com/locate/healthplace.

Pettit, T., Irga, P. J., & Torpy, F. R. (2020). The botanical biofiltration of elevated air pollution concentrations associated the Black Summer wildfire natural disaster. *Journal of Hazardous Materials Letters*, *1*. Available from https://doi.org/10.1016/j.hazl.2020.100003, https://www.journals.elsevier.com/journal-of-hazardous-materials-letters.

Priya, U. K., & Senthil, R. (2021). A review of the impact of the green landscape interventions on the urban microclimate of tropical areas. *Building and Environment*, *205*. Available from https://doi.org/10.1016/j.buildenv.2021.108190, http://www.elsevier.com/inca/publications/store/2/9/6/index.htt.

Ramakreshnan, L., Aghamohammadi, N., Fong, C. S., & Sulaiman, N. M. (2021). A comprehensive bibliometrics of 'walkability' research landscape: Visualization of the scientific progress and future prospects. *Environmental Science and Pollution Research*, *28*(2), 1357−1369. Available from https://doi.org/10.1007/s11356-020-11305-x, https://link.springer.com/journal/11356.

Ramakreshnan, L., & Aghamohammadi, N. (2024). The application of nature-Based solutions for Urban Heat Island mitigation in Asia: Progress, challenges, and recommendations. *Current Environmental Health Reports*, 1–14. Available from https://link.springer.com/article/10.1007/s40572-023-00427-2.

Ren, Z., He, X., Pu, R., & Zheng, H. (2018). The impact of urban forest structure and its spatial location on urban cool island intensity. *Urban Ecosystems*, *21*(5), 863−874. Available from https://doi.org/10.1007/s11252-018-0776-4.

Ren, Z., Zhao, H., Fu, Y., Xiao, L., & Dong, Y. (2022). Effects of urban street trees on human thermal comfort and physiological indices: A case study in Changchun city, China. *Journal of Forestry Research*, *33*(3), 911−922. Available from https://doi.org/10.1007/s11676-021-01361-5.

Reyes, R., Bustamante, W., Gironás, J., Pastén, P. A., Rojas, V., Suárez, F., Vera, S., Victorero, F., & Bonilla, C. A. (2016). Effect of substrate depth and roof layers on green roof temperature and water requirements in a semi-arid climate. *Ecological Engineering*, *97*, 624−632. Available from https://doi.org/10.1016/j.ecoleng.2016.10.025, http://www.elsevier.com/inca/publications/store/5/2/2/7/5/1.

Saaroni, H., Amorim, J. H., Hiemstra, J. A., & Pearlmutter, D. (2018). Urban Green Infrastructure as a tool for urban heat mitigation: Survey of research methodologies and findings across different climatic regions. *Urban Climate*, *24*, 94−110. Available from https://doi.org/10.1016/j.uclim.2018.02.001.

Santamouris, M. (2013). Using cool pavements as a mitigation strategy to fight urban heat island—A review of the actual developments. *Renewable and Sustainable Energy Reviews*, *26*, 224−240. Available from https://doi.org/10.1016/j.rser.2013.05.047.

Seddon, N., Sengupta, S., García-Espinosa, M., Hauler, I., Herr, A.R. & Rizvi. (2019). Nature-based solutions in nationally determined contributions: Synthesis and recommendations for enhancing climate ambition and action by 2020.

Shahmohammad, M., Hosseinzadeh, M., Dvorak, B., Bordbar, F., Shahmohammadmirab, H., & Aghamohammadi, N. (2022). Sustainable green roofs: A comprehensive review of influential factors. *Environmental Science and Pollution Research*, *29*(52), 78228−78254. Available from https://doi.org/10.1007/s11356-022-23405-x.

Shirinde, J., & Wichmann, J. (2022). Temperature modifies the association between air pollution and respiratory disease mortality in Cape Town, South Africa. *International Journal of Environmental Health Research*, 1–10. Available from https://doi.org/10.1080/09603123. 2022.2076813.

Siah, C. J. R., Kua, E. H., & Goh, Y. S. S. (2022). The impact of restorative green environment on mental health of big cities and the role of mental health professionals. *Current Opinion in Psychiatry*, *35*(3), 186–191. Available from https://doi.org/10.1097/YCO.0000000000000778.

Thorsson, S., Lindberg, F., Björklund, J., Holmer, B., & David, R. (2011). Potential changes in outdoor thermal comfort conditions in Gothenburg, Sweden due to climate change: The influence of urban geometry. *International Journal of Climatology*, *31*(2), 324–335. Available from https://doi.org/10.1002/joc.2231.

Solà, M. D. M., De Ayala, A., Galarraga, I., & Escapa, M. (2021). Promoting energy efficiency at household level: A literature review. *Energy Efficiency*, *14*(1). Available from https://doi. org/10.1007/s12053-020-09918-9, http://www.springer.com/environment/journal/12053.

Solomon, C. G., Salas, R. N., Perera, F., & Nadeau, K. (2022). Climate change, fossil-fuel pollution, and children's health. *New England Journal of Medicine*, *386*(24), 2303–2314. Available from https://doi.org/10.1056/nejmra2117706.

Sousa, W.P., Sousa, P.P., & Sousa, C.P. (2022). Literature review: Influence of green areas for physical and mental health. 13.

Srbinovska, M., Andova, V., Mateska, A. K., & Krstevska, M. C. (2021). The effect of small green walls on reduction of particulate matter concentration in open areas. *Journal of Cleaner Production*, *279*. Available from https://doi.org/10.1016/j.jclepro.2020.123306, https://www.journals.elsevier.com/journal-of-cleaner-production.

Swamy, G. S. N. V. K. S. N., Nagendra, S. M., & Schlink, Uwe (2020). Impact of urban heat island on meteorology and air quality at microenvironments. *Journal of the Air & Waste Management Association*, *70*(9), 876–891. Available from https://doi.org/10.1080/10962247. 2020.1783390.

Tomlinson, C. J., Chapman, L., Thornes, J. E., & Baker, C. J. (2011). Including the urban heat island in spatial heat health risk assessment strategies: A case study for Birmingham, UK. *International Journal of Health Geographics*, *10*. Available from https://doi.org/10.1186/ 1476-072X-10-42, http://www.ij-healthgeographics.com/content/10/1/42, United Kingdom.

Ulpiani, G. (2021). On the linkage between urban heat island and urban pollution island: Three-decade literature review towards a conceptual framework. *Science of The Total Environment*, *751*. Available from https://doi.org/10.1016/j.scitotenv.2020.141727.

Unger J., (2009). Connection between urban heat island and sky view factor approximated by a software tool on a 3D urban database. International Journal of Environment and Pollution. 36 (1-3), 59–80, http://www.inderscience.com/ijep. doi: 10.1504/ijep.2009.021817.

Van Renterghem, T., Hornikx, M., Forssen, J., & Botteldooren, D. (2013). The potential of building envelope greening to achieve quietness. *Building and Environment*, *61*, 34–44. Available from https://doi.org/10.1016/j.buildenv.2012.12.001.

Vox, G., Blanco, I., Convertino, F., & Schettini, E. (2022). Heat transfer reduction in building envelope with green façade system: A year-round balance in Mediterranean climate conditions. *Energy and Buildings*, *274*. Available from https://doi.org/10.1016/j.enbuild.2022.112439.

Wagner, F., Nusrat, F., Thiem, L., & Akanda, A. S. (2022). Assessment of urban water-energy interactions and heat island signatures in rhode island. *Energy Nexus*, *7*. Available from https://doi.org/10.1016/j.nexus.2022.100093.

Wang, X. H., Wu, Y., Gong, J., Li, B., & Zhao, J. J. (2019). Urban planning design and sustainable development of forest based on heat island effect. *Applied Ecology and Environmental*

Research, *17*(4), 9121−9129. Available from https://doi.org/10.15666/aeer/1704_91219129, http://aloki.hu/pdf/1704_91219129.pdf.

Wang, Zj, Song, H., Liu, H., & Ye, J. (2020). Coupling of solar energy and thermal energy for carbon dioxide reduction: Status and prospects. *Angewandte Chemie - International Edition*, *59*(21), 8016−8035. Available from https://doi.org/10.1002/anie.201907443, http://onlinelibrary.wiley.com/journal/10.1002/(ISSN)1521-377, 3.

William, R., Goodwell, A., Richardson, M., Le, P. V. V., Kumar, P., & Stillwell, A. S. (2016). An environmental cost-benefit analysis of alternative green roofing strategies. *Ecological Engineering*, *95*, 1−9. Available from https://doi.org/10.1016/j.ecoleng.2016.06.091, http://www.elsevier.com/inca/publications/store/5/2/2/7/5/1.

Wong, N. H., Tan, C. L., Kolokotsa, D. D., & Takebayashi, H. (2021). Greenery as a mitigation and adaptation strategy to urban heat. *Nature Reviews Earth and Environment*, *2*(3), 166−181. Available from https://doi.org/10.1038/s43017-020-00129-5, nature.com/natrevearthenviron/.

Yang, B., Yang, X., Leung, L. R., Zhong, S., Qian, Y., Zhao, C., Chen, F., Zhang, Y., & Qi, J. (2019). Modeling the impacts of urbanization on summer thermal comfort: The role of urban land use and anthropogenic heat. *Journal of Geophysical Research: Atmospheres*, *124*(13), 6681−6697. Available from https://doi.org/10.1029/2018JD029829, http://agupubs.onlinelibrary.wiley.com/hub/jgr/journal/10.1002/(ISSN)2169-8996/.

Yang, J., Mohan Kumar, Dl, Pyrgou, A., Chong, A., Santamouris, M., Kolokotsa, D., & Lee, S. E. (2018). Green and cool roofs' urban heat island mitigation potential in tropical climate. *Solar Energy*, *173*, 597−609. Available from https://doi.org/10.1016/j.solener.2018.08.006, http://www.elsevier.com/inca/publications/store/3/2/9/index.htt.

Yang, J., Wang, Z. H., & Kaloush, K. E. (2015). Environmental impacts of reflective materials: Is high albedo a 'silver bullet' for mitigating urban heat island? *Renewable and Sustainable Energy Reviews*, *47*, 830−843. Available from https://doi.org/10.1016/j.rser.2015.03.092, https://www.journals.elsevier.com/renewable-and-sustainable-energy-reviews.

Zafirah, Y., Lin, Y. K., Andhikaputra, G., Deng, L. W., Sung, F. C., & Wang, Y. C. (2021). Mortality and morbidity of asthma and chronic obstructive pulmonary disease associated with ambient environment in metropolitans in taiwan. *PLoS One*, *16*(7). Available from https://doi.org/10.1371/journal.pone.0253814, https://journals.plos.org/plosone/article/file?id = 10.1371/journal.pone.0253814&type = printable.

Zander, K. K., Botzen, W. J. W., Oppermann, E., Kjellstrom, T., & Garnett, S. T. (2015). Heat stress causes substantial labour productivity loss in Australia. *Nature Climate Change*, *5*(7), 647−651. Available from https://doi.org/10.1038/nclimate2623, http://www.nature.com/nclimate/index.html.

Zhang, Y., Zhang, L., & Meng, Q. (2022). Dynamic heat transfer model of vertical green façades and its co-simulation with a building energy modelling program in hot-summer/warm-winter zones. *Journal of Building Engineering*, *58*.

Zinia, N. J., & McShane, P. (2018). Ecosystem services management: An evaluation of green adaptations for urban development in Dhaka, Bangladesh. *Landscape and Urban Planning*, *173*, 23−32. Available from https://doi.org/10.1016/j.landurbplan.2018.01.008, http://www.elsevier.com/inca/publications/store/5/0/3/3/4/7.

Zinzi, M., & Carnielo, E. (2017). Impact of urban temperatures on energy performance and thermal comfort in residential buildings. The case of Rome, Italy. *Energy and Buildings*, *157*, 20−29. Available from https://doi.org/10.1016/j.enbuild.2017.05.021.

Chapter 6

The impact of heat mitigation on energy demand

Synnefa Afroditi

School of Built Environment, Faculty of Arts, Design and Architecture, University of New South Wales, Sydney, NSW, Australia

6.1 Introduction

Overheating of urban areas is caused by the urban heat island (UHI) effect, in which urban areas exhibit higher temperatures than the surrounding rural areas as a result of positive thermal balance of cities caused mainly by the increased absorption of solar radiation and heat storage, high anthropogenic heat, and reduced heat losses. Global climate change further exacerbates urban overheating, increasing the magnitude of the ambient temperature and the frequency of extreme heat events. Urban overheating has been documented in hundreds of large cities in the world, and its intensity is quite high, especially during the summer period, and on average, it may exceed 5°C (Santamouris, 2015a, 2016a,b). Increased urban temperatures have a significant impact on the energy consumption of buildings and peak electricity demand, indoor and outdoor thermal comfort, air quality and pollution, heat-related mortality and morbidity, and the economy, affecting negatively the sustainability and livability of cities (Santamouris, 2015b, 2020).

A number of UHI mitigation techniques and technologies are developed and implemented in several large-scale projects to counterbalance the negative impacts of urban overheating. Mitigation strategies involve the increase of urban albedo by the use advanced materials for buildings and the urban fabric, the use of additional green infrastructure, evaporative technologies, solar control systems, and heat dissipation systems to reject the excess heat in atmospheric sinks of low temperature (Akbari et al., 2016). A significant number of large-scale projects implementing UHI mitigation strategies have been carried out worldwide demonstrating the potential of such solutions to reduce the ambient temperature, by a maximum of up to 3°C (Santamouris et al., 2017). Using experimental and theoretical approaches, the impact of mitigation technologies on urban climate, energy consumption, outdoor

Mitigation and Adaptation of Urban Overheating. DOI: https://doi.org/10.1016/B978-0-443-13502-6.00006-3

environmental quality, air quality, heat-related mortality and morbidity, and so forth is estimated, and their performance is assessed.

This chapter aims at presenting the existing information and knowledge on the energy impact of UHI mitigation strategies. As most available existing studies report performance data on increased albedo and vegetation, the analysis will focus on the potential reduction of the energy consumption resulting from increased albedo technologies and additional vegetation in cities, including studies assessing the impact of a combination of both mitigation strategies. The most commonly used methods by researchers to quantify the energy impact of heat mitigation strategies at building and urban scale are presented. A number of representative case studies estimating and reporting the reduction of energy consumption from the implementation of UHI mitigation strategies are analyzed, and the main factors affecting their performance are discussed.

6.2 Urban heat island mitigation technologies to decrease energy demand

Urban overheating may increase the energy consumption for cooling and peak electricity demand, forcing power utilities to build additional power plants and increase the cost of electricity supply. The increase of the electricity energy consumption places under stress both the consumers and the electricity networks. This increased demand for electricity and peak electricity especially during heat weaves can potentially lead to systems overload and blackouts when utility plants cannot meet the increased demand. Existing studies show that the potential increase of the electricity demand per degree of temperature rise varies between 0.5% and 8.5%, while the peak electricity demand increases from 0.45% to 4.6% per degree of ambient temperature rise (Santamouris et al., 2015).

Among the UHI mitigation technologies that have been proposed, increased albedo and increased vegetation technologies present a significant potential in reducing urban overheating and decreasing energy demand of buildings and cities. These technologies have been analyzed in detail in a previous chapter, and in this section, we describe their main characteristics with focus on their energy impacts.

6.3 Increased albedo technologies

Increasing the albedo of cities is one of the most efficient strategies to mitigate the negative effects of UHIs. This can be achieved by implementing highly reflective materials (i.e., cool materials) on the building envelope (roofs and walls) and pavements such as roads, sidewalks, parking lots, and so forth. A significant number of large-scale projects involving the use of reflective materials have been implemented around the world, and the analysis of performance data indicates that the use of reflective materials installed on building roofs or pavements presents a significant mitigation potential with an average peak temperature drop for all projects close to 1.3 C (Santamouris et al., 2017).

Cool materials are characterized by high solar reflectance (SR) and high infrared emittance (e). These two properties result in affecting the temperature of a surface (Synnefa et al., 2006). A surface with high SR and infrared emittance exposed to solar radiation will exhibit lower surface temperature compared to a similar surface with lower SR and e values. If the cool surface is on the building envelope, this would result in decreasing the heat penetrating into the building, reducing the need for electricity for space cooling in conditioned buildings. A cool surface in the urban fabric (roof or pavement) would contribute to decreasing the temperature of the ambient air, as less heat would be transferred from the cooler surface, reducing indirectly the cooling needs. Since building heat gains through the roof peaks in late afternoon, when summer electricity use is highest, cool roofs can also reduce peak electricity demand. Lower peak demand saves on total electricity use but also can reduce demand fees that some utilities charge commercial and industrial building owners. In addition, it can contribute to reduce the strain on the electric grid and reduce the likelihood of power failures on extremely hot days Moreover, lower building cooling needs because of cool building envelopes present the possibility to downsize new or replacement air conditioning equipment, potentially increasing cooling efficiency and producing corresponding monetary savings (Levinson, Akbari et al., 2005; Synnefa & Santamouris, 2012). Cool roofs may present an additional cooling benefit, involving the influence of a large roof surface area such as that of a commercial or industrial building on local air temperatures 0.5−1.5 m above the roofs and thereby on the performance of rooftop HVAC equipment since it will result in decreasing the temperature lift between the source and the output resulting in decreased HVAC energy consumption (Carter, 2011; Green et al., 2020; Leonard & Stay, 2006; Pisello et al., 2013; Wray & Akbari, 2008). Moreover, combining solar panels with a cool roof can increase the output of a photovoltaic system, as reduced heat will be transferred from the cooler roof to the PV module, increasing its performance (Vasilakopoulou et al., 2023).

A large number of cool materials with high SR and infrared emittance values are commercially available representing different surface options. Nowadays, there is a cool option for almost every type of roof or other surface of the urban fabric (European Cool Roofs Council ECRC, 2021; Santamouris, 2013; U.S. Environmental Protection Agency, 2012; U.S. Cool Roof Rating Council (CRRC), n.d.). In addition, research in the field of advanced materials for heat mitigation in the built environment has been ongoing for more than 15 years and a number of innovative materials with advanced radiative and thermal properties aiming to reduce solar and heat gains and increase heat losses have been developed and tested, demonstrating significant cooling potential (Santamouris & Yun, 2020; Santamouris et al., 2011).

6.4 Increased vegetation

Urban vegetation provides environmental, social, and economic ecosystem service benefits (European Commission, The Multifunctionality of Green Infrastructure, Science for Environmental Policy in-Depth Report European Commission/DG Environment, Brussels, 2012). It improves the urban climate and contributes to mitigate the UHI effect, while in parallel, it offers other well-known benefits related to thermal comfort and human health, energy, air quality/ pollution/ GHG emissions, carbon storage and sequestration, water management and quality, as well as other quality of life benefits (Santamouris, Ban-Weiss et al., 2018; Santamouris & Osmond, 2020; U.S. Environmental Protection Agency, 2008). It represents one of the most considered urban heat mitigation measures, and many cities around the world, recognizing the multiple undeniable benefits it offers, have adopted policies and programs aiming to increase urban greenery such as Sydney (Greening Our City Program, n.d.) or New York (Milion Trees NYC, n.d.) tree planting programs. Urban greenery may be part of the city landscape, parks, street, and open space greenery, or it may be integrated into the exterior envelope of buildings like green roofs and vertical green facades. A large number of studies have tried to quantify the mitigation potential of various types of increased greenery in cities. A recent review has found that a reasonable increment of green infrastructure fraction by 20% can result in a peak temperature drop of up to 0.3°C, while with an increase of up to 100%, a maximum drop of average daily peak temperatures of up to 1.8°C is expected. Most studies reported higher cooling effects at night compared to daytime conditions; so a maximum nocturnal temperature decrease of up to 2.3°C may be reached by increasing green infrastructure by 80% (Santamouris & Osmond, 2020).

Urban vegetation may reduce energy consumption mainly through the following processes: (1) shading building and other surfaces of the urban fabric, (2) lowering ambient temperature via evapotranspiration, that is, using heat from the air to evaporate water, (3) altering wind patterns, speed, and advection. For example, appropriately selecting and placing vegetation on a surface or around a structure can provide cooling shade in summer by intercepting direct sunlight, while allowing for sunlight penetration for passive solar heating of the structure during winter. Vegetation may also decrease cooling and heating demands by slowing wind speeds and reducing the infiltration of hot or cold air into a structure. Finally, evapotranspiration decreases ambient air temperatures and thereby reduces energy demand for air conditioning. Green roofs contribute to increasing building energy savings also by acting as insulators, decreasing heat gains and heat losses through the roof in summer and winter, respectively (Akbari et al., 2016; Bowler et al., 2010; Huang et al., 1987; Hwang et al., 2015; Santamouris, Ban-Weiss et al., 2018; Santamouris & Osmond, 2020; U.S. Environmental Protection Agency, 2008; U.S. Environmental Protection Agency, 2008). Many factors have been found to affect the mitigation potential of increased urban greenery, such as the specific climatic conditions, the type and characteristics of vegetation, the way it is distributed in the landscape, the availability of soil moisture, and so forth. These factors should be considered in each specific project in order achieve the desired outcome and avoid potential negative effects such as increased humidity that could be undesirable especially in tropical cities or a potential warming effect due to the entrapment by the canopy of the upwelling long-wave heat reflected or emitted from horizontal and vertical surfaces (Aflaki et al., 2017; Akbari et al., 2016; Morakinyo et al., 2017; Santamouris, Ban-Weiss et al., 2018; Santamouris & Osmond, 2020; U.S. Environmental Protection Agency, 2008).

6.5   Direct and indirect effects of urban heat island mitigation strategies

Studies aiming to evaluate the energy impact of UHI mitigation strategies consider their direct effects on building energy requirements, their indirect effects, and/ or their combination. Planting trees around a building or installing a green roof or reflective materials on the building envelope directly changes the energy balance and therefore the cooling needs of that particular building. Such direct effects provide immediate benefits to the building that applies them. For large-scale application of UHI mitigation measures such as city-wide increases in urban greenery or albedo, the energy balance of the whole city is modified, resulting in city-wide changes of the urban climate. City-wide changes in climate will have an indirect impact on the energy use of an individual building and are therefore considered as indirect effects. So, in order to achieve energy benefits from indirect effects, large-scale application of UHI mitigation measures is required. Fig. 6.1 describes the impact of heat-island mitigation measures on energy use.

6.6   Methods to estimate the energy impact of heat mitigation strategies

Many approaches have been proposed by researchers aiming to provide quantification of the energy impact of heat mitigation strategies depending on the scale examined by the study, ranging from building scale for investigation of the impact of the mitigation measures when applied on a single building on its energy demand to urban scale for exploring the energy impact of large scale/ city-wide application of heat mitigation measures. Although

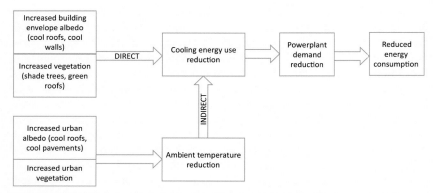

FIGURE 6.1 Impact of urban heat island (UHI) mitigation strategies on energy use Impact of UHI mitigation strategies on energy use. *Adapted from Akbari, H, Pomerantz, M., & Taha, H. (2001). Cool surfaces and shade trees to reduce energy use and improve air quality in urban areas. Solar Energy, 70(3), 295–310. https://doi.org/10.1016/s0038–092x(00)00089-x.*

there are a lot of variations in the employed methodologies depending on the scope of the research, available means, project limitations, and so forth, there are some general actions that can be defined. This section reports the most commonly used methods for estimating the energy impact of heat mitigation measures when implemented at building or urban scale.

6.7 Building scale

Many different approaches have been proposed by researchers aiming to quantify the impact of UHI mitigation strategies at building scale (direct effect) involving experimental or numerical modeling and simulation methods. Many studies involving the increase of envelope albedo (Akbari et al., 2005, 2011; Bozonnet et al., 2011; Kolokotsa et al., 2012; Kolokotroni et al., 2013, 2018; Revel et al., 2014; Romeo & Zinzi, 2013; Rosado & Levinson, 2019; Suehrcke et al., 2008, Synnefa et al., 2012; Zinzi, 2016a,b; Zinzi & Agnoli, 2012) and green roofs and/or increased vegetation (shade trees) case studies (Akbari et al., 1997; Berardi, 2016; Foustalieraki et al., 2017; Karachaliou et al., 2016; Konopacki & Akbari, 2000; Niachou et al., 2001; Olivieri et al., 2013; Santamouris et al., 2007; Silva et al., 2016) have been conducted in residential and nonresidential buildings. The experimental assessment consists of the monitoring activities performed in the existing buildings involving the building with (1) the conventional envelope (the reference case) and (2) the cool or green envelope or increased vegetation after installation. The following actions are usually taken: (1) Audit of the building(s) and data collection to assess the impact of the cool or green envelope, including weather conditions in the area, air temperature inside the building, surface temperatures, AC and total building power consumption, operation schedules, and other building configuration and use data. Ideally, the experimental period should cover one full year for pre- and postintervention periods. (2) Identification and estimation/measurement of the parameters affecting the performance of the envelope before and after the intervention. (3) Statistical analysis of collected data in order to determine the energy impact of the cool or green envelope at building scale including calculation of AC energy savings and demand reduction. (4) A cost—benefit analysis may be performed to quantify if installing a cool/green envelope solution is beneficial for the specific case on a monetary basis.

With the aim to assess the impacts of a cool or green envelope in terms of energy consumption, some experimental studies have utilized test cells (Coma et al., 2016; La Roche & Berardi, 2014; Levinson et al., 2007; Simpson & McPherson, 1997; Suman & Verma, 2003). This method consists of using at least two identical test cells, usually without openings so as to only account for the envelope influence, one of which is covered by a conventional material and is used as a reference and the other is covered by the cool envelope material or the green roof configuration.

Building energy models involving whole building energy simulation methods for a whole year have been used for quantifying the energy impacts of increased envelope albedo and/ or increased vegetation at building scale (Ascione et al., 2013; Gao et al., 2014; Levinson & Akbari, 2010; Synnefa et al., 2007; Taha et al., 1996; Wang et al., 2008; Wong et al., 2003). Various building energy model tools (e.g., EnergyPlus, DOE-2, ESP-r, and TRNSYS) are used to investigate the response of the building envelope against the installation of mitigation technologies such as increased albedo or vegetation that require detailed input and give hourly values of the selected output parameters with short computational times. This is a particularly useful methodology when an experimental campaign cannot be carried out, for example, when a sensitivity or parametric analysis must be performed or in cases where the benefits of mitigation measures must be assessed under different climatic conditions, with different building typologies and characteristics, for comparisons of different mitigation strategies with other energy saving measures and so forth. In order to perform such an analysis, all available data on the building (plans, building configuration, building use, systems, bills etc.) must be collected or defined; if it is a theoretical case, data on the parameters necessary to simulate the initial and final envelope roof configuration are to be used as inputs to the model as well as appropriate climatic data files (involving hourly values). Output data must be analyzed in order to estimate energy savings or other desired output parameters.

There are several web-based calculators allowing for a quick estimation of annual energy and monetary savings associated with choosing a cool roof or a green roof instead of a conventional roof for a large number of cities in the United States (ASU Green Roof Calculator, n.d; U.S.

Department of Energy's Oak Ridge National Laboratory Cool Roof Calculator (Version 1.2)., n.d) and one for cool roofs in Europe (EU Cool Roofs Toolkit Available on, 2009).

Many studies have employed methods combining both experimental and simulation activities in order to assess the energy impacts of heat mitigation measures at building scale (Foustalieraki et al., 2017; Synnefa et al., 2012). This is particularly useful in cases where some monitoring activities can be performed but, for example, the pre- and postcool/green roof installation monitoring periods are short and correspond to different outdoor conditions that cannot be directly compared. In such cases, collected experimental data may be employed to calibrate and validate the developed building energy model.

6.8 Urban scale

In order to estimate the energy impact of the large-scale implementation of UHI mitigation strategies, the effects on local climate must first be calculated. This analysis is done mostly by the use of mesoscale meteorological models such as WRF, CSUMM, and MM5. These models are based on the governing equations of fluid dynamics, and models such as radiation, cloud cover, and soil are integrated into the calculations. Meso-scale models are applied on very coarse cells, indicating a weak resolution on the surface layer to observe interactions between buildings and their environment. To avoid this limitation, such models can be coupled with land surface and urban canopy models. The methodology can be summarized in the following steps: (1) development and validation of the mesoscale model in order to represent the climatic characteristics in the area under investigation, (2) design the UHI mitigation strategies/scenarios, (3) estimation of the cooling potential of the UHI mitigation scenarios, (4) development of modified hourly weather data files to account for the mitigated climatic conditions that would prevail in the area after the application of the UHI mitigation strategies, applying appropriate statistical techniques, (5) calculation of the energy performance of representative building typologies using building energy simulation programs (Energy Plus, TRNSYS, DOE-2, etc.) and the base case (unmitigated) and modified (mitigated) weather files, and (6) estimation of city-scale energy impact from the building energy simulation results and the use of statistical data (Garshasbi et al., 2023; Konopacki & Akbari, 2000, 2002, 2005; Touchaei et al., 2016).

A variant of this methodology, used to investigate the impact of UHI mitigation strategies at neighborhood scale (microscale), on building energy performance, involves combining building energy simulation models with microclimate models, CFD models, that account for the interaction of a building with its surrounding environment in the surface (Aboelata, 2021; Adilkhanova et al., 2023; Castaldo et al., 2018; Haddad et al., 2020;

Santamouris, Haddad et al., 2018; Sedaghat & Sharif, 2022). It should be noted that CFD simulations require a high-resolution representation of the urban geometry, knowledge of the boundary conditions, and adequate computational resources (Mirzaei, 2015). Envimet (n.d) is a 3D urban climate modeling tool that simulates the microclimatic effects of buildings, vegetation, and other objects and represents one popular option for such studies, raising the possibility of investigating different UHI mitigation scenarios compared to the reference case. The climatic variables extracted from the microclimate model results for the different UHI mitigation scenarios can be used to develop modified weather files utilized as inputs to the building model. The building energy performance is then calculated taking into account the modified climate due to the adoption of UHI mitigation strategies. Performing field measurements to validate the model establish confidence on the results (Oke et al., 2017).

It is evident that these methodologies may take into consideration both direct and indirect effects of UHI mitigation strategies.

6.9 The energy impact of urban heat island mitigation strategies

This section compiles a number of representative research studies that estimate and report the reduction of energy consumption from the implementation of UHI mitigation strategies at (1) building scale and (2) urban scale. The reported studies have employed methodologies that include experimental and/or simulation activities, various building types (residential, commercial, etc.), UHI mitigation characteristics, climatic conditions, and so forth. The first part includes studies that aim to quantify building energy savings from the application of cool roofs and walls and / or green roofs and shade trees (direct effect), while the second part considers studies that estimate the energy conservation potential from the urban-scale implementation of increased albedo and vegetation technologies (indirect and direct effects).

6.10 Building scale

6.10.1 Increased envelope albedo

A large number of studies have attempted to quantify the potential energy savings attributed to the increase of envelope albedo in residential and non-residential buildings.

The installation of a cool roof on a new or existing building can significantly improve the energy efficiency during the cooling season and throughout the year, as reported by annual HVAC or cost savings. A literature review on cool roof studies reports cooling energy savings varying from 2% to 44% and averaged about 20% (Haberl & Cho, 2004). The same study reports peak cooling energy savings from cool roofs between 3% and 35%, which depend on ceiling insulation levels, duct placement, and attic configuration. Lower peak electricity demand not only saves on total electrical use but also can reduce demand fees that some utilities charge commercial and industrial building owners and assist in preventing unwanted electricity shutdowns on hot summer days Moreover, most HVAC systems are designed based on peak summer cooling loads; therefore reduced peak electricity loads will lead to the downsizing of HVAC systems, which can operate more efficiently throughout the year, including the heating season (Akbari et al., 1997; Levinson, Akbari et al., 2005). In a more recent literature review, including more than 100 international studies employing simulation and experimental methods, regarding the thermal performance of reflective materials applied to building components, it was reported that daily cooling energy decrease varies between 1% and 80% depending on the climate and the building construction characteristics (Hernández-Pérez et al., 2014).

We present below some examples of either simulation-based or experimental studies for residential and nonresidential building types and different climatic conditions focusing on the impact of increased envelope albedo on cooling energy savings.

A simulation study aiming to investigate the impact of using cool roofs on the energy loads of residential buildings for various climatic conditions (27 cities around the world found that the decrease in the cooling loads for an increase in roof SR by 0.4 varies between 6.8 and 29 kWh/m^2 and for a higher increase by 0.65 between 8.4 and 48 kWh/m^2). Regarding peak cooling loads, it was shown that increasing the SR by 0.65 can achieve savings that vary between 10.7% and 27%. Synnefa et al. (2007) and Levinson et al. (2007) simulated the energy impact of increased roof albedo (by 0.35) on commercial building prototypes testing new/old construction with roof insulation levels R-19 and R-7, respectively, for 236 US locations. The savings for all evaluated locations and building types were then scaled up to a national level using US building stock and building density data. It was found that cooling energy use savings per unit conditioned roof area varied from 3.30 kWh/m^2 in Alaska to 7.69 kWh/m^2 in Arizona with an average of 5.02 kWh/m^2 nationwide. A study for California houses has been carried out aiming to estimate the impact of cool NIR reflective coatings for tiles on annual cooling energy savings (Levinson et al., 2007). For a typical 139 m^2 house, increasing the SR of the roof by 0.3 by the application of cool colored coatings on tiles was done to present whole-house peak power savings of 230 W in Fresno, 210 W in San Bernardino, and 210 W in San Diego. The corresponding absolute and fractional cooling energy savings are 92 kWh/year (5%), 67 kWh/year (6%), and 8 kWh/year (1%), respectively. The cooling load savings when increasing the SR to 0.6 were estimated for a hospital and an office building during the summer conditions in the hot and humid areas of Iran. Simulation results show that a light-colored roof reduced the cooling load by 10% for both the hospital and the office building (Hatamipour et al., 2007; Wang et al., 2008) compared to the electricity consumption of a single story retail shed with different coatings for six locations around the world (Durban, South Africa, Kuala Lumpur, Malaysia, Lisbon, Portugal, Miami, and Phoenix, United States, and Shanghai, China). The highly reflective coatings significantly reduced the energy consumption in hot climates in the range of 25%−38%.

The roof albedo was increased by 0.4 on a standard-compliant Chinese office and residential building prototype in seven Chinese cities (Harbin, Changchun, Beijing, Chongqing, Shanghai, Wuhan, and Guangzhou), and it was found that for the office building, the annual cooling load was reduced by 2.3 kWh/m^2 (Harbin) to 12.4 kWh/m^2 (Guangzhou). The annual energy load savings ranged from −1.5 kWh/m^2 (Changchun) to 10.5 kWh/m^2 (Guangzhou) and were positive everywhere but Harbin and Changchun. For the residential building, the annual cooling load reduction ranged from 0 (in Harbin and Changchun, where the residential building was not cooled) to 10.9 kWh/m^2 (Guangzhou) (Gao et al., 2014).

In the framework of an EU-funded project called Cool Roofs, some case studies have been implemented, aiming to demonstrate cool roof potential in real buildings with different typologies and different climatic conditions, in terms of improving the thermal conditions in non-air-conditioned buildings and reducing the energy consumption in air-conditioned buildings. The methodology followed in all cases includes building monitoring activities under free-floating conditions pre- and postcool roof application and use of experimental data to perform calibrated simulation and estimate annual energy and thermal performances. In Athens, Greece, increasing the roof albedo from 0.2 to 0.89 on a noninsulated 410 m^2 school building resulted in annual cooling energy load reduction of 40% and a heating penalty of 10%. Lower reductions of 35% and 4%, respectively, were estimated when considering the building to be insulated (Synnefa et al., 2012). In Heraklion, Crete, Greece, increasing the SR of a 50 m^2 one-floor well-insulated laboratory and office building by 0.69 resulted in energy conservation equal to 19.8% for the whole year and 27% for the summer period (Kolokotsa et al., 2012). Increasing the roof albedo by 0.54 of a 700 m^2 roof of a single-story office/laboratory building with no insulation in Trapani, Sicily, Italy, reduced the cooling energy demand by 54% and by 24% considering an insulated variant (Romeo & Zinzi, 2013). Applying a cool colored coating ($\Delta SR = 0.5$) on a 137 m^2 flat roof in a naturally ventilated office university building in the area of London, United Kingdom, resulted in an a heating load increase and a cooling load decrease with an overall energy demand reduction between 1% and 8.5% (Kolokotroni et al., 2013).

Akbari et al. (2005) monitored the cooling energy use savings from increasing roof SR of a retail store in Sacramento, an elementary school in San Marcos, and a cold storage building in Reedley, California. The increase in SR was 0.61, 0.54, and 0.61, respectively. Results showed that the average daily savings in cooling energy use for the retail store was estimated to be 72 W h/m^2 (52% savings), and the peak demand reduction was found to be 10 W/m^2 (50% savings). The estimated daily savings in cooling energy use and peak demand for the school were about 42−48 Wh/m^2 (17%−18% savings) and 5 W/m^2 (12% savings), respectively. The cold storage facility had daily cooling savings of 69 Wh/m^2 (4% savings) and peak demand reduction of 5−6 W/m^2 (6% savings).

A study has been performed aiming to quantify the impact of a 1004 m^2 cool roof of a single-story "warehouse" style retail building in subtropical Brisbane, Australia (Suehrcke et al., 2008). The radiative properties of the initial roof were SR = 0.2 and $e = 0.25$, and after the application of the cool roof, the corresponding values were SR = 0.875 and $e = 0.9$. Experimental monitoring results before and after the cool roof installation were used to perform a calibrated simulation. Regarding cooling energy consumption, a 13%, or 2.84 MWh, reduction was achieved because of the cool roof, and an energy reduction was calculated for every month, confirming that the cool roof does not introduce a heating penalty for the specific case. In addition, the energy efficiency potential of cool roof technology applied to similar retail buildings across Australia was examined for seven different climates from cool temperate (Canberra) to hot humid summer and warm winter climate (Darwin). In all cases, cooling energy savings have been found, with the greatest reduction occurring in tropical, subtropical, and dessert environments, where an energy saving of 2.8−8.4 kWh/year/m^2 was estimated.

The impact of a cool roof (SR- 0.82) on a low-income single-story semidetached house in Jamaica is examined by Kolokotroni et al. (2018). A cooling load reduction of 188 kWh/m^2/year corresponding to about of 38% was estimated.

A field study involving a 700 m^2 commercial building with increased roof albedo by 0.6 in Hyderabad, India, found that the annual energy saving achieved was 20−22 kWh/m^2; the air-conditioning energy use reduction was 14%−26% (Akbari et al., 2011).

There are fewer studies aiming to estimate the impact of cool walls on building energy demand. A simulation study found cooling energy savings of 10%−20% when cool walls were applied to a three-story residential building in the Italian cities of Palermo, Rome, and Milan (Zinzi, 2016). Cooling energy savings from the installation of cool walls on residential buildings (prototypes) were found to range between 0.2 and 2.9 kWh/m^2 wall per 0.1 increase in wall albedo for the Mediterranean cities of Marseille, Athens, and Cairo. Zinzi (2016) and Revel et al. (2014) simulated the effect of cool-tile walls on a five-story building and on a large single-story industrial building in three European cities Madrid, Rome, and Palermo and found yearly total energy reductions up to 3.5 kWh/m^2. A study (Rosado & Levinson, 2019) investigating the energy impact of wall albedo used building energy simulations methods considering 10 different building categories, 3 building vintages, 16 California climate zones, and 15 U.S. climate zones. In California, annual whole-building HVAC energy cost savings were 4.0%–27% in single-family homes, 0.5%–3.8% in medium office buildings, and 0.0%–8.5% in standalone retail stores. In warm U.S. climates, annual HVAC energy cost savings were 1.8%–8.3% in single-family homes, 0.3%–4.6% in medium office buildings, and 0.5%–11% in standalone retail stores. The magnitude of savings and penalties from cool walls depends on key factors including climate, wall construction characteristics, wall orientation, building orientation, and HVAC efficiency.

6.11   Increased vegetation

Several studies have demonstrated that strategically planted shade trees do reduce energy demand. A recent review study (Yekang Ko, 2018) reports that building energy savings from trees and vegetation show an extremely wide range of values, from 2.3% to 90% for cooling and from less than 1% to 20% for heating. It is explained that this large variation is in part due to the different climate regions studied, but also because of the different methods and assumptions used. Parker (1981) found that cooling-energy savings from properly located trees and shrubs around a mobile trailer in Florida reduced the daily air-conditioning electricity use by as much as 50%. Simpson and McPherson (1996), using simulation methods, indicated that two trees shading the west-facing exposure of a house and one tree shading the east-facing exposure reduced annual energy use for cooling by 10%–50% and peak electrical use up to 23%. Akbari et al. (1997) used experimental methods to investigate the impact of shade trees on two houses in Sacramento, California. They monitored the cooling-energy use of both houses to establish a base case relationship between the energy use of the houses. Eight large and eight small shade trees were then installed at one of the sites for a period of four weeks, and then they were moved from one site to the other. The results showed seasonal cooling-energy savings of about 30% (about 4 kWh/day). Akbari and Konopacki (2005) used simulation methods to estimate the potential of heat-island reduction strategies (here, we report the results for shade trees) to reduce cooling-energy use in three building types, residences, offices, and retail stores (old and new vintage), considering adding 4, 8, and 10 shade trees, respectively. Results are presented by ranges of heating degree days. For pre-1980, gas-heated residential buildings, shade trees savings potentials ranged from about 298 kWh/1000 ft^2 (4%) (HDD < 500) to about 185 kWh/1000 ft^2 (9%) (5500 < HDD < 6000). The heating energy penalties ranged from 0 to 19 therms (0%–2%). For post-1980 stock of residential buildings, shade trees savings potentials ranged from about 192 kWh (4%) (HDD < 500) to about 98 kWh (11%) (5500 < HDD < 6000). The heating energy penalties ranged from 0 to 9 therms (0%−2%). The peak demand electricity savings ranged from 0.08 to 0.14 kW/1000 ft^2 for pre-1980 stock and 0.09−0.08 kW/1000 ft^2 for post-1980 stock. For stock of pre-1980 gas-heated office buildings, the shade trees savings potentials ranged from about 317 kWh (2%) (HDD < 500) to about 594 kWh (7%) (5500 < HDD < 6000). The heating energy penalties ranged from 0 to 6 therms (0%−2%). For post-1980 stock of office buildings, the shade trees savings potentials ranged from about 202 kWh (5%) (HDD < 500) to about 244 kWh (6%) (5500 < HDD < 6000). The heating energy penalties ranged from 0 to 4 therms (0%−3%). The peak demand electricity savings ranged from 0.12 to 0.34 kW/1000 ft^2 for pre-1980 stock and 0.06 to 0.22 kW/1000 ft^2 for post-1980 stock. For stock of pre-1980 gas-heated retail store buildings, the shade trees potentials ranged from about 320 kWh (2%) (HDD < 500) to about 423 kWh (5%) (5500 < HDD < 6000). The heating energy penalties ranged from 0 to 1 therms. For post-1980 stock of retail store buildings, the shade trees savings potentials ranged from about 195 kWh (2%) (HDD < 500) to about 174 kWh (5%) (5500 < HDD < 6000). The heating energy penalties was 0 therms. The peak demand electricity savings ranged from 0.09 to 0.17 kW/1000 ft^2 for pre-1980 stock and 0.07−0.08 kW/1000 ft^2 for post-1980 stock.

A large number of experimental and simulation studies have investigated and reported the energy benefits of green roofs over the past decades. Green roofs are highly efficient in reducing the level of building energy consumption both in warm and cold climates. A recent review of green roof studies (Mihalakakou et al., 2023) highlights that green roofs contribute to a substantial decrease in the cooling load and an increase in annual energy savings and reports that cooling load reduction by up to 70% was achieved, while annual energy savings ranged from 10% to 60%. Below, we present some green roof studies covering a range of building types and climatic conditions.

Niachou et al. (2001) used experimental and numerical methods and found that a green roof on an uninsulated building in Athens, Greece, may result in energy savings of 37% for the entire year, which reached 48% when night ventilation was also applied. Santamouris et al. (2007) examined the energy impact of a green roof in a nursery school building in Athens, Greece, and reported a cooling load reduction of 6%−49% for the entire building and 12%−87% for the last floor. Spala et al. (2008) found a 40% reduction of the cooling load from a green roof on an office building in the greater Athens area. Jaffal et al. (2012) simulated the impact of a green roof in three cities with different climates and found a reduction of the total annual energy demand of 32% for Athens, 6% for La Rochelle, and 8% for the cold climate of Stockholm. A field experiment comparing the effect of an extensive green roof compared to a conventional roof on a lightweight building in Shanghai, China, reports a maximum heat flux difference between the green roof and a conventional roof up to 15 W/m^2 for the summer period He et al. (2016) and Yang et al. (2018) used simulation methods and reported a reduction of heat gain by 13.14 (31%) KWh/m^2 for the whole of a summer design day, considering a green roof on an office building in the tropical climate of Singapore. Karachaliou et al. (2016) investigated the energy impact of a large-scale intensive green roof system installed on an office building using simulation and experimental methods and reported a significant reduction of the annual energy consumption, up to 19% for the cooling and up to 11% for the heating load. In Toronto, Canada, the adoption of a green roof retrofit on a university building resulted in an energy demand reduction by 3% for an LAI equal to 2 and a soil depth of 30 cm. Berardi (2016) and Morakinyo et al. (2017) used simulation methods to evaluate the impact of different green roofs on energy demand reduction under four different climates (Hong Kong, Paris, Cairo, and Tokyo) and urban densities. In terms of cooling demand reduction, 5.2% was observed in hot-dry climate on the hottest day of the year with full-intensive green roof, while the least saving of 0.1% was found with semiextensive green roof in temperate climate. Ascione et al. (2013) performed a parametric analysis comparing green roof typologies to traditional roofing technology, considering a well-insulated office building and different European climatic conditions. In warm climates, green roofs can contribute to a reduction of the cooling load without a significant increase of the heating load, offering an annual energy demand reduction up to 11%. For cold climates, green roofs offered a decrease in both heating and cooling energy demand, contributing to energy savings up to 7%. Foustalieraki et al. (2017) used numerical and experimental methods to evaluate a medium-scale green roof system installed on a commercial building, in Athens, Greece, and found that the overall annual energy reduction during the cold and hot periods reached 15%, with the reduction of the cooling load reaching 18.7% and that of the heating load reaching 11.4%. Zhang et al. (2022) investigated eight green roof types in commercial and residential buildings in the subtropical monsoon climate of Nanning, China. A reduction of the annual energy consumption ranging from 30% to 55% was demonstrated for various soil thicknesses.

6.12 Urban scale

Many studies have estimated the potential of UHI mitigation strategies in reducing the ambient temperature and mitigating the UHI effect when implemented at urban scale. However, few studies have attempted to quantify the energy impact of large-scale implementation of UHI mitigation strategies. This section compiles research studies that aim to estimate the energy saving potential of UHI mitigation strategies, mainly increased albedo and/or increased vegetation, considering that they have been implemented at urban scale, taking into account the indirect and/or both direct and indirect benefits.

The impact of large-scale tree-planting programs in 10 US metropolitan areas (Atlanta, GA, Chicago, IL, Dallas, TX, Houston, TX, Los Angeles, CA, Miami, FL, New York, NY, Philadelphia, PA, Phoenix, AZ, and Washington DC) have been analyzed considering both direct and indirect effects on air-conditioning energy use on residential and office buildings, using a combination of a building simulation program for energy calculations and a mesoscale simulation model for meteorological calculations. The meteorological simulations showed that trees could decrease ambient temperature on average by about 0.3K−1K at 1400 hours. For most cities, the energy analysis found total (direct and indirect) annual energy savings to be 5−35 $/1000 ft roof area for residential buildings and 2−18$/1000 ft of roof area of office buildings (Taha et al., 1996).

Another study performed by Rosenfeld et al. (1998) has estimated both direct and indirect energy savings from increasing roof and pavement albedo by 0.35 and 0.25, respectively, and increasing vegetation by adding 11 M trees, for Los Angeles Basin. Using a combination of mesoscale modeling (Taha, 1996, 1997) and building energy simulations, they found for residential buildings net direct savings of 24$/year (15% reduction compared to base case) from cooler roofs, 18$/year (12%) from shade trees, and 36$/year (23%) from the lower ambient temperature by 3 C resulting from the UHI mitigation measures applied, amounting to a total of 78 $/year (50%). Total energy savings from commercial buildings were considered equivalent to 25% of residential savings. Extrapolating these savings to all of L.A., the total (direct and indirect) annual benefits are $171 M for roofs, $273 M for trees, and $91 M for pavements, amounting to about $535 M/year, if all mitigation strategies are combined, while the avoided peak power for air conditioning can reach about 1.5 GW (more than 15% of the city's air conditioning).

Konopacki and Akbari (2000, 2002) have estimated the direct and indirect energy effects of heat island reduction measures involving cool roofs and pavements, shade trees, and urban vegetation in five U.S. metropolitan areas: Baton Rouge, Chicago, Houston, Sacramento, and Salt Lake City. They have considered three building types, single-family residence, office and retail store, and old and new construction. Direct savings were determined for buildings with eight shade trees (retail store: four) and a high-albedo roof (residential: 0.5, commercial: 0.6, base case: 0.2) The analysis indicated potential net annual energy savings of $15 M (79% residential, 6% office, and 15% retail store), $30 M (37% residential, 27% office, and 36% retail store), $82 M (79% residence, 7% office, and 14% retail store), $30 M (51% residence, 16% office, and 32%retail store), and $4 M (11% residence, 31% office, and 58% retail store) and peak power avoidance of 130, 400, 730, 490, and 85 MW, respectively, for the five cities. Of the overall annual energy savings for Baton Rouge, Chicago, Houston, Sacramento, and Salt Lake City, the indirect impacts (cooler ambient air temperature) of heat island reduction strategies were 15%, 18%, 19%, 19%, and 22%, respectively.

Another study aiming to estimate the energy impact of UHI mitigation strategies, taking into consideration the effect of (1) cool roofs, that is, increased albedo by 0.3 for residential and 0.4 for office and retail store buildings (direct effect), (2) strategic placement of shade trees near the building, that is, 4, 8, and 10 trees for the residence, office, and retail store, respectively (direct effect), and (3) urban vegetation and cool roofs and pavements (indirect effect), has been conducted in 240 regions in the United States (Akbari & Konopacki, 2005). The study considers residences, offices, and retail stores, each characterized by pre-1980 (old) or post-1980 (new) construction vintage and with natural gas or electricity as heating fuel. Using building energy simulation, the authors estimate energy use and peak power demand. The indirect savings were calculated using modified weather data representing the UHI mitigation strategies, that is, modified urban fabric, derived via statistical analysis of previously completed simulations (Akbari, 2002; Konopacki & Akbari, 2000, 2002). For residential buildings, the total energy savings from UHI mitigation strategies were found to range from 12% to 25% (with a corresponding heating penalty of 0%−5%) and peak demand electricity savings ranged from 0.2 to 0.6 kW/1000 ft^2. For office buildings, the total energy savings from UHI mitigation strategies were found to range from 5% to 18% (with a corresponding heating penalty of 0%−7%) and peak demand electricity savings ranged from 0.2 to 1 kW/ 1000 ft^2. For retail stores, the total energy savings ranged from 7% to 17% (with heating penalty ranging from 0% to 13%), while the peak demand electricity savings ranged from 0.2 to 0.7 kW/1000 ft^2. Finally, it was highlighted that for all building types, over 75% of the total savings were from direct effects of cool roofs and shade trees.

Some studies have been performed for Canadian cities, where the climate is dominated by high heating demands during the winter and moderate cooling needs during the summer.

Akbari and Taha (1992) used simulation methods to investigate the effect of moderate increases in surface SR and vegetative cover (trees) in four Canadian cities (Toronto, Edmonton, Montreal, and Vancouver). The simulations indicated that by increasing the vegetative cover by 30%, the heating energy use in Toronto can be reduced by 10% in urban houses and 20% in houses located in open suburban areas (mostly because of the wind-shielding effect of trees). Results also showed that by increasing the albedo of houses by 0.2 (from moderate-dark to medium-light color), the cooling energy use can be reduced by about 30%−40%. In urban residential neighborhoods of Edmonton, Montreal, and Vancouver, average savings in heating energy use are 8%, 11%, and 10%, respectively. Cooling energy can be totally offset in Edmonton and Vancouver and average savings of 35% can be achieved in Montreal.

Akbari and Konopacki (2004) used simulation methods (PSU/NCAR MM5 and DOE -2) to estimate the potential of heat island reduction strategies (i.e., solar-reflective roofs, shade trees, wind-shielding, reflective pavements, and urban vegetation) to reduce cooling energy use in buildings in the Greater Toronto Area, Canada (GTA). They have considered both direct effect (reducing heat gain through the building shell) and indirect effect (reducing the ambient air temperature). A total of nine building prototypes including residential, office, and retail stores, old and new construction, were modeled. The roof albedo was increased by 0.3 for residential roofs and by 0.4 for commercial roofs, and the number of shade trees modeled was 4, 8, and 10 for the residence, office, and retail store, respectively. A 20% increase in vegetation cover around buildings was considered for the wind-shielding effect. For gas-heated residential prototypes, the simulations predicted annual total energy savings of about 3%−5% from combined direct and indirect effects for residential buildings, 10% for offices and 12% for retail stores. Electrically heated units showed less reductions because the electric heating penalty is more expensive than that of gas. An annual natural gas deficit was found for all building types and all UHI mitigation strategies with the exception of wind-shielding amounting to $2−6/100 m^2 for residences, $11−12/100 m^2 for offices, and only $0−3/100 m^2 for retail stores. Peak power reduction in cooling electricity for both direct and indirect effects was 21%−23% in residences and 13%−16% in offices and retail stores. Considering the entire GTA, annual energy savings of over $11 M may be achieved (with uniform residential and commercial electricity and gas prices of $0.084/kWh and $5.54/GJ, respectively) from the combined direct and indirect effects of UHI mitigation strategies. The direct effects account for about 88% of that total amount and the indirect effects for 12%. The residential sector accounts for over half (about 59%) of the total savings, offices 13%, and retail stores 27%. Savings from cool roofs were about 20%, shade trees 30%, wind-shielding by trees 37%, and indirect effects 12%.

Touchaei et al. (2016) investigated the impact of increasing the albedo of roofs, walls, and pavements (by 0.45, 0.4, and 0.25, respectively) on the energy consumption of commercial building prototypes (small, medium, large office, and retail store), old and new construction, and different HVAC systems in Montreal, Canada, using a combination of mesoscale modeling (WRF) and building energy simulation (DOE-2) in order to account for both direct and indirect effects. The maximum air temperature reduction was 0.6°C. They found that the combined effect of decreased solar heat gain by building surfaces and decreased air temperature reduces the energy consumption of HVAC systems by 2% (~ 0.1 W/m^2) on average with a maximum decrease of 0.4 W/m^2.

Adilkhanova et al. (2023) performed a study to evaluate the urban cooling effect from the city-scale application of high albedo materials on urban buildings' energy consumption in Seoul. The methodology involves a multidisciplinary approach that combines (1) urban climate modeling (WRF) for considering increased albedo (by 0.5) impacts on the urban microclimate and developing weather files, (2) urban building energy modeling (CitySim software) for urban-scale building energy simulations, and statistical analysis of actual energy data to develop an energy prediction model. To evaluate the effect of increasing surface albedo on the energy performance of the buildings, a building stock consisting of 1000 buildings with various construction characteristics, located in the area under investigation, was selected. The results showed that high albedo materials could reduce UHI and its adverse effects on energy consumption, reaching a monthly temperature drop of up to 2.08°C and significant urban-scale monthly cooling energy use reductions of up to 2.91 kWh/m^2.

Garshasbi et al. (2023) carried out a study that uses a holistic methodology combining mesoscale modeling (WRF + SLUCM) and building energy simulations (Energy Plus via Design Builder) to quantify the magnitude and spatial variation of the cooling load savings and peak electricity demand reductions from the building-scale (direct effect) and combined building-scale and urban-scale (direct and indirect effect) implementation of cool roofs (0.65 increase of roof albedo) under various climatic conditions in Australia, ranging from high humidity summer and warm winter conditions (Darwin) to cool temperate climate (Hobart). The study involves 17 different building types including existing and new residential, office, school, and commercial, with various building construction and use characteristics. For warm/mild temperate climatic conditions (Sydney), cool roofs were found to reduce the cooling loads of a typical noninsulated low-rise office building by 10.2−13.8 kWh/m^2 (37.6%−42.0%) during the summer months of January and February, while taking into account both direct and indirect effects; the cooling load conservation potential increases to 14.9−17.4 kWh/m^2 (50.3%−63.7%). The corresponding peak electricity demand reduction is 53%−70%. Accordingly, for all 17 different building types and major Australian cities, cooling load reduction due to cool roof application during summer was found to range between 0.5% and 63.6% and between 13.2% and 72.4%, when both direct and indirect effects are considered. Residential buildings showed the greatest cooling load saving potential of 1.8%−63.6% (building-scale application of cool roofs) and 29.8%−72.4% (combined building-scale and urban-scale implementation of cool roofs), followed by office buildings achieving 0.8%−51.3% and 22.4%−62.5% and commercial buildings achieving 0.5%−16.7% and 13.2%−29.1%, respectively. Annual cooling load savings of cool roofs for all building types and climate zones, ranging between 0 and 52.7 kWh/m^2, were found to be significantly higher than corresponding heating load penalties (0−8.8 kWh/m^2), except for noncommercial buildings in Hobart with harsh winter climate conditions. It was also estimated that cool roofs can generate an additional cooling load savings of around 3%−20% due to the higher efficiency of AC systems operating at lower ambient temperatures.

Santamouris, Haddad et al. (2018) applied a variant of the previous methodology. They have used advanced 3D microclimate modeling (ENVIMET) combined with building energy simulations in an effort to quantify the local impacts and energy benefits of different urban heat mitigation scenarios for a selected area in Chippendale, Sydney, that represents the typical pattern of a compact urban site in Sydney's central area and is characterized by a mild climate. They investigated the impact of several UHI mitigation scenarios including the increase of global albedo (roofs, streets, and pavements), cool roofs, cool pavements, increased greenery, and so forth. They estimated that increasing the global albedo by 0.6 on all outdoor surfaces including roofs, pavements, and streets can reduce the peak ambient temperature by 3°C and peak cooling demand of residential buildings by up to 20%. Increasing pavement greenery by 55% can reduce the peak ambient temperature by 1.5°C and peak cooling demand of residential buildings by up to 2.3%.

A holistic study that aims to estimate the benefits of city-scale heat mitigation technologies to energy consumption and peak electricity demand has been conducted for the tropical city of Darwin, NT, Australia (Haddad et al., 2020). The authors have used microclimate modeling supported by detailed measurements of meteorological data, combined with building energy simulation in order to assess the impact on cooling energy needs of mitigation technologies involving the use of (1) cool roofs and pavements (increase of global albedo from 0.2 to 0.6), (2) increased greenery to occupy 30% of the open spaces, and (3) the combination of global increase in albedo to 0.6, application of shading on streets and parking lots, and 30% increased greenery. The best performing mitigation scenario, combining cool materials, shading, and greenery, was found to reduce the peak ambient temperature by 2.7°C and consequently decrease the annual cooling load of residential buildings by 7.2% (by 31.9 kWh/m^2 from 441 kWh/m^2 in the unmitigated condition) and that of office buildings by 5.1% (by 30 kWh/m^2 from 585 kWh/m^2 in the unmitigated condition). Cool roofs and pavements were found to reduce the peak ambient temperature by 2.8°C and the annual cooling load by 5.8% (by 25.8 kWh/m^2) and by 4.1% (by 24.3 kWh/m^2) for the residential and office buildings respectively. Urban greenery may decrease the ambient temperature by a maximum of 2.6°C and the annual cooling load of residential and office buildings by 2.6% (by 11.6 kWh/m^2) and by 1.4% (by 8.3 kWh/m^2), respectively. Cooling load savings in the wet season were estimated to be higher (by 22%−28% for the residential building and 37%− 67% for the office building) than those in the dry weather regime, consistent with the higher air temperature reduction observed during the wet season. The annual total cooling load savings for residential and commercial buildings at Local Government Area (LGA) level in the city of Darwin, based on statistical data of the total floor area, resulting from the application of greenery, cool materials, and the combination of strategies, are estimated to be 88.4, 214 GWh, and 265 GWh/y^{-1}, respectively. Finally, a peak electricity demand reduction of 2% was estimated for the combined scenario compared to the unmitigated scenario, using a model developed from measured electricity data for the unmitigated condition deriving the relation between ambient conditions and electricity demand and thus the impact of local climate mitigation strategies.

Table 6.1 summarizes the main characteristics of the studies assessing the energy impact of large-scale implementation of increased albedo and vegetation.

6.13 Discussion on factors affecting the energy performance of urban heat island mitigation strategies

The implementation of UHI mitigation strategies such as the use of cool materials on buildings and the urban fabric and increased vegetation at building and/ or urban scale is found to reduce cooling energy needs and peak electricity demand. However, the research studies previously analyzed present a significant variability in the energy performance of buildings due to increased albedo or vegetation, depending on the specific boundary conditions in terms of local climate, UHI mitigation solution properties and characteristics, building construction details, type and use, HVAC systems, location specifications, and so forth. Some important factors that affect the energy performance of the UHI mitigation strategies are discussed below.

Increasing the urban albedo (including roofs and pavements) can potentially lead to an increase of heating loads and annual heating energy use in climates that have a heating season. Studies have shown that in most cases, this increase is far less important than the corresponding cooling energy savings, resulting in positive net savings for warm/moderate climatic conditions. This is explained by the fact that during winter, the sun is much lower in the sky and solar radiation arriving to a horizontal surface less intense. There is a higher probability of overcast skies, and there is less solar availability (fewer hours of sunshine), so less total energy arrives on a surface to be absorbed or reflected over the same period of time as during the summer (Gao et al., 2014; Kolokotsa et al., 2012; Kolokotroni et al., 2013; Levinson, Akbari et al., 2005; Synnefa et al., 2007, 2012; Zinzi & Agnoli, 2012), for example, in the framework of the simulation study previously mentioned by Synnefa et al. (2007). It was demonstrated that the heating penalty $(0.2-17 \text{ kWh/m}^2 \text{ year})$ is less important compared to the cooling load reduction $(9-48 \text{ kWh/m}^2 \text{ year})$ for the climates studied. The building use plays an important role, and in some cases, internal gains might be so high that air conditioning may be required throughout the year and the increased albedo will be beneficial even in colder climatic conditions. Mastrapostoli et al. (2016) and Hosseini and Akbari (2016) showed that increased heating penalties are overestimated in studies that do not consider the effect of snow that raises the albedo of conventional surfaces.

TABLE 6.1 Summary of methods, urban heat island (UHI) mitigation scenarios and main findings of studies investigating the energy impact from the large scale implementation of UHI mitigation strategies.

Study	Location	UHI mitigation strategy	ΔTmax	Cooling load reduction	City-wide energy savings	Building type	Modeling tools
Taha et al. (1996)	Atlanta GA, Chicago IL, Dallas TX, Houston TX, Los Angeles CA, Miami FL, New York NY, Philadelphia PA, Phoenix AZ, Washington DC	Additional trees in the metropolitan area ranging from 1.4 M to 5 MThree trees (a,b)	1°C–3°C	≈5−35$/1000 ft (a) ≈2−18$/1000 ft (b)	–	(a) Residential (b) Office	DOE 2, CSUMM
Rosenfeld et al. (1998)	Los Angeles	Increased roof albedo by 0.35 Increased pavements albedo by 0.25	1.5°C	24 $/year (18%) (a) –	$171 M $91 M	(a) Residential (b) Commercial (energy savings from commercial buildings equivalent to 25% of residential savings)	CSUMM, DOE 2
		Increased vegetation 11 M trees	1.5°C	18 $/year (12%) (a)	$273 M		
		Combined	3°C	78 $/year (50%) (a)	$535 M		

Reference	Location	Measures	Temperature change	Energy use	Cost savings	Building type	Model
Konopacki and Akbari (2000), Konopacki and Akbari (2002/2002)	Baton Rouge, Chicago, Houston, Sacramento Salt Lake City	Shade trees (a,b:8, c:4) Cool roofs (ΔSR: a.0.3,b,c.0.4) urban reforestation and high-albedo surfaces	2°F, 0°F, 2°F, 3°F, 3°F,		$15 M $30 M $82 M $30 M $4 M (net annual energy savings)	(a) Residential (b) Office building (c) Retail	CSUMM DOE-2
Akbari and Konopacki (2005)	240 US metropolitan regions	Cool roofs (ΔSR: a.0.3,b,c.0.4) Trees (a.4,b.8,c.10) urban vegetation, and high-albedo surfaces	–	12%–25% (heating penalty 0%–5%) (b) 5%–8% (heating penalty) 0%–7% (c) 7%–17% (heating penalty 0%–13%)		(a) Residential (b) Office building (c) Retail	CSUMM DOE-2
Akbari and Taha (1992)	Toronto, Edmonton, Montreal, Vancouver	Increased roof and wall albedo by 20% (only for direct savings) Increased vegetation: 30%	–	Toronto: 10%–20% heating and 30%–40% cooling energy use reduction Edmonton: 8% heating and 100% cooling energy use reduction Montreal: 11% heating and 35% cooling energy use reduction Vancouver: 10% heating and 100% cooling energy use reduction		Residential	URBMET, DOE-2

(Continued)

TABLE 6.1 (Continued)

Study	Location	UHI mitigation strategy	$\Delta Tmax$	Cooling load reduction	City-wide energy savings	Building type	Modeling tools
Akbari and Konopacki (2004)	Toronto, Canada	Cool roofs (ΔSR: a.0.3,b.c.0.4) Increased vegetation: 20%	1.6°C	(a) 3%–5% (b) 10% (c) 12%	$11 M (annual energy savings)	(a) Residential (b) Office (c) Retail stores	MM5, DOE-2
Touchaei et al. (2016)	Montreal, Canada	Increased albedo: roofs by 0.45 Walls by 0.4 pavements by 0.25	0.6°C	Reduction of HVAC energy consumption $\approx 2\%$ $(0.1\ W/m^2)$		(a) Office (b) Retail store	WRF, DOE-2
Adilkhanova et al. (2023)	Seoul, Republic of Korea	Increased albedo of roofs, walls and streets by 0.5	2.08°C	—	monthly cooling energy use reductions: 2.91 kWh/ m^2 (August)	(a) Building stock consisting of 1000 buildings located in the examined area with various characteristics	WRF, CitySim software,
Garshasbi et al. (2023)	Major Australian cities	Increased roof albedo by 0.65	—	Cooling load reduction during summer: 13.2%–72.4% (all building types) 29.8%–72.4% (residential)		Residential, office, school, and commercial	WRF, SLUCM, Energy Plus via Design Builder

Reference	Location	Intervention	Temperature reduction	Energy savings results	Annual total cooling load savings	Building type	Software
Santamouris, Haddad et al. (2018)	Chipendale, Sydney, AU	Increase of global albedo by 0.6; Increase of pavement greenery by 55%; green roofs	3°C; 1.5°C	22.4%–62.5% (office); 13.2%–29.1% (commercial); All building types: Annual cooling load savings: 0–52.7 kWh/m²; Heating penalty: 0–8.8 kWh/m²; (from 2.35 to 1.89) (20%); from 2.35 to 2.29 (2.3%)	—	Residential	Envimet, Energy plus
Haddad et al. (2020)	Darwin, NT, Australia	Urban greenery 30%; Increase of global albedo by 0.4; Combination of increased greenery by 30%, cool roofs and pavements (ΔSR = 0.4) and urban shading (on top of main streets of Darwin)	2.6°C; 2.8°C; 2.7°C	(a) 11.6 kWh/m² (2.6%); (b) 8.3 kWh/m² (1.4%); (a) 25.8 kWh/m² (5.8%); (b) 24.3 kWh/m² (4.1%); (a) 31.9 kWh/m² (7.2%).; (b) 30 kWh/m² (5.1%)	88.4 GWh; 214 GWh; 265 GWh (annual total cooling load savings for residential and commercial buildings)	(a) Residential; (b) Office building	Envimet, Energy plus

Another factor that significantly affects the performance and mitigation potential of increased albedo technologies is aging, that is, the decrease of cool materials SR overtime due to weathering, soiling, and biological growth. The aging effect mainly depends on the type of cool material, the characteristics of the local climate, and the initial value of SR (Berdahl et al., 2008; Mastrapostoli et al., 2016; Sleiman et al., 2011; Takebayashi et al., 2016). Washing and cleaning practices were found to be able to restore the initial albedo or a percentage of it in some cases (Levinson, Berdahl et al., 2005). This loss in SR results in a decrease in the cool materials mitigation and energy performance. For example, Paolini et al. (2014) reported that aging by 0.14 and 0.24 reduces cool roof cooling load savings of 14%−23% in Roma or 20%−34% in Milano. Aging was found to reduce the cool roof energy savings of 8.8% in Xiamen or 15.8% in Chengdu, China (Shi et al., 2019). Finally, Paolini et al. (2020) performed a 4-year natural exposure study on vertical concrete slates with initial albedo 0.75 and 0.46 in Milan, Italy; after 4 years, the aged albedos were 0.55 and 0.38, respectively. They then simulated a 10-story residential building using these new and aged wall albedos. The 0.20 drop in wall albedo increased annual cooling energy use by 5%−11% and reduced annual heating energy use by 2%−4%.

Intensive research activities aiming to overcome the abovementioned limitations have led to the development of a range of innovative advanced materials with enhanced properties and performance (Santamouris & Yun, 2020).

Shading of buildings by trees and vegetation can potentially increase the heating energy use during the winter. Four studies that have evaluated the energy impact of shade trees on residential buildings in Sacramento, California, report annual AC savings per tree ranging from 3.2% to 8.0% for newer buildings, and 5.2% to 7.7% for older buildings, while the corresponding annual heating penalty from shade was about 1%−2% per tree for old and new construction (Akbari et al., 1993; Huang et al., 1987; Simpson & McPherson, 1998; Thayer & Maeda, 1985). Other simulation studies considering a variety of climatic conditions and building types report heating penalty values lower compared to corresponding energy savings in most cases (Akbari & Konopacki, 2005). Trees can reduce energy consumption in a city, and such savings depend on climatic conditions. Tree location and species play an important role in energy savings. Planting medium to large solar-friendly trees around 9−10 m away from the west wall of the building has been found to show the largest cooling energy savings, followed by the east orientation (Akbari, 2002; Ko et al., 2015). Several studies have highlighted the parameters that influence energy savings from increased vegetation such as species, vegetation fraction, leaf area index (LAI), trunk and tree height and crown diameter, foliage density, and so forth (Morakinyo et al., 2018). In choosing the appropriate vegetation species, irrigation requirements should also be considered (Santamouris, Ban-Weiss et al., 2018). In terms of achieving long-term energy benefits, the tree survival rate is an important factor (Ko et al., 2015). Green spaces placement and vegetation species selection must be considered in the full context of the landscape conditions so that potential energy benefits are being maximized.

The energy impact of green roofs depends on the climate, building, and the system and vegetation characteristics. Many studies have attempted to evaluate the impact on the energy performance of parameters such as solar radiation, ambient air temperature, air relative humidity and wind speed, LAI, foliage height and density, plant coverage, albedo, soil layer thickness, plant transpiration rate, irrigation, and others. For instance, increases in LAI values are found to significantly reduce energy consumption during summer having a potentially a negative effect during winter (Ascione et al., 2013; Jaffal et al., 2012; Sailor, 2008). Increasing soil thickness is considered as an additional layer contributing to the U-value of the roof and was found to increase energy savings (Theodosiou, 2003). Increasing plant density in a green roof was found to reduce cooling consumption even in well-insulated roofs (Olivieri et al., 2013). Relative humidity of the atmosphere and wind speed affect significantly the cooling capacity of the green roof system, as they influence the evapotranspiration process (Theodosiou, 2003). A comprehensive review of the parameters affecting the energy performance of green roofs is included in Berardi (2016) and Mihalakakou et al. (2023).

Energy benefits of all considered UHI mitigation strategies strongly depend on building characteristics and heat transfer processes, which are primarily determined by the buildings' thermal resistance or U-value. High values of thermal resistance reduce the conduction of heat through the envelope and therefore decrease the effect of increased albedo or increased vegetation technologies and a lower envelope temperature on heat transfer to the indoor space. Many research studies show that benefits of increased albedo or increased vegetation technologies are more important in low or noninsulated buildings, as it is the case for old construction buildings (Akbari & Konopacki, 2005; Garshasbi et al., 2023; Synnefa et al., 2007; Zinzi & Agnoli, 2012). In parallel, the characteristics of the energy load of the buildings (more cooling or heating load) contribute to understanding the possible contribution of increased albedo and vegetation technologies (Jaffal et al., 2012; Mastrapostoli et al., 2016).

Finally, taking into consideration that climate is expected to significantly increase building cooling demand (Santamouris, 2016a, 2019), the importance of considering possible future climate scenarios when estimating the energy impact of UHI mitigation strategies is highlighted, and increased albedo and vegetation technologies present an attractive solution, as cooling savings are expected to be even more important in future climatic conditions (Hosseini et al., 2018). However, taking into consideration a changing climate, with more frequent and severe droughts and heatwaves, careful selection of appropriate tree/ vegetation species will become more important in order to obtain the desired energy savings and other ecosystem services (Santamouris, Ban-Weiss et al., 2018).

6.14   Conclusions

Developing high-performance heat mitigation strategies and optimally integrating them in urban areas is absolutely necessary in order to counterbalance urban overheating and its negative impacts on energy, environmental quality, air quality, health, and economy and thus increase the sustainability and livability of cities. This chapter has presented case studies that have quantified the impact on building energy consumption from the implementation of UHI mitigation strategies, namely, increased albedo and vegetation, at building and/ or urban scale and has discussed the most commonly used methodologies for such analyses. The existing literature provides concrete evidence that increasing the urban albedo and green infrastructure in urban areas contributes to the reduction of cooling energy use and peak electricity demand, offering important energy, environmental, and socioeconomic benefits. It should be highlighted though that there is a limited number of available studies evaluating the impact of large scale increases of urban albedo and vegetation. In addition, due to the limited real-life implementation of UHI mitigation strategies at urban scale, most existing data come from studies using advanced simulation techniques and tools. The existing studies cover a wide range of boundary conditions, input parameters, methods, and assumptions, and therefore, data, solutions, and conclusions are difficult to be homogenized and may appear confusing or even conflicting. It is evident that more research is needed to improve knowledge on the topic, including detailed experimental studies and assessment of the potential benefits and drawbacks with the aim to provide decision-makers, stakeholders, and urban planners with the necessary information and knowledge to successfully implement UHI mitigation strategies, such as increased albedo and green infrastructure, in order to achieve optimal benefits. The existing and ongoing research activities and programs show that there is significant drive and potential for large-scale mitigation projects in the future, facilitated by efforts made by local authorities and governments to establish incentives, standards, and legislation to further promote implementation of UHI mitigation strategies.

References

Aboelata, A. (2021). Assessment of green roof benefits on buildings' energy-saving by cooling outdoor spaces in different urban densities in arid cities. *Energy, 219*. Available from https://doi.org/10.1016/j.energy.2020.119514.

Adilkhanova, I., Santamouris, M., & Yun, G. Y. (2023). Coupling urban climate modeling and city-scale building energy simulations with the statistical analysis: Climate and energy implications of high albedo materials in Seoul. *Energy and Buildings, 290*. Available from https://doi.org/10.1016/j.enbuild.2023.113092.

Aflaki, A., Mirnezhad, M., Ghaffarianhoseini, A., Ghaffarianhoseini, A., Omrany, H., Wang, Z. H., & Akbari, H. (2017). Urban heat island mitigation strategies: A state-of-the-art review on

Kuala Lumpur, Singapore and Hong Kong. *Cities*, *62*, 131–145. Available from https://doi.org/10.1016/j.cities.2016.09.003, http://www.elsevier.com/inca/publications/store/3/0/3/9/6/.

Akbari, H., Levinson, R., & Rainer, L. (2005). Monitoring the energy-use effects of cool roofs on California commercial buildings. *Energy and Buildings*, *37*(10), 1007–1016. Available from https://doi.org/10.1016/j.enbuild.2004.11.013.

Akbari, H., Bretz, S., Kurn, D. M., & Hanford, J. (1997). Peak power and cooling energy savings of high-albedo roofs. *Energy and Buildings*, *25*(2), 117–126. Available from https://doi.org/10.1016/s0378-7788(96)01001-8.

Akbari, H., Bretz, S.E., Hanford, J.W., Kurn, D.M., Fishman, B.L., Taha, H.G., & Bos, W. (1993). Report LBL-3441 Lawrence Berkeley Laboratory, University of California Berkeley, CA 94720 Unpublished content Monitoring peak power and cooling energy savings of shade trees and white surfaces in the Sacramento Municipal Utility District (SMUD) service area. https://heatisland.lbl.gov/publications/monitoring-peak-power-and-cooling.

Akbari, H., Cartalis, C., Kolokotsa, D., Muscio, A., Pisello, A. L., Rossi, F., Santamouris, M., Synnefa, A., Wong, N. H., & Zinzi, M. (2016). Local climate change and urban heat island mitigation techniques - The state of the art. *Journal of Civil Engineering and Management*, *22*(1), 1–16. Available from https://doi.org/10.3846/13923730.2015.1111934, http://www.tandfonline.com/loi/tcem20.

Akbari, H., Kurn, D. M., Bretz, S. E., & Hanford, J. W. (1997). Peak power and cooling energy savings of shade trees. *Energy and Buildings*, *25*(2), 139–148. Available from https://doi.org/10.1016/s0378-7788(96)01003-1, https://www.journals.elsevier.com/energy-and-buildings.

Akbari, H., & Taha, H. (1992). The impact of trees and white surfaces on residential heating and cooling energy use in four Canadian cities. *Energy*, *17*(2), 141–149. Available from https://doi.org/10.1016/0360-5442(92)90063-6.

Akbari, H., & Konopacki, S. (2005). Calculating energy-saving potentials of heat-island reduction strategies. *Energy Policy*, *33*(6), 721–756. Available from https://doi.org/10.1016/j.enpol.2003.10.001.

Akbari, H., & Konopacki, S. (2004). Energy effects of heat-island reduction strategies in Toronto, Canada. *Energy*, *29*(2), 191–210. Available from https://doi.org/10.1016/j.energy.2003.09.004, http://www.elsevier.com/inca/publications/store/4/8/3/.

Akbari, H. (2002). Shade trees reduce building energy use and CO_2 emissions from power plants. *Environmental Pollution*, *116*(1), S119–S126. Available from https://doi.org/10.1016/s0269-7491(01)00264-0.

Akbari, H., Xu, T., Taha, H., Wray, C., Sathaye, J., Garg, V., & Reddy, K.N. (2011). Using cool roofs to reduce energy use, greenhouse gas emissions, and urban heat-island effects: Findings from an India experiment (No. LBNL-4746E). Available from https://doi.org/10.2172/1026804.

Ascione, F., Bianco, N., de' Rossi, F., Turni, G., & Vanoli, G. P. (2013). Green roofs in European climates. Are effective solutions for the energy savings in air-conditioning? *Applied Energy*, *104*, 845–859. Available from https://doi.org/10.1016/j.apenergy.2012.11.068, http://www.elsevier.com/inca/publications/store/4/0/5/8/9/1/index.htt.

ASU Green Roof Calculator, https://sustainability-innovation.asu.edu/urban-climate/green-roof-calculator/.

Berardi, U. (2016). The outdoor microclimate benefits and energy saving resulting from green roofs retrofits. *Energy and Buildings*, *121*, 217–229. Available from https://doi.org/10.1016/j.enbuild.2016.03.021.

Berdahl, P., Akbari, H., Levinson, R., & Miller, W. A. (2008). Weathering of roofing materials - An overview. *Construction and Building Materials*, *22*(4), 423–433. Available from https://doi.org/10.1016/j.conbuildmat.2006.10.015.

Bowler, D. E., Buyung-Ali, L., Knight, T. M., & Pullin, A. S. (2010). Urban greening to cool towns and cities: A systematic review of the empirical evidence. *Landscape and Urban Planning, 97*(3), 147−155. Available from https://doi.org/10.1016/j.landurbplan.2010.05.006, http://www.elsevier.com/inca/publications/store/5/0/3/3/4/7.

Bozonnet, E., Doya, M., & Allard, F. (2011). Cool roofs impact on building thermal response: A French case study. *Energy and Buildings, 43*(11), 3006−3012. Available from https://doi.org/10.1016/j.enbuild.2011.07.017.

Carter, T. G. (2011). Proceedings of Building Simulation 2011: 12th Conference of International Building Performance Simulation Association 2911−2918 International Building Performance Simulation Association Australia Issues and solutions to more realistically simulate conventional and cool roofs. http://www.ibpsa.org/proceedings/BS2011/P_1927.pdf.

Castaldo, V. L., Pisello, A. L., Piselli, C., Fabiani, C., Cotana, F., & Santamouris, M. (2018). How outdoor microclimate mitigation affects building thermal-energy performance: A new design-stage method for energy saving in residential near-zero energy settlements in Italy. *Renewable Energy, 127*, 920−935. Available from https://doi.org/10.1016/j.renene.2018.04.090.

Coma, J., Pérez, G., Solé, C., Castell, A., & Cabeza, L. F. (2016). Thermal assessment of extensive green roofs as passive tool for energy savings in buildings. *Renewable Energy, 85*, 1106−1115. Available from https://doi.org/10.1016/j.renene.2015.07.074.

Envimet. https://www.envi-met.com/Accessed.

EU Cool Roofs Toolkit 2009. Available on http://pouliezos.dpem.tuc.gr/coolroof/coolcalcenergy_eu.html.

European Commission, The Multifunctionality of Green Infrastructure, Science for Environmental Policy in-Depth Report European Commission/DG Environment, Brussels. (2012). EC.

European Cool Roofs Council (ECRC), Rated Products database. (2021). 2023 5 0. https://coolroofcouncil.eu/product-rating-database/.

Foustalieraki, M., Assimakopoulos, M. N., Santamouris, M., & Pangalou, H. (2017). Energy performance of a medium scale green roof system installed on a commercial building using numerical and experimental data recorded during the cold period of the year. *Energy and Buildings, 135*, 33−38. Available from https://doi.org/10.1016/j.enbuild.2016.10.056.

Gao, Y., Xu, J., Yang, S., Tang, X., Zhou, Q., Ge, J., Xu, T., & Levinson, R. (2014). Cool roofs in China: Policy review, building simulations, and proof-of-concept experiments. *Energy Policy, 74*(C), 190−214. Available from https://doi.org/10.1016/j.enpol.2014.05.036, http://www.journals.elsevier.com/energy-policy/.

Garshasbi, S., Feng, J., Paolini, R., Jonathan Duverge, J., Bartesaghi-Koc, C., Arasteh, S., Khan, A., & Santamouris, M. (2023). On the energy impact of cool roofs in Australia. *Energy and Buildings, 278*. Available from https://doi.org/10.1016/j.enbuild.2022.112577, https://www.journals.elsevier.com/energy-and-buildings.

Green, A., Gomis, L. L., Paolini, R., Haddad, S., Kokogiannakis, G., Cooper, P., Ma, Z., Kosasih, B., & Santamouris, M. (2020). Above-roof air temperature effects on HVAC and cool roof performance: Experiments and development of a predictive model. *Energy and Buildings, 222*. Available from https://doi.org/10.1016/j.enbuild.2020.110071.

Greening our City Program. https://www.dpie.nsw.gov.au/premiers-priorities/greening-our-city/greening-our-city-grant.

Haberl, J. S., & Cho, P. E. S. (2004), Literature review of uncertainty of analysis methods.

Haddad, S., Paolini, R., Ulpiani, G., Synnefa, A., Hatvani-Kovacs, G., Garshasbi, S., Fox, J., Vasilakopoulou, K., Nield, L., & Santamouris, M. (2020). Holistic approach to assess co-benefits of local climate mitigation in a hot humid region of Australia. *Scientific Reports, 10*(1). Available from https://doi.org/10.1038/s41598-020-71148-x.

Hatamipour, M. S., Mahiyar, H., & Taheri, M. (2007). Evaluation of existing cooling systems for reducing cooling power consumption. *Energy and Buildings*, *39*(1), 105−112. Available from https://doi.org/10.1016/j.enbuild.2006.05.007.

He, Y., Yu, H., Dong, N., & Ye, H. (2016). Thermal and energy performance assessment of extensive green roof in summer: A case study of a lightweight building in Shanghai. *Energy and Buildings*, *127*, 762−773. Available from https://doi.org/10.1016/j.enbuild.2016.06.016.

Hernández-Pérez, I., Álvarez, G., Xamán, J., Zavala-Guillén, I., Arce, J., & Simá, E. (2014). Thermal performance of reflective materials applied to exterior building components-A review. *Energy and Buildings*, *80*, 81−105. Available from https://doi.org/10.1016/j.enbuild.2014.05.008.

Hosseini, M., Tardy, F., & Lee, B. (2018). Cooling and heating energy performance of a building with a variety of roof designs; the effects of future weather data in a cold climate. *Journal of Building Engineering*, *17*, 107−114. Available from https://doi.org/10.1016/j.jobe.2018.02.001.

Hosseini, M., & Akbari, H. (2016). Effect of cool roofs on commercial buildings energy use in cold climates. *Energy and Buildings*, *114*, 143−155. Available from https://doi.org/10.1016/j.enbuild.2015.05.050.

Huang, Y. J., Akbari, H., Taha, H., & Rosenfeld, A. H. (1987). The potential of vegetation in reducing summer cooling loads in residential buildings. *Journal of Climate and Applied Meteorology*, *26*(9), 1103−1116. 10.1175/1520-0450(1987)026 < 1103:tpovir > 2.0.co;2.

Hwang, W. H., Wiseman, P. E., & Thomas, V. A. (2015). Tree planting configuration influences shade on residential structures in four U.S. cities. *Arboriculture and Urban Forestry*, *41*(4), 208−222. Available from http://auf.isa-arbor.com/request.asp?JournalID = 1&ArticleID = 3363&Type = 2.

Jaffal, I., Ouldboukhitine, S. E., & Belarbi, R. (2012). A comprehensive study of the impact of green roofs on building energy performance. *Renewable Energy*, *43*, 157−164. Available from https://doi.org/10.1016/j.renene.2011.12.004.

Karachaliou, P., Santamouris, M., & Pangalou, H. (2016). Experimental and numerical analysis of the energy performance of a large scale intensive green roof system installed on an office building in Athens. *Energy and Buildings*, *114*, 256−264. Available from https://doi.org/10.1016/j.enbuild.2015.04.055.

Ko, Y., Lee, J. H., McPherson, E. G., & Roman, L. A. (2015). Long-term monitoring of Sacramento Shade program trees: Tree survival, growth and energy-saving performance. *Landscape and Urban Planning*, *143*, 183−191. Available from https://doi.org/10.1016/j.landurbplan.2015.07.017, http://www.elsevier.com/inca/publications/store/5/0/3/3/4/7.

Kolokotroni, M., Gowreesunker, B. L., & Giridharan, R. (2013). Cool roof technology in London: An experimental and modelling study. *Energy and Buildings*, *67*, 658−667. Available from https://doi.org/10.1016/j.enbuild.2011.07.011.

Kolokotroni, M., Shittu, E., Santos, T., Ramowski, L., Mollard, A., Rowe, K., Wilson, E., Filho, J. Pd. B., & Novieto, D. (2018). Cool roofs: High tech low cost solution for energy efficiency and thermal comfort in low rise low income houses in high solar radiation countries. *Energy and Buildings*, *176*, 58−70. Available from https://doi.org/10.1016/j.enbuild.2018.07.005.

Kolokotsa, D., Diakaki, C., Papantoniou, S., & Vlissidis, A. (2012). Numerical and experimental analysis of cool roofs application on a laboratory building in Iraklion, Crete, Greece. *Energy and Buildings*, *55*, 85−93. Available from https://doi.org/10.1016/j.enbuild.2011.09.011.

Konopacki, S., & Akbari, H. (2002). Lawrence Berkeley National Laboratory Report LBNL-49638 Energy savings of heat island reduction strategies in Chicago and Houston (including updates for Baton Rouge).

Konopacki, S., & Akbari, H. (2000). Energy savings calculations for heat island reduction strategies in Baton Rouge, Sacramento and Salt Lake City. Proceedings ACEEE Summer Study on Energy Efficiency in Buildings. 9.

La Roche, P., & Berardi, U. (2014). Comfort and energy savings with active green roofs. *Energy and Buildings*, *82*, 492−504. Available from https://doi.org/10.1016/j.enbuild.2014.07.055.

Leonard, T., & Stay, T. L. (2006). Stay cool: A roof system on a Minnesota building demonstrates energy-saving technology. https://www.professionalroofing.net/Articles/Stay-cool-04-01-2006/835.

Levinson, R., & Akbari, H. (2010). Potential benefits of cool roofs on commercial buildings: Conserving energy, saving money, and reducing emission of greenhouse gases and air pollutants. *Energy Efficiency*, *3*(1), 53−109. Available from https://doi.org/10.1007/s12053-008-9038-2.

Levinson, R., Akbari, H., Konopacki, S., & Bretz, S. (2005). Inclusion of cool roofs in nonresidential title 24 prescriptive requirements. *Energy Policy*, *33*(2), 151−170. Available from https://doi.org/10.1016/S0301-4215(03)00206-4.

Levinson, R., Berdahl, P., Asefaw Berhe, A., & Akbari, H. (2005). Effects of soiling and cleaning on the reflectance and solar heat gain of a light-colored roofing membrane. *Atmospheric Environment*, *39*(40), 7807−7824. Available from https://doi.org/10.1016/j.atmosenv.2005.08.037.

Levinson, R., Berdahl, P., Akbari, H., Miller, W., Joedicke, I., Reilly, J., Suzuki, Y., & Vondran, M. (2007). Methods of creating solar-reflective nonwhite surfaces and their application to residential roofing materials. *Solar Energy Materials and Solar Cells*, *91*(4), 304−314. Available from https://doi.org/10.1016/j.solmat.2006.06.062.

Mastrapostoli, E., Santamouris, M., Kolokotsa, D., Vassilis, P., Venieri, D., & Gompakis, K. (2016). On the ageing of cool roofs: Measure of the optical degradation, chemical and biological analysis and assessment of the energy impact. *Energy and Buildings*, *114*, 191−199. Available from https://doi.org/10.1016/j.enbuild.2015.05.030.

Mihalakakou, G., Souliotis, M., Papadaki, M., Menounou, P., Dimopoulos, P., Kolokotsa, D., Paravantis, J. A., Tsangrassoulis, A., Panaras, G., Giannakopoulos, E., & Papaefthimiou, S. (2023). Green roofs as a nature-based solution for improving urban sustainability: Progress and perspectives. *Renewable and Sustainable Energy Reviews*, *180*, 113306. Available from https://doi.org/10.1016/j.rser.2023.113306.

Milion Trees NYC. https://www.milliontreesnyc.org/.

Mirzaei, P. A. (2015). Recent challenges in modeling of urban heat island. *Sustainable Cities and Society*, *19*, 200−206. Available from https://doi.org/10.1016/j.scs.2015.04.001, http://www.elsevier.com/wps/find/journaldescription.cws_home/724360/description#description.

Morakinyo, T. E., Lau, K. K. L., Ren, C., & Ng, E. (2018). Performance of Hong Kong's common trees species for outdoor temperature regulation, thermal comfort and energy saving. *Building and Environment*, *137*, 157−170. Available from https://doi.org/10.1016/j.buildenv.2018.04.012, http://www.elsevier.com/inca/publications/store/2/9/6/index.htt.

Morakinyo, T. E., Kong, L., Lau, K. K. L., Yuan, C., & Ng, E. (2017). A study on the impact of shadow-cast and tree species on in-canyon and neighborhood's thermal comfort. *Building and Environment*, *115*, 1−17. Available from https://doi.org/10.1016/j.buildenv.2017.01.005, http://www.elsevier.com/inca/publications/store/2/9/6/index.htt.

Niachou, A., Papakonstantinou, K., Santamouris, M., Tsangrassoulis, A., & Mihalakakou, G. (2001). Analysis of the green roof thermal properties and investigation of its energy performance. *Energy and Buildings*, *33*(7), 719−729. Available from https://doi.org/10.1016/s0378-7788(01)00062-7.

Oke, T. R., Mills, G., Christen, A., & Voogt, J. A. (2017). Urban climates. Cambridge University\n8 Press. Available from https://doi.org/10.1017/9781139016476.

Olivieri, F., Di Perna, C., D'Orazio, M., Olivieri, L., & Neila, J. (2013). Experimental measurements and numerical model for the summer performance assessment of extensive green roofs in a Mediterranean coastal climate. *Energy and Buildings, 63*, 1−14. Available from https://doi.org/10.1016/j.enbuild.2013.03.054.

Paolini, R., Terraneo, G., Ferrari, C., Sleiman, M., Muscio, A., Metrangolo, P., Poli, T., Destaillats, H., Zinzi, M., & Levinson, R. (2020). Effects of soiling and weathering on the albedo of building envelope materials: Lessons learned from natural exposure in two European cities and tuning of a laboratory simulation practice. *Solar Energy Materials and Solar Cells, 205*. Available from https://doi.org/10.1016/j.solmat.2019.110264, http://www.sciencedirect.com/science/journal/09270248/100.

Paolini, R., Zinzi, M., Poli, T., Carnielo, E., & Mainini, A. G. (2014). Effect of ageing on solar spectral reflectance of roofing membranes: Natural exposure in Roma and Milano and the impact on the energy needs of commercial buildings. *Energy and Buildings, 84*, 333−343. Available from https://doi.org/10.1016/j.enbuild.2014.08.008.

Parker, J. H. (1981). Department of Physical Sciences, lorida International University, Miami, FL Use of Landscaping for Energy Conservation.

Pisello, A. L., Santamouris, M., & Cotana, F. (2013). Active cool roof effect: Impact of cool roofs on cooling system efficiency. *Advances in Building Energy Research, 7*(2), 209−221. Available from https://doi.org/10.1080/17512549.2013.865560.

Revel, G. M., Martarelli, M., Emiliani, M., Celotti, L., Nadalini, R., Ferrari, A. D., Hermanns, S., & Beckers, E. (2014). Cool products for building envelope - Part II: Experimental and numerical evaluation of thermal performances. *Solar Energy, 105*, 780−791. Available from https://doi.org/10.1016/j.solener.2014.02.035, http://www.elsevier.com/inca/publications/store/3/2/9/index.htt.

Romeo, C., & Zinzi, M. (2013). Impact of a cool roof application on the energy and comfort performance in an existing non-residential building. A Sicilian case study. *Energy and Buildings, 67*, 647−657. Available from https://doi.org/10.1016/j.enbuild.2011.07.023.

Rosado, P. J., & Levinson, R. (2019). Potential benefits of cool walls on residential and commercial buildings across California and the United States: Conserving energy, saving money, and reducing emission of greenhouse gases and air pollutants. *Energy and Buildings, 199*, 588−607. Available from https://doi.org/10.1016/j.enbuild.2019.02.028, https://www.journals.elsevier.com/energy-and-buildings.

Rosenfeld, A. H., Akbari, H., Romm, J. J., & Pomerantz, M. (1998). Cool communities: Strategies for heat island mitigation and smog reduction. *Energy and Buildings, 28*(1), 51−62. Available from https://doi.org/10.1016/S0378-7788(97)00063-7.

Sailor, D. J. (2008). A green roof model for building energy simulation programs. *Energy and Buildings, 40*(8), 1466−1478. Available from https://doi.org/10.1016/j.enbuild.2008.02.001.

Santamouris, M. (2016a). Cooling the buildings − Past, present and future. *Energy and Buildings, 128*, 617−638. Available from https://doi.org/10.1016/j.enbuild.2016.07.034.

Santamouris, M. (2016b). Innovating to zero the building sector in Europe: Minimising the energy consumption, eradication of the energy poverty and mitigating the local climate change. *Solar Energy, 128*, 61−94. Available from https://doi.org/10.1016/j.solener.2016.01.021.

Santamouris, M., & Osmond, P. (2020). Increasing green infrastructure in cities: Impact on ambient temperature, air quality and heat-related mortality and morbidity. *Buildings, 10*(12). Available from https://doi.org/10.3390/buildings10120233.

Santamouris, M., Synnefa, A., & Karlessi, T. (2011). Using advanced cool materials in the urban built environment to mitigate heat islands and improve thermal comfort conditions. *Solar Energy, 85*(12), 3085−3102. Available from https://doi.org/10.1016/j.solener.2010.12.023.

Santamouris, M. (2015a). Analyzing the heat island magnitude and characteristics in one hundred Asian and Australian cities and regions. *Science of The Total Environment, 512−513*, 582−598. Available from https://doi.org/10.1016/j.scitotenv.2015.01.060.

Santamouris, M. (2015b). Regulating the damaged thermostat of the cities—Status, impacts and mitigation challenges. *Energy and Buildings, 91*, 43−56. Available from https://doi.org/10.1016/j.enbuild.2015.01.027.

Santamouris, M. (2019). *Minimizing energy consumption, energy poverty and global and local climate change in the built environment: Innovating to zero, Copyright ©*. Elsevier Inc. Available from http://doi.org/10.1016/C2016-0-01024-0.

Santamouris, M. (2020). Recent progress on urban overheating and heat island research. Integrated assessment of the energy, environmental, vulnerability and health impact. *Synergies with the Global Climate Change. Energy and Buildings, 207*. Available from https://doi.org/10.1016/j.enbuild.2019.109482.

Santamouris, M., Cartalis, C., Synnefa, A., & Kolokotsa, D. (2015). On the impact of urban heat island and global warming on the power demand and electricity consumption of buildings - A review. *Energy and Buildings, 98*, 119−124. Available from https://doi.org/10.1016/j.enbuild.2014.09.052.

Santamouris, M., Pavlou, C., Doukas, P., Mihalakakou, G., Synnefa, A., Hatzibiros, A., & Patargias, P. (2007). Investigating and analysing the energy and environmental performance of an experimental green roof system installed in a nursery school building in Athens, Greece. *Energy, 32*(9), 1781−1788. Available from https://doi.org/10.1016/j.energy.2006.11.011.

Santamouris, M., Ban-Weiss, G., Osmond, P., Paolini, R., Synnefa, A., Cartalis, C., Muscio, A., Zinzi, M., Morakinyo, T. E., Ng, E., Tan, Z., Takebayashi, H., Sailor, D., Crank, P., Taha, H., Pisello, A. L., Rossi, F., Zhang, J., & Kolokotsa, D. (2018). Progress in urban greenery mitigation science − Assessment methodologies advanced technologies and impact on cities. *Journal of Civil Engineering and Management, 24*(8), 638−671. Available from https://doi.org/10.3846/jcem.2018.6604, http://journals.vgtu.lt/.

Santamouris, M., & Yun, G. Y. (2020). Recent development and research priorities on cool and super cool materials to mitigate urban heat island. *Renewable Energy, 161*, 792−807. Available from https://doi.org/10.1016/j.renene.2020.07.109, http://www.journals.elsevier.com/renewable-and-sustainable-energy-reviews/.

Santamouris, M., Ding, L., Fiorito, F., Oldfield, P., Osmond, P., Paolini, R., Prasad, D., & Synnefa, A. (2017). Passive and active cooling for the outdoor built environment − Analysis and assessment of the cooling potential of mitigation technologies using performance data from 220 large scale projects. *Solar Energy, 154*, 14−33. Available from https://doi.org/10.1016/j.solener.2016.12.006.

Santamouris, M., Haddad, S., Saliari, M., Vasilakopoulou, K., Synnefa, A., Paolini, R., Ulpiani, G., Garshasbi, S., & Fiorito, F. (2018). On the energy impact of urban heat island in Sydney: Climate and energy potential of mitigation technologies. *Energy and Buildings, 166*, 154−164. Available from https://doi.org/10.1016/j.enbuild.2018.02.007.

Santamouris, M. (2013). Using cool pavements as a mitigation strategy to fight urban heat island—A review of the actual developments. *Renewable and Sustainable Energy Reviews, 26*, 224−240. Available from https://doi.org/10.1016/j.rser.2013.05.047.

Sedaghat, A., & Sharif, M. (2022). Mitigation of the impacts of heat islands on energy consumption in buildings: A case study of the city of Tehran, Iran. *Sustainable Cities and Society, 76*. Available from https://doi.org/10.1016/j.scs.2021.103435, http://www.elsevier.com/wps/find/journaldescription.cws_home/724360/description#description.

Shi, D., Zhuang, C., Lin, C., Zhao, X., Chen, D., Gao, Y., & Levinson, R. (2019). Effects of natural soiling and weathering on cool roof energy savings for dormitory buildings in Chinese cities with hot summers. *Solar Energy Materials and Solar Cells*, *200*. Available from https://doi.org/10.1016/j.solmat.2019.110016, http://www.sciencedirect.com/science/journal/09270248/100.

Silva, C. M., Gomes, M. G., & Silva, M. (2016). Green roofs energy performance in Mediterranean climate. *Energy and Buildings*, *116*, 318−325. Available from https://doi.org/10.1016/j.enbuild.2016.01.012.

Simpson, J. R., & McPherson, E. G. (1996). Potential of tree shade for reducing residential energy use in California. *Journal of Arboriculture*, *22*(1), 10−18.

Simpson, J. R., & McPherson, E. G. (1998). Simulation of tree shade impacts on residential energy use for space conditioning in Sacramento. *Atmospheric Environment*, *32*(1), 69−74. Available from https://doi.org/10.1016/s1352-2310(97)00181-7.

Simpson, J. R., & McPherson, E. G. (1997). The effects of roof albedo modification on cooling loads of scale model residences in Tucson, Arizona. *Energy and Buildings*, *25*(2), 127−137. Available from https://doi.org/10.1016/s0378-7788(96)01002-x.

Sleiman, M., Ban-Weiss, G., Gilbert, H. E., François, D., Berdahl, P., Kirchstetter, T. W., Destaillats, H., & Levinson, R. (2011). Soiling of building envelope surfaces and its effect on solar reflectance - Part I: Analysis of roofing product databases. *Solar Energy Materials and Solar Cells*, *95*(12), 3385−3399. Available from https://doi.org/10.1016/j.solmat.2011.08.002.

Spala, A., Bagiorgas, H. S., Assimakopoulos, M. N., Kalavrouziotis, J., Matthopoulos, D., & Mihalakakou, G. (2008). On the green roof system. Selection, state of the art and energy potential investigation of a system installed in an office building in Athens, Greece. *Renewable Energy*, *33*(1), 173−177. Available from https://doi.org/10.1016/j.renene.2007.03.022.

Suehrcke, H., Peterson, E. L., & Selby, N. (2008). Effect of roof solar reflectance on the building heat gain in a hot climate. *Energy and Buildings*, *40*(12), 2224−2235. Available from https://doi.org/10.1016/j.enbuild.2008.06.015.

Suman, B. M., & Verma, V. V. (2003). Measured performance of a reflective thermal coating in experimental rooms. *Journal of Scientific and Industrial Research*, *62*(12), 1152−1157.

Synnefa, A., Saliari, M., & Santamouris, M. (2012). Experimental and numerical assessment of the impact of increased roof reflectance on a school building in Athens. *Energy and Buildings*, *55*, 7−15. Available from https://doi.org/10.1016/j.enbuild.2012.01.044.

Synnefa, A., Santamouris, M., & Akbari, H. (2007). Estimating the effect of using cool coatings on energy loads and thermal comfort in residential buildings in various climatic conditions. *Energy and Buildings*, *39*(11), 1167−1174. Available from https://doi.org/10.1016/j.enbuild.2007.01.004.

Synnefa, A., Santamouris, M., & Livada, I. (2006). A study of the thermal performance of reflective coatings for the urban environment. *Solar Energy*, *80*(8), 968−981. Available from https://doi.org/10.1016/j.solener.2005.08.005.

Synnefa, A., & Santamouris, M. (2012). Advances on technical, policy and market aspects of cool roof technology in Europe: The cool roofs project. *Energy and Buildings*, *55*, 35−41. Available from https://doi.org/10.1016/j.enbuild.2011.11.051, Greece.

Taha, H. (1997). Modeling the impacts of large-scale albedo changes on ozone air quality in the South Coast Air Basin. *Atmospheric Environment*, *31*(11), 1667−1676. Available from https://doi.org/10.1016/s1352-2310(96)00336-6.

Taha, H. (1996). Modeling impacts of increased urban vegetation on ozone air quality in the South Coast Air Basin. *Atmospheric Environment*, *30*(20), 3423−3430. Available from https://doi.org/10.1016/1352-2310(96)00035-0, http://www.elsevier.com/locate/atmosenv.

Taha, H., Konopacki, S., & Gabersek, S. (1996). Modeling the meteorological and energy effects of urban heat islands and their mitigation: A 10-region study.

Takebayashi, H., Miki, K., Sakai, K., Murata, Y., Matsumoto, T., Wada, S., & Aoyama, T. (2016). Experimental examination of solar reflectance of high-reflectance paint in Japan with natural and accelerated aging. *Energy and Buildings*, *114*, 173−179. Available from https://doi.org/10.1016/j.enbuild.2015.06.019.

Thayer, R., & Maeda, B. (1985). Measuring street tree impact on solar performance: A five-climate computer modeling study. *Arboriculture & Urban Forestry*, *11*(1), 1−12. Available from https://doi.org/10.48044/jauf.1985.001.

Theodosiou, T. G. (2003). Summer period analysis of the performance of a planted roof as a passive cooling technique. *Energy and Buildings*, *35*(9), 909−917. Available from https://doi.org/10.1016/S0378-7788(03)00023-9.

Touchaei, A. G., Hosseini, M., & Akbari, H. (2016). Energy savings potentials of commercial buildings by urban heat island reduction strategies in Montreal (Canada). *Energy and Buildings*, *110*, 41−48. Available from https://doi.org/10.1016/j.enbuild.2015.10.018.

U.S. Cool Roof Rating Council (CRRC), U.S. Cool Roof Rating Council (CRRC), Rated Products Directory 2023. https://coolroofs.org/directory.

U.S. Department of Energy's Oak Ridge National Laboratory Cool Roof Calculator (Version 1.2). https://web.ornl.gov/sci/buildings/tools/cool-roof/.

U.S. Environmental Protection Agency, Reducing urban heat islands: Compendium of strategiesTrees and Vegetation. (2008). U.S. Environmental Protection Agency. 2008. "Green Roofs." In *Reducing urban heat islands*: Available from https://www.epa.gov/heat-islands/heat-island-compendium.

U.S. Environmental Protection Agency. (2008). "Green Roofs." In *Reducing urban heat islands: Compendium of strategies*. https://www.epa.gov/heat-islands/heat-island-compendium.

U.S. Environmental Protection Agency. (2012). "Cool Pavements." In Reducing urban heat islands: Compendium of strategies. https://www.epa.gov/heat-islands/heat-island-compendium.

Vasilakopoulou, K., Ulpiani, G., Khan, A., Synnefa, A., & Santamouris, M. (2023). Cool roofs boost the energy production of photovoltaics- Investigating the impact of roof albedo on the energy performance of monofacial and bifacial photovoltaic modules. *Solar Energy*.

Wang, X., Kendrick, C., Ogden, R., & Maxted, J. (2008). Dynamic thermal simulation of a retail shed with solar reflective coatings. *Applied Thermal Engineering.*, *28*(8−9), 1066−1073. Available from https://doi.org/10.1016/j.applthermaleng.2007.06.011.

Wong, N. H., Cheong, D. K. W., Yan, H., Soh, J., Ong, C. L., & Sia, A. (2003). The effects of rooftop garden on energy consumption of a commercial building in Singapore. *Energy and Buildings*, *35*(4), 353−364. Available from https://doi.org/10.1016/s0378-7788(02)00108-1.

Wray, C., & Akbari, H. (2008). The effects of roof reflectance on air temperatures surrounding a rooftop condensing unit. *Energy and Buildings*, *40*(1), 11−28. Available from https://doi.org/10.1016/j.enbuild.2007.01.005.

Yang, J., Mohan Kumar, Dl, Pyrgou, A., Chong, A., Santamouris, M., Kolokotsa, D., & Lee, S. E. (2018). Green and cool roofs' urban heat island mitigation potential in tropical climate. *Solar Energy*, *173*, 597−609. Available from https://doi.org/10.1016/j.solener.2018.08.006, http://www.elsevier.com/inca/publications/store/3/2/9/index.htt.

Yekang, K. (2018). Trees and vegetation for residential energy conservation: A critical review for evidence-based urban greening in North America. *Urban Forestry & Urban Greening*, *34*, 318−335. Available from https://doi.org/10.1016/j.ufug.2018.07.021.

Zhang, K., Garg, A., Mei, G., Jiang, M., Wang, H., Huang, S., & Gan, L. (2022). Thermal performance and energy consumption analysis of eight types of extensive green roofs in subtropical monsoon climate. *Building and Environment, 216*. Available from https://doi.org/10.1016/j.buildenv.2022.108982.

Zinzi, M. (2016a). Characterisation and assessment of near infrared reflective paintings for building facade applications. *Energy and Buildings, 114*, 206–213. Available from https://doi.org/10.1016/j.enbuild.2015.05.048.

Zinzi, M. (2016b). Exploring the potentialities of cool facades to improve the thermal response of Mediterranean residential buildings. *Solar Energy, 135*, 386–397. Available from https://doi.org/10.1016/j.solener.2016.06.021, http://www.elsevier.com/inca/publications/store/3/2/9/index.htt.

Zinzi, M., & Agnoli, S. (2012). Cool and green roofs. An energy and comfort comparison between passive cooling and mitigation urban heat island techniques for residential buildings in the Mediterranean region. *Energy and Buildings, 55*, 66–76. Available from https://doi.org/10.1016/j.enbuild.2011.09.024.

Chapter 7

The impact of heat mitigation on urban environmental quality

M.E. González-Trevizo, K.E. Martínez-Torres and J.C. Rincón-Martínez
Facultad de Ingeniería, Arquitectura y Diseño, Universidad Autónoma de Baja California, Ensenada, Mexico

7.1 Introduction

Recently, planetary urbanization managed to concentrate a global urban population of about 55% and is projected to reach 68% by 2050. In 2020, the population in Latin America and the Caribbean was concentrated two-thirds in cities, leaving just 18.8% of inhabitants dwelling in rural areas, and data reported by the United Nations (UN) Population Division just 2 years before indicated that 14.2% of the population resides in one of the six megacities with at least 10 million inhabitants, such as Mexico City, Buenos Aires, or Rio de Janeiro (UNDESA, 2018). It is also anticipated that 53 million more will be added by 2035, which would result in the consolidation of around 50 new metropolises to the final list of about 429 new metropolitan areas worldwide (UN-HABITAT, 2020). According to the projections made by the UN Department of Economic and Social Affairs, it is anticipated that the worldwide population would see a growth from 8.5 billion in the year 2030 to 9.9 billion by 2050, and further rise to 10.9 billion by the year 2100 (UN Department of Economic & social affairs, 2019). In the foreseeble future, it is expected that all levels of governance will have to promote climate adaptation and global warming mitigation actions with the greatest and positive health, environmental, and socioeconomic externalities to design locally relevant policies for building climate-resilient and long-term sustainable ecosystems (World Health Organization, 2021). Thus, the sustainable development perspective on resilience for the Anthropocene era is lately addressed under social-ecological systems insights (Reyers et al., 2018). The undisputable human imprint responsible for the land cover replacement modified the natural thermal balance and surface properties with a dense concentration of pavement, construction materials, and other man-made infrastructure that alters the sensible and latent heat transfers and drives, at the same time, urban overheating (U.S. Environmental Protection Agency, 2023).

Mitigation and Adaptation of Urban Overheating. DOI: https://doi.org/10.1016/B978-0-443-13502-6.00007-5

In light of the abovementioned anthropogenic activities, scientists have emphasized the significance of generating adaptation plans to address the resulting multiple climate change narratives and adverse storylines (Roberts et al., 2019). In fact, since 2014, the worst IPCC representative concentration pathway (RCP 8.5) calculates an atmospheric carbon concentration beyond 100 $GtCO_2$/year emissions from fossil fuels and cement and >1000 $GtCO_2$/yr by 2100 as well as human-induced warming between 3.5°C and 4.5°C in terms of a median climate response model (Intergovernmental Panel on Climate Change, 2014). Therefore, through the assessment reports for limiting global warming, decision-making policies toward climate-resilient development for managing adaptation gaps and risks to terrestrial, freshwater, and coastal ecosystems, global water cycle, and projected water systems must be implemented. In the meantime, through impact and risk management plans for migration and displacement and key infrastructure of the urban metabolism. Global surface temperature will increase constantly, and the global warming limit of 5°C will be surpassed within this century unless noteworthy reductions in atmospheric carbon and other greenhouse gas emissions occur in the coming decades (Boggia et al., 2018).

In this respect, overpopulation and dense urban settlements with closed fabrics are particularly susceptible since they experience higher land surface temperature levels than the surrounding suburban and rural areas due to human-induced activities, climatic factors, and urban structural features (Nwakaire et al., 2020). The phenomenon known as urban heat island (UHI) is originated by dramatic land cover alterations, the evaporative surface inhibited by construction materials that modify the storage, and emissive heat transfers, as well as the modification of sensible and latent heat fractions of the environment (Thanvisitthapon et al., 2023), and has synergistic interactions related to urban overheating effects, such as the intensity and duration of extreme events like heat waves, especially during hot summers (Park et al., 2023).

Urban overheating, defined by Nazarian et al. as *"the exceedance of locally defined thermal thresholds that lead to negative impacts on people and urban systems,"* represents a newly introduced definition (Nazarian et al., 2022) to assess urban heat as a multidimensional hazard to the human health, productivity, and also to the energy conservation for sustainable societies. It is affected by intricate connections between climate and the urban environment at different scales, which, in the presence of tropospheric precursors and sunlight are major sources of ground-level ozone. Thus, deposition of air pollutants such as nitrogen dioxide (NO_2), Ozone (O_3), carbon monoxide (CO), and sulfur dioxide (SO_3), as well as aerodynamic particulate matter ($PM_{2.5}$), and (PM_{10}) have a significant impact on heat-related mortality and morbidity manifested as dehydration, cerebrovascular, and ischemic heart diseases (Pyrgou & Santamouris, 2018).

Land use and land cover (LULC) adequate policies, urban design, and urban planning are notable for reducing the different thermal stress externalities in the built environment, according to research conducted in the last decade, which gathered the most remarkable set of actions within the urban heat mitigation approach (UHMA). Consequently, the purpose of this research is to incorporate a critical evaluation of the most implemented adaptation and mitigation measures to reduce environmental quality degradation impacts. Beyond a bibliometric analysis of a comprehensive repository comprised of 159 manuscripts as baseline collection (Gonzalez-Trevizo et al., 2021), this chapter explores, via the implementation of data cluster mapping analysis, those derivative studies related to prominent topics, as well as the resulting research interactions among their key factors. The collaborative work between the leading researchers in the field of knowledge and the multidimensional factors that are influencing the agendas for the prevention of the synergistic effects of environmental risk and uneven distribution of vulnerability is mentioned.

7.2 General perspective

The urbanization process, population growth, geographic location, and climate have generated changes in LULC associated with the UHI, affecting environmental quality and quality of life. In this context, to mitigate the effects of urban overheating, researchers are particularly interested in the impacts of heat mitigation on urban environmental quality. The literature identifies the following primary concerns regarding environmental quality and overheating: (1) thermal comfort; (2) air quality; (3) energy; and (4) health.

As indicated previously, it is relevant to identify the environmental challenges linked to overheating and the trends of heat mitigation impacts for establishing a framework to improve urban environmental performance. In other words, to conduct an advanced literature search with a clear scope, the combination of keywords, field codes, and proximity operators considered the concepts "UHI," "mitigation strategies," and "environmental quality" with the use of Boolean operators to identify literature in Scopus and Web of Science. The protocol guidelines provided by the "CEE evidence synthesis methodology" (Collaboration for Environmental Evidence, 2013) were used to identify the relevant articles. For this Chapter, only articles published from 2001 to 2022 were considered.

7.2.1 Prevailing topics

A bibliographic network was generated to identify the most relevant concepts associated with urban overheating and environmental quality.

The network was derived by the use of a clustering approach, namely by using bibliographic coupling and co-citation features with the aid of the VosViewer tool (van Eck & Waltman, 2017). During the analysis, only terms with link strength were considered; acronyms and abbreviations were omitted to ascertain the genuine influence of the primary keywords. The concepts with the most occurrences on the topic and their interconnections are illustrated in Fig. 7.1. The network shows 447 keywords occurrences contained in 159 documents, and the colors represent a collection of words that compose a cluster.

The most relevant keyword was "urban heat island" with 94 occurrences and 405 interactions with other related concepts, preceded by "LST" with 25 occurrences and 119 interactions, "cool material" with 13 occurrences and 63 interactions, "UHI mitigation" with 13 occurrences and 50 interactions, and "WRF model" with eight occurrences and 25 interactions. The network map also shows other associated topics such as climate change, albedo, LULC, urbanization, the importance of impervious surfaces, UHI intensity, heat wave, urban resilience, outdoor thermal comfort, and energy consumption.

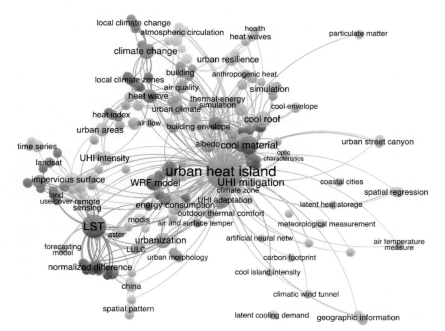

FIGURE 7.1 Network visualization of thematic cluster studies associated with urban heat and Urban Environmental Quality instead (UEQ).

7.2.2 Leading authors in the field of knowledge

The following waterfall chart describes the synthesis and evaluation of a repository comprised of 158 investigations carried out to address urban heat mitigation and adaptation strategies, their leading authors, and the resulting perspective literature. The search queries used an original collection reported previously as an input parameter, and for the purpose of this critical evaluation study, the literature review mapping tool was extensively utilized, as it permits the filtering of search results by author relevance and publication date and also displays the relationships between the articles in the collection as connecting lines that trace all the multiple interactions (Litmaps tool, 2023). As shown in Fig. 7.2, each research from 2001 to 2022 has a distinctive set of colors based on the heat mitigation approach and related strategies involved regarding the impacts on the urban environmental quality that will be discussed in greater detail in Section 7.3: (1) LULC, (2) urban design, and (3) urban planning.

The relevance is illustrated on the y-axis by the number of citations between highly cited authors and their associated interactions with multiple studies by different authors from around the globe, while the x-axis depicts the evolution over time. The most relevant reference in terms of citations and association with other studies addressed the influence of urban climate on the energy consumption of buildings (Santamouris et al., 2001), and as a result of its crucial contribution, it is shown as "seed publication" in Fig. 7.2; it is cited by more than a thousand authors. It is followed by the same leading author (Santamouris et al., 2011, 2012), 703 citations,

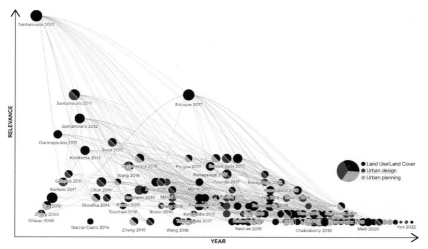

FIGURE 7.2 Leading authors in terms of citation relevance (vertical axis) and date of publication (horizontal axis).

Estoque et al., (2017), 638, Rossi et al. (2015), 151, Kolokotsa et al. (2013), and Giannopoulou et al. (2011), 99 citations. The number of interactions and citation influence between past and recent studies is expected to decline steadily in compliance with the evolution in time, while highly cited authors define trends, thereby creating a common framework and worldwide vision of the studied phenomenon.

In the beginning, the characterization of the UHI intensity during summer was the umbrella to cover the general interest; however, the understanding of building heat transfer reduction led to experimental tests and developments of different cool materials and urban technologies like envelopes with green facade systems, and more recently, the use of remote sensing with low and high-resolution data and imagery from different satellites such as Landsat, ASTER, IKONOS-1, Terra or SPOT-6 to study spatiotemporal change patterns in land cover or to compare effects of multiple mitigation strategies aimed at mitigating the impact of the built environment. The latest studies encompass important findings in Greece, Sydney, Italy, Germany, Sri Lanka, and China, among others. In Africa, Simwanda et al. reported the urban footprint effect in growing cities like Lagos or Nairobi (Simwanda et al., 2019). In this regard, the important studies addressed the impact of urban form on wind patterns and intraurban surface in relation to temperature variability (Agathangelidis et al., 2019) as well as the use of extensive green roof designs for the same purpose (López-Silva et al., 2022).

7.3 Approaches on heat mitigation impacts and urban environmental quality

7.3.1 Climate parameters, mitigation, and adaptation

The atmospheric parameters result in increased daytime temperatures, attenuated nocturnal ventilation, and elevated air pollution (U.S. Environmental Protection Agency, 2023). This situation affects human health and human thermal comfort, which are particularly compromised in Hot semi-arid (BSh), Warm-summer Mediterranean (Csb), Mediterranean hot summer (Csa), and Cold semi-arid climates (BSk), which tend to be the most vulnerable climates because of inhabitants' adaptability thresholds. Although the broad use of the climate parameters is consistently equated with the nature of the approach, scale, and strategies to face the impact of urban overheating, as described before and shown in Table 7.1, the common UHMA implied different spatial complexities to perform mitigation strategies, which are dependent on the scale of intervention: mesoscale, local scale, and microscale.

TABLE 7.1 Mitigation strategies applied in semiarid and Mediterranean climates.

Author/year	City and country	Scale	Köppen-Geiger climate zone	Mitigation and adaptation strategies							
				▲	◧	■	▢	◉	☀	⇅	▣
Rodríguez et al. (2022)	Sevilla, Spain	Micro	Csa							✓	✓
López-Silva et al. (2022)	Ensenada, Mexico	Micro	BSk		✓	✓					
Núñez-Peiró et al. (2021)	Madrid, Spain	Meso	BSk								✓
García and Díaz (2021)	Granada, Spain	Meso	Csa								✓
Pyrgou et al. (2020)	Nicosia, Cyprus	Meso	BSh				✓				
Agathangelidis et al. (2019)	Athens, Greece	Meso	Csa								✓
Duan et al. (2019)	Athens, Greece	Micro	Csa						✓	✓	
Jato-Espino (2019)	Alicante; Castellon; Valencia, Spain	Meso	BSk			✓					
Shirani-bidabadi et al. (2019)	Isfahan, Iran	Meso	BSk	✓							
Yang and Bou-Zeid (2019)	Los Angeles, USA	Micro	Csb		✓					✓	

(Continued)

TABLE 7.1 (Continued)

Author/year	City and country	Scale	Köppen-Geiger climate zone	Mitigation and adaptation strategies							
				▲	◨	■	▭	◉	☀	⇅	▣
Jandaghian and Akbari (2018a)	Sacramento, USA	Meso	Csa								✓
Kim et al. (2018)	Boise, USA	Micro	Csb	✓		✓	✓				✓
Kyriakodis and Santamouris (2018)	Athens, Greece	Multiple	Csa				✓				
Liu and Weng (2018)	Los Angeles; Long Beach; Anaheim; Santa Ana, USA	Meso	Csb, BSk, BSk, Csa	✓							
Lontorfos et al. (2018)	Athens, Greece	Micro	Csa			✓					
Morini et al. (2018)	Rome, Italy	Meso	Csa			✓	✓				
Pyrgou and Santamouris (2018)	Nicosia, Cyprus	Meso	BSh	✓			✓				
Rosso et al. (2018)	Rome, Italy	Micro	Csa	✓	✓	✓	✓	✓	✓	✓	✓
Rousta et al. (2018)	Teheran, Iran	Meso	Csa	✓	✓		✓	✓	✓	✓	
Zheng and Weng (2018)	Los Angeles, USA	Meso	Csb		✓		✓	✓	✓	✓	
Castellani et al. (2017)	Los Angeles, USA	Micro	Csb	✓			✓	✓	✓	✓	
Founda and Santamouris (2017)	Athens, Greece	Meso	Csa	✓			✓	✓			

Study	Location	Scale	Climate							
Foustalieraki et al. (2017)	Athens, Greece	Micro	Csa		✓					✓
Georgakis and Santamouris (2017)	Athens, Greece	Micro	Csa				✓	✓		
Morini et al. (2017)	Athens, Greece	Micro	Csa,	✓			✓	✓		
Paravantis et al. (2017)	Athens, Greece	Meso	Csa	✓			✓	✓		
Efthymiou et al. (2016)	Athens, Greece	Micro	Csa		✓		✓			✓
Haashemi et al. (2016)	Teheran, Iran	Meso	Csa		✓		✓			✓
Karachaliou et al. (2016)	Athens, Greece	Micro	Csa		✓		✓			✓
Morini and Touchaei (2016)	Terni, Italy	Meso	Csa		✓		✓	✓		
Rossi, Castellani, et al. (2016)	Terni, Italy	Micro	Csa		✓		✓	✓		
Rossi, Bonamente, et al. (2016)	Rome, Italy	Meso	Csa		✓		✓			✓
Santamouris et al. (2015)	Athens, Greece	Multiple	Csa						✓	
García-Cueto et al. (2013)	Tijuana; Ensenada, Mexico	Meso	BSk		✓	✓			✓	

(Continued)

TABLE 7.1 (Continued)

Author/year	City and country	Scale	Köppen–Geiger climate zone	Mitigation and adaptation strategies							
				▲	◧	■	▭	◉	☀	↕	▣
Georgakis et al. (2014)	Athens, Greece	Micro	Csa		✓	✓					
Skoulika et al. (2014)	Athens, Greece	Micro	Csa	✓							
Kolokotsa et al. (2013)	Chania, Greece	Micro	Csa	✓	✓		✓	✓			
Santamouris et al. (2012)	Athens, Greece	Micro	Csa		✓						✓
Giannopoulou et al. (2011)	Athens, Greece	Meso	Csa	✓							
Gobakis et al. (2011)	Athens, Greece	Meso	Csa		✓		✓				
Santamouris et al. (2011)	Athens, Greece	Micro	Csa		✓						
Ghiaus et al. (2006)	Athens, Greece.	Micro	Csa			✓	✓				
Santamouris et al. (2001)	Athens, Greece	Meso	Csa	✓		✓	✓		✓	✓	

Scale: Micro (microscale: canopy UHI, urban object); Meso (mesoscale: boundary/surface UHI, 1–10,000 km²). Köppen–Geiger climate zone: hot semiarid climate (BSh); cold semiarid climate (BSk); warm summer Mediterranean climates (CSa); Mediterranean hot summer Mediterranean climates (CSb). Mitigation strategy: Parks and green areas (▲), green roofs or facades (◧), albedo, reflectance, and retroreflectance (■), pavements (▭), urban designs and water systems (◉), urban design and air systems (↕), and urban morphology (▣).

7.3.2 Prevailing mitigation and adaptation measures

7.3.2.1 Land use and land cover

A critical aspect of understanding urban overheating is identifying the land use dynamics as an important component of land cover alterations (SSCIPO-LUCC, 1999). In this respect, other studies define land cover as the surface where the land's vegetation, soil, water, and other physical features are located (Sultana & Satyanarayana, 2022). Land use corresponds to the human-made arrangements, actions, and inputs in a particular land cover type that generates, alters, or preserves it (FAO, 2000).

LULC is associated with urban sprawl and the loss of rural land (Patel et al., 2019). Furthermore, the replacement of natural areas with artificial surfaces that have limited porosity and reflectance, as well as high thermal conductivity and heat capacity, as a result of accelerated urbanization, generates surfaces that store heat, limit evaporation, and reduce long-wave radiation, thereby increasing temperatures (Sultana & Satyanarayana, 2022).

Open and green spaces tend to be transformed into built-up areas and impervious surfaces, contributing to UHI (Estoque et al., 2017) and detrimental impacts on urban environmental quality (UEQ). Different LULC strategies are implemented address the urban overheating and its consequences on environmental:

- Parks and greenering
- Green roofs/Cool roofs or Green facades/Cool facades
- Albedo, reflectance, retroreflectance, etc.
- Pavements
- Water bodies.

To understand recent studies that analyze the implementation of LULC strategies, we summarize four benefits in terms of impacts on the following: (1) air quality (e.g., pollutant particulate matter such as $PM_{2.5}$ and PM_{10}); (2) cooling energy consumption; (3) habitability (thermal comfort); and (4) health (mortality and morbidity rates). Table 7.2 shows some studies that implement mitigation strategies.

Indeed, modifying LULC constitutes the main strategies for mitigating the effects of UHI. Researchers in Ohio conducted a study in 2018 that shows that increased greenery reduces temperatures in summer (Chun & Guldmann, 2018). Also, other research have demonstrated the UHI mitigation potential of cool and green roofs (Chatterjee et al., 2019) and also indicated the influence of high albedo pavements on human thermal perception (Rosso et al., 2016). Fig. 7.3 shows the interaction between 159 leading studies pertaining to research on LULC actions (inbound multicolor dots based on baseline literature collection) and 20 prospective researches that address the heat mitigation impacts on UEQ, as indicated on outbound crossed dots.

TABLE 7.2 Implementation and impacts of land use and land cover strategies.

Author	LULC strategy	Air quality	Cooling energy consumption	Habitability	Health
López-Silva et al. (2022)	◨	–	Reducing buildings' thermal load	Improved thermal comfort	–
Martinez et al. (2021)	◨	–	Energy savings	Reduced outdoor overheating	–
Cardinali et al. (2020)	▲, ■	–	Annual energy saving up to 12%	–	–
Castaldo et al. (2018)	▲, ◨	–	Reduced 10% of the energy (HVAC)	–	–
Jandaghian and Akbari (2018a)	▲, ■, ▢	Decreased PM2.5 and O3 concentrations	–	–	–
Jandaghian and Akbari (2018b)	■	–	–	The discomfort index improved by 3%	Heat-related mortality decreased by 3.2% (heat wave periods)
Rousta et al. (2018)	▲, ▢, ◉	–	–	Use of strategies to minimize SUHI	–
Rosso et al. (2016)	■, ▢	–	–	Improved thermal perception	–
Wang et al. (2016)	◨, ▢	–	–	PET could be reduced by average 1.8°C.	–

LULC strategy: Parks and green areas (▲), green roofs or facades (◨), albedo, reflectance and retroreflectance (■), pavements (▢), and urban designs and water systems (◉).

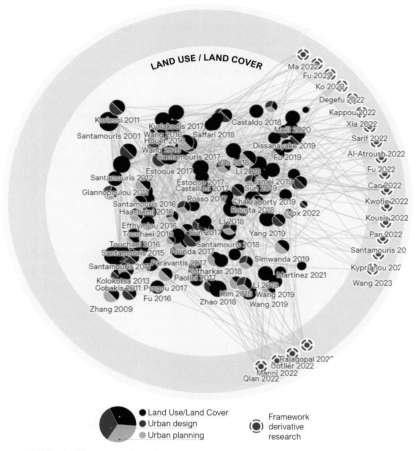

FIGURE 7.3 Incident research on land use and land cover actions and prospective research to address the heat mitigation impacts on UEQ.

Prospective research could be summed up as follows: (1) impacts of land classification through land use and LCZ in UHI formation and intensity, the study of material's surface properties, and green and gray infrastructure (Rajagopal et al., 2023); (2) the definition of standardized monitoring protocols must be addressed for monitoring efficiently urban microclimates (Kousis et al., 2022; Degefu et al., 2022); (3) applications of new technologies of cool pavements to mitigate UHI (Kappou et al., 2022) and established protocols for field measurements to identify the direct impacts (Ko et al., 2022); (4) novel approaches to decrease pavement surface temperature and air temperature to improve sustainable mobility (Al-Atroush et al., 2022); (5) understand the urbanization impacts on UHI (Qian et al., 2022) and the connection between urban development and surface temperature (Kwofie et al., 2022); (6) monitoring the land surface to identify driving factors with urban morphological parameters (Cao et al., 2022), analyzing the urban cool island phenomenon (Sarif et al., 2022), understanding the UHI effect across time series (Fu, Yao, et al., 2022), analyzing surface UHIs (SUHI) using zoning techniques of local climate (Xia et al., 2022), and identifying extreme heat waves and land surface temperature (Cotlier & Jimenez, 2022); (7) LULC modifications and their impacts on ecological quality (Pan et al., 2022), analyzing the spatiotemporal SUHI effect (Ma & Peng, 2022; Wang et al., 2023); (8) impacts of an urban park on microclimate (Fu, Wang, et al., 2022); and (9) mitigation and adaptation strategies to improve urban health: (Kyprianou et al., 2023; Santamouris, 2023).

7.3.2.2 Urban design

The urban design aims to safeguard the integrity of the design of buildings, places, spaces, and networks to ensure the development of all levels of government, the private sector, and nonprofit organizations by promoting coherent strategies and guaranteeing decent housing, expanding economic opportunities, and a suitable living environment, the last of which is the most vulnerable; all these aspects are disturbed and reevaluated when constraints are imposed in transport, mobility, and accessibility by exceptional scenarios like COVID-19 pandemic or heat wave intense events (Faedda et al., 2022). In North America, multiple climate action plans have been implemented to combat the current climate crisis, promote climate resilience and adaptation, and reduce carbon emissions by reducing building energy demand and transitioning to low-carbon or zero-carbon sources to reduce disparities and promote environmental justice; guidance and targeted climate and environmental justice policies for disadvantaged and low-income communities are essential to achieve these objectives (U.S. Department of Housing & Human Services, 2016). From Medellin green corridors and Paris *"cool islands"* to tackle heat waves to Sydney's tree plan or LA's white paint experiment to mitigate sensible balance distortions, according to the UN Environment Programme, almost 90 cities declared heat alerts regarding heat waves throughout the summer's extreme weather (UN Environment Programme, 2022).

The long-term view in urban design, its robust spatial dimension, and key aspects such as the built context, cultural character, people's choice, network city connections, creative environments, buildings and infrastructure custodianship, and decision-makers' collaboration are important in design protocols (Pirrit et al., 2005). Different entities are extended to the communities' livability environmental concept to manage synergies before the negative impacts defined thermal threshold of urban overheating. More than two decades ago, these protocols emerged as powerful platforms for creating urban infrastructure and the diverse connection system between places (English Partnership & The Housing Corporation, 2000); to better comprehend them, we must distinguish between the various approaches that incorporate urban design elements and have been incorporated in recent studies to reduce residents' exposure to heat. The framework defined by the street network and urban grid composition that affects the energy and resource efficiency in terms of solar radiation has been well documented; this is exemplified in the work undertaken by different authors: Agathangelidis et al. studied the "Integration of Urban Form, Function, and Energy Fluxes in a Heat Exposure Indicator" named *"UHeatEx"* that integrates the physical processes integrating remote sensing techniques to reflect the variability of the thermal environment regarding the intra-UHI assessment and climate change adaptation (Agathangelidis et al., 2019). In addition, another study conducted an environmental data clustering analysis involving CO_2 concentration (ppm), global solar radiation (W/m^2), and apparent temperature (°C) measures to identify intraurban granular morphologies from pedestrian transects, highlighting compelling findings during daytime with the greatest temperature difference of 9.7°C between the different case studies reported (Pigliautile & Pisello, 2020).

Therefore, the causal role of urban densification as a physical dynamic for thermal interactions between buildings and their surrounding environment has been the subject of intense analysis and parametric studies involving energy, latent cooling demand, and boundary layers (Shi et al., 2019). Other forms of addressing urban density consider compatible uses by mixing centers, edges, neighborhood units, transition zones, and significant landscape aspects that involve urban forestry, open and public access networks, and microclimate management considering a framework focused on urban design variables. Within this context, several studies have developed and applied techniques, methods, or numerical models for solar analyses in urban canyon configurations (Manni et al., 2020) to simulate microscale thermal interactions in different building environments (Chatterjee et al., 2019), define the city-scale morphological influence on diurnal urban air temperature (Wang et al., 2020), or enhance energy and environmental performance of near-zero energy settlements (Cardinali et al., 2020). In this regard, urban technologies tend to be more concerned with the passive cooling potential of construction materials and radiative interactions between surfaces (specular, Lambertian, or retroreflective) to assess solar irradiance and LST (Fabiani et al., 2019; Manni et al., 2019), whereas urban systems have a very definite awareness of the driving forces behind convective heat transfers related to urban elements and systems such as green surfaces, fountains, pools, ponds, evaporative towers, sprinklers, and purificators (Fan et al., 2019; Li et al., 2019).

Urban design strategies to reduce urban overheating and detrimental consequences are mainly listed as follows:

- Urban design and solar radiation
- Urban design and fluids
  - Urban heterogeneity and airflows
  - Urban morphology and circulations
  - Urban design and air or water systems.

In detail, Fig. 7.4 demonstrates that influential authors have in recent years advocated for continuing research on urban design-related aspects to address the impact of warmer cities on global environmental quality.

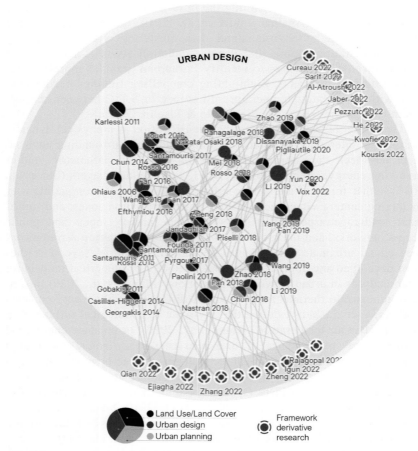

FIGURE 7.4 Incident research on urban design actions and prospective research to address the heat mitigation impacts on UEQ.

For instance, in the Arab world, countries like Egypt, Saudi Arabia, or the United Aran Emirates are most interested in assessing the alterations in biological (flora and fauna), physical (land, water, and air), and human (social, economic, health, political, cultural, and visual) components of UHIs to build a consistent narrative (Jaber, 2022). However, Rajagopal et al. reviewed the latest advances evaluating environmental measures on UHI; the study revealed a significant research gap in connectivity between green, blue, and gray infrastructures where a vast amount of measuring variables and indexes are remarkable: NDBI, SVF, ISA, IBI, mean building height, building footprint, patch density, building coverage ratio, aggregation index, building surface fraction, H/W ratio, surface/volume ratio, building orientation, gravitational urban morphology index, and others (Rajagopal et al., 2023). If we now turn to similar investigations, Mirabi and Davies provided evidence to correlate the urban linear infrastructure (ULI) effects and its interaction with different typologies with subsurface, surface, boundary, and canyon UHI. The ULI types can broadly be defined by utility networks (power and water) and transportation networks, with categories involving flyover, at-grade, and underground contributing factors (Mirabi & Davies, 2022).

In broad terms, to enhance outdoor thermal comfort and building energy efficiency, Zheng et al. implemented a parametric simulation framework for urban streets; the advanced BEM model in conjunction with an urban canopy model at a neighborhood scale for six Chinese cities resulted in 31,104 simulations with peak cooling load (ΔCL_{peak}) thresholds of about 3.2 kWh/m^2, revealing street orientation, window type, and window-to-wall ratio as the most significant factors (Zheng et al., 2023). For Wang et al., the numerous urban forms and natural environment elements modifies the urban climate, and predicting the ambient temperature demands a multidimensional approach; they proposed the urban thermal environment index, which implies a trained gradient-boosted regression trees mode and nine different influence pathways. The results indicated that urban form variables contribute more substantially to increasing the ambient temperature during the winter, but natural components perform better during the summer, with greater relative humidity and solar radiation contribution rates (Wang et al., 2022). In addition, Yoa et al. evaluated the impacts of urban configurations on outdoor thermal perceptions in two empirical sites using on-site measurements and carried out simulations via ENVI-met. Interestingly, the enclosed urban geometries were observed to enhance short and long-wave radiation, whereas air velocity minimally impacts the WBGT (Yola et al., 2022).

Turning now to the evidence on climate importance, Qian et al. investigated the influence of urban development on regional climate and severe climate conditions in datasets and methods to summarize the scientific insights on more than 500 studies for research priorities (Qian et al., 2022), while other researchers estimated driving factors within seasonal and diurnal SUHI intensities through 3D urban morphological parameters and moderate-resolution imaging spectroradiometer (Cao et al., 2022), compared optimization strategies to evaluate the effectiveness of various thermal environment actions involving the use of cool pavements and facades as well as grass, trees, and green roofs (Zhang et al., 2022), studied the nighttime heat wave enhancement over urban clusters due to dry conditions generated by large scale circulations, particularly aggravated in temperate climates (Igun et al., 2023), and showed that not only low windspeed and evapotranspiration, but also population, built-up size, and landscape pattern play a relevant role in warmer nighttime LST (Ejiagha et al., 2022).

Finally, the framework derivative research shows the continuity of energy balance models (Abunnasr et al., 2022), sophisticated outdoor monitoring systems (Cureau et al., 2022), material science relevance, urban cooling technologies, as well as the heat transfer modification process associated with a building's physical properties; walls, roofs, and ground coverage could be seized under the green building approach to offer a comprehensive solution for local and global urban overheating by incorporating an appropriate adaptation through adequate site planning, built form, land use, energy waste management, water saving systems, and material selection (He, 2022). On this basis, comparing the evidence from different authors, it can be seen that cool pavements as natural heat radiators must consider the role of solar reflectance, thermal emittance, thermal capacity, and permeability in their heat-exchange processes to represent an urban cooling technology with potential in high and low building densities since albedo effectiveness is comparable across climates and varies with urban morphology, where H/W aspect ratio and geometric features such as roughness, thermal inertia, and SVF are essential, but anthropogenic heat transfer is vital for microclimatic forecasting models (Kappou et al., 2022; Pezzuto et al., 2022).

7.3.2.3   Urban planning

Urban climate and environmental quality studies are supported by meteorological data and thermally driven phenomena to address the adverse effects concerning global warming, which are evident in low aerodynamics, such as extreme heat and poor air quality under calm synoptic conditions (Schau-Noppel et al., 2020). A holistic approach of multidisciplinary perspective and public/private practices between urban design and urban planning should consider a sociopolitical framework including environmental policies, infrastructure, socioeconomic features, and multilevel governance aspects, where the participation of users, decision-makers, and stakeholders is essential (Abd Elrahman & Asaad, 2021).

Notable examples of urban planning influencing the thermal environment involve interactions between city-scale morphological aspects (Wang et al., 2020) or spatiotemporal green infrastructure variability as heat stress modulators (Chakraborty & Lee, 2019). These cases demonstrate the significance of urban heterogeneity patterns in gray, green, and blue infrastructure concerning the environmental advantages provided by ecosystem services, as well as in the convective and radiative mechanisms linked to sensible and latent energy fluxes given by the physical properties of cities. As previously stated, it is clear that urban environmental quality is dramatically affected by rising temperatures; a great variety of authors worldwide have tracked and assessed the cooling effects of several mitigation initiatives.

Thus, the interaction between mitigation strategies proposed by prominent authors to address urban overheating impacts via applying key principles that integrate environmental assessment into urban planning is shown in Fig. 7.5. Thus, three general urban planning categories are implemented to alleviate local, city-regional, and global burdens and are listed as follows:

- Urban policies and community awareness.
- City-scale, urban tissue, or city grid.
- Infrastructure, traffic intensity, mobility.

Thus far, recent literature identifies clear UHI and overheating conditions under the effect of unplanned urban growth (Bek et al., 2018) in relation to the peak energy demand and the cooling energy penalty leading to urban citizens' vulnerability. Significant climate change adaptation actions in terms of urban planning have been taken; from urban form integration in intra-UHI scales (Agathangelidis et al., 2019) to the energy performance evaluation of green roof systems (Foustalieraki et al., 2017), different BEP-BEM and three-dimensional microclimatic simulation tools such as ENVI-met have gained ground in investigations by modeling and analyzing particulate matter $PM_{(0.25;10)}$, and gas concentrations, evapotranspiration and sensible heat fluxes, water and heat exchanges, as well as biometereological metrics in complex urban scenarios and local climate zones composed of buildings, vegetation, and other objects in the fields of urban design are increasingly considered (Cardinali et al., 2020; Chatterjee et al., 2019).

The framework derivative research could be summarized as follows: (1) analyzing urban warming in most populated cities (Rajagopal et al., 2023); (2) ULI systems contribution to UHI effects (Mirabi & Davies, 2022); (3) using green infrastructure to mitigate UHI (Shao & Kim, 2022; Zheng et al., 2022) and green cover to generate the cooling effect (Huang et al., 2022) and analyzing sustainable green roofs (Shahmohammad et al., 2022); (4) critical analysis of UHI mitigation strategies (Cakmakli & Rashed-Ali, 2022); (5) analyzing LST and environmental factors (Soltanifard et al., 2022), LULC patterns (Dong et al., 2022), spatial patterning of LST (Liu et al., 2023), and spatiotemporal variations of LST (Chen et al., 2022; Wu et al., 2022; Yang et al., 2022; Zhou et al., 2022); (6) evaluating urban morphology (Han et al., 2022) and urban street design to improve thermal comfort (Zheng et al., 2023); (7) multidimensional factors to predict thermal conditions (Wang et al., 2022); and (8) analyzing UHMA and urban health (Kyprianou et al., 2023).

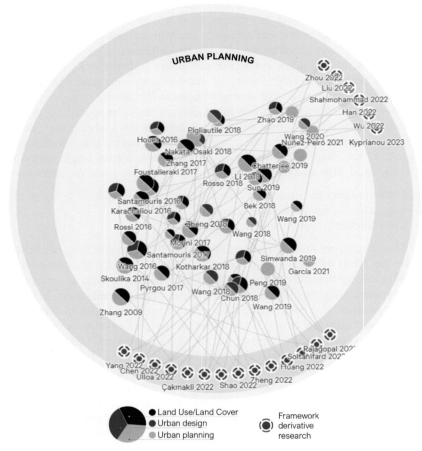

FIGURE 7.5 Incident research on urban planning actions and prospective research to address the heat mitigation impacts on UEQ.

7.4 Present challenges and future pathways

The conceptual framework of an urbanized world of agglomerations and megacities faces unprecedented challenges; sophisticated growing urban surfaces are responsible for complex phenomena that transform the environment in detrimental scenarios; they degrade sensitive ecosystems and demand a large amount of renewable and nonrenewable resources, thereby diminishing the likelihood of social equity and planning practices based on environmental justice toward sustainable development. As a consequence of this circumstance, regional leaders and decision-makers must make crucial decisions under the current climate change scenarios. As a consequence, scientific research plays a significant role in this regard by combining energy, health, and environmental science to design human life-improving measures.

In this regard, Table 7.3 represents a summary of the most studied impacts to be addressed by the various urban overheating countermeasures over the last few decades of research, one aspect to consider is the presence of health vulnerabilities and related hazards that arise as a result of climate change, and urban pollution and energy-sensitive aspects on the other hand. As described previously in this chapter, the emerging trends addressed in the derivative research referred to in different sections encompass a multidisciplinary and comprehensive perspective; they are indicated in the left column and the headings of the remaining columns. There is a direct interaction between the various phenomena associated with global warming, urban environmental quality, and energy demand and supply.

Moreover, the current understanding of interlinkages between urbanization prospects and the least developed populations advocates for preferential modalities for rural immigration inclusion, city management, urban mobility, quantitative commitments, market access for technology, and risk/health services to prevent heat-related diseases during extreme wheatear events, as detailed later in this section. Emerging trends to develop LULC policies include using logical methods with big data predictions or applied geography frameworks, geographic information system (GIS) science linked to personal wearable systems, and remote sensing techniques, which serve the diagnostic process of geospatial driving factors to monitor new definitions, such as local climate zones and the land surface temperature seasonality, including multitemporal influencing factors, and implications on the built environment microclimates; to prevent intraurban thermal inversion anomalies, the implementation of advanced technologies is a game changer.

A well-known example of this is the microscale and mesoscale computational models used to comprehend urban overheating. For instance, the ANSYS Fluent and Open FOAM were used to evaluate fluid mechanics at the surface layer environment. Also, microclimate tools like ENVI-met simulated urban surface—plant—air interactions. Computer-aided design and energy simulation tools (e.g., Energy Plus, DOE-2, and BLAST) are also mentioned in the literature. Some coupled model techniques, such as WRF (WRF, WRF-Chem), are useful to study the urban climate, even on a regional scale. Different tools and techniques for acquiring, analyzing, and displaying geographic data and imagery permit analyzing the impacts of overheating on the environment. For example, literature reports use GIS (e.g., ArcGIS, ArcMap, QGIS, and others) and remote sensing to measure urban expansion, characterize urban LULC change, and identify urban growth. Also, machine learning techniques improve LULC classification and permit modeling future land use patterns and their environmental implications (Saha et al., 2022). R Software, SPSS, and Stata were also used to study the UHI influence on the microscale and mesoscale models.

TABLE 7.3 Prevailing impacts on vulnerable populations.

Urban Overheating			Impacts		
•		•	Health vulnerabilities and risks		
	•	•		Urban pollution	
•		•			Sensitive energy aspects
Global warming	Urban environmental quality	Energy demand and supply	**a) Heat related health vulnerabilities** • Comfort and heat streets in low-income population. • Cardiovascular, cerebrovascular, respiratory, renal and infectious diseases on morbidity rates. • Cardiovascular and respiratory mortality on extreme heat events. **b) Risks** • Socioeconomic and risks. • Biodiversity alterations and affectations. • Agriculture and ecosystem damages.	• Tropospheric ozone concentration (NOx y VOC). • Higher particulate matter presence ($PM_{2.5}$, PM_{10}). • Intraurban thermal inversion anomalies. • Power plant greenhouse gas emissions. • Air/water pollution influence on urban vegetation. • Environmental risk on generation network plants.	• Underestimation of cooling demand of buildings. • Higher peak electricity demand. • Greater load duration variability and curve peaks. • Insufficient power generation and storage capacity. • Transmission infrastructure affectation and lifetime. • Generation curtailments and power disruption risks. • Heat wave impacts on renewable technologies.

In terms of pollution research, there is robust evidence addressing the importance of a holistic outlook for the urban design configuration and the articulation of green buildings and traditional infrastructure with ecologically efficient surroundings, the blue infrastructure, as well as air, water, and vegetation configurations as part of the essential set of elements from the urban tissue to redefine context dependency, landscapes, parks, integration of passive cooling technologies, pervious concretes, and cool roofs and albedo prediction models to decrease tropospheric ozone concentration (NOx and VOC) and the presence of inconvenient particulate matter ($PM_{2.5}$ and PM_{10}), which is important to consider the surface energy balance or the urban morphological parameters. Climate-based critical thinking considers that simulation tools will have a significant role in the future in optimizing early-stage urban planning to stimulate energy performance and the resulting greenhouse gas emissions in modern cities.

Additionally, significant research was conducted on different strategies to reduce overheating to improve urban environmental quality. The literature shows that the studies discuss the impacts of LULC, urban planning, and urban design in different contexts, representing challenges due to the spatial-temporal variations in the study areas, which makes it difficult to contrast the application impacts of strategies to mitigate overheating. In addition, it is important to highlight the research deficit in those countries with a low-income population. In these countries, there is a vulnerability tendency to heat-related risks, so it is essential to evaluate mitigation strategies that integrate adequate solutions. Considering the continuous urbanization process and the temperature rise, it is imperative to investigate and develop appropriate solutions and technologies for the built environment to improve comfort and well-being further.

Heat waves represent a potential risk due to their severe health implications, associated, in their most severe form, with heat stroke deaths and can also lead to circulatory or cardiac problems in vulnerable groups (e.g., children and elderly adults). The heat waves are increasingly frequent and longer lasting in certain regions. Though this phenomenon was studied, it was identified that, in those cities with vulnerable populations, it is necessary to carry out studies that correlate the overheating impact on mortality and morbidity to define mitigation and adaptation strategies that contribute to urban environmental quality. It is also relevant to identify socioeconomic risks, alterations in biodiversity, and ecosystem damage in the context of overheating in the urban environment, where artificial materials predominate and have replaced natural surfaces, contributing to more significant impacts due to heat waves. However, the literature has identified the impacts caused by vegetation and the use of appropriate materials as mitigation strategies to face climate change. There is still a need for studies to select the appropriate green infrastructure for an urban area considering other aspects such as availability of water, climatic context, use, and access to technologies. Research carried out by Kyprianou et al. (2023) notes a future perspective related to investigating overheating adaptation and mitigation strategies applicable to urban revitalization initiatives. Despite the primary focus on tangible actions and the influence of climate change on urban health, important proposals promote social participation and community approaches for urban climate and decision-making dynamics.

References

Abd Elrahman, A. S., & Asaad, M. (2021). Urban design & urban planning: A critical analysis to the theoretical relationship gap. *Ain Shams Engineering Journal*, *12*(1), 1163−1173. Available from https://doi.org/10.1016/j.asej.2020.04.020.

Abunnasr, Y., Mhawej, M., & Chrysoulakis, N. (2022). SEBU: A novel fully automated Google Earth Engine surface energy balance model for urban areas. *Urban Climate*, *44*. Available from https://doi.org/10.1016/j.uclim.2022.101187.

Agathangelidis, I., Cartalis, C., & Santamouris, M. (2019). Integrating urban form, function, and energy fluxes in a heat exposure indicator in view of intra-urban heat island assessment and climate change adaptation. *Climate*, *7*(6). Available from https://doi.org/10.3390/cli7060075.

Al-Atroush, M. E., Mustaffa, Z., & Sebeay, T. A. (2022). Emerging trends in overcoming the weather barrier to sustainable mobility in gulf and tropical cities. *IOP Conference Series: Earth and Environmental Science*, *1026*(1). Available from https://doi.org/10.1088/1755-1315/1026/1/012040.

Bek, M. A., Azmy, N., & Elkafrawy, S. (2018). The effect of unplanned growth of urban areas on heat island phenomena. *Ain Shams Engineering Journal*, *9*(4), 3169−3177. Available from https://doi.org/10.1016/j.asej.2017.11.001.

Boggia, A., Massei, G., Pace, E., Rocchi, L., Paolotti, L., & Attard, M. (2018). Land use policy spatial multicriteria analysis for sustainability assessment: A new model for decision making. *Land Use Policy*, *71*(December 2017), 281−292. Available from https://doi.org/10.1016/j.landusepol.2017.11.036.

Cakmakli, A. B., & Rashed-Ali, H. (2022). A climate-based critical analysis of urban heat island assessment methods and mitigation strategies. *Journal of Green Building*, *17*(4), 129−149. Available from https://doi.org/10.3992/jgb.17.4.129.

Cao, S., Cai, Y., Du, M., Weng, Q., & Lu, L. (2022). Seasonal and diurnal surface urban heat islands in China: An investigation of driving factors with three-dimensional urban morphological parameters. *GIScience & Remote Sensing, 59*(1), 1121−1142. Available from https://doi.org/10.1080/15481603.2022.2100100.

Cardinali, M., Pisello, A. L., Piselli, C., Pigliautile, I., & Cotana, F. (2020). Microclimate mitigation for enhancing energy and environmental performance of near zero energy settlements in Italy. *Sustainable Cities and Society, 53.* Available from https://doi.org/10.1016/j.scs.2019.101964.

Castaldo, V. L., Pisello, A. L., Piselli, C., Fabiani, C., Cotana, F., & Santamouris, M. (2018). How outdoor microclimate mitigation affects building thermal-energy performance: A new design-stage method for energy saving in residential near-zero energy settlements in Italy. *Renewable Energy, 127*(1), 920−935. Available from https://doi.org/10.1016/j.renene.2018.04.090.

Castellani, B., Morini, E., Anderini, E., Filipponi, M., & Rossi, F. (2017). Development and characterization of retro-reflective colored tiles for advanced building skins. *Energy and Buildings, 154,* 513−522. Available from https://doi.org/10.1016/j.enbuild.2017.08.078.

Chakraborty, T., & Lee, X. (2019). A simplified urban-extent algorithm to characterize surface urban heat islands on a global scale and examine vegetation control on their spatiotemporal variability. *International Journal of Applied Earth Observation and Geoinformation, 74,* 269−280. Available from https://doi.org/10.1016/j.jag.2018.09.015.

Chatterjee, S., Khan, A., Dinda, A., Mithun, S., Khatun, R., Akbari, H., Kusaka, H., Mitra, C., BhattI, S., Saleem, D., Quang, V., & Wang, Y. (2019). Simulating micro-scale thermal interactions in different building environments for mitigating urban heat islands. *Science of The Total Environment,* 610−631. Available from https://doi.org/10.1016/j.scitotenv.2019.01.299.

Chen, Y., Wang, Y., Zhou, D., Gu, Z., & Meng, X. (2022). Summer urban heat island mitigation strategy development for high-anthropogenic-heat-emission blocks. *Sustainable Cities and Society, 87.* Available from https://doi.org/10.1016/j.scs.2022.104197.

Chun, B., & Guldmann, J. M. (2018). Impact of greening on the urban heat island: Seasonal variations and mitigation strategies. *Computers, Environment and Urban Systems.* Available from https://doi.org/10.1016/J.COMPENVURBSYS.2018.05.006.

Collaboration for Environmental Evidence. (2013). Guidelines for systematic reviews in environmental management. March Guidelines for Systematic Reviews in Environmental Management. http://www.environmentalevidence.org/Documents/Guidelines/Guidelines4.2.pdf.

Cotlier, G., & Jimenez, J. C. (2022). The extreme Heat Wave over western north america in 2021: An assessment by means of Land Surface Temperature. *Remote Sensing, 14*(3). Available from https://doi.org/10.3390/rs14030561.

Cureau, R., Pigliautile, I., & Pisello, A. L. (2022). A new wearable system for sensing outdoor environmental conditions for monitoring hyper-microclimate. *Sensors, 22*(2). Available from https://doi.org/10.3390/s22020502.

Degefu, M. A., Argaw, M., Feyisa, G. L., & Degefa, S. (2022). Regional and urban heat island studies in megacities: A systematic analysis of research methodology. *Indoor and Built Environment,* 1775−1786. Available from https://doi.org/10.1177/1420326X211061491.

Dong, R., Wurm, M., & Taubenböck, H. (2022). Seasonal and diurnal variation of land surface temperature distribution and its relation to land use/land cover patterns. *International Journal of Environmental Research and Public Health, 19*(19). Available from https://doi.org/10.3390/ijerph191912738.

Duan, S., Luo, Z., Yang, X., & Li, Y. (2019). The impact of building operations on urban heat/cool islands under urban densification: A comparison between naturally-ventilated and air-conditioned buildings. *Applied Energy, 235*(October 2018), 129−138. Available from https://doi.org/10.1016/j.apenergy.2018.10.108.

Efthymiou, C., Santamouris, M., Kolokotsa, D., & Koras, A. (2016). Development and testing of photovoltaic pavement for heat island mitigation. *Solar Energy*, *130*, 148−160. Available from https://doi.org/10.1016/j.solener.2016.01.054.

Ejiagha, I. R., Ahmed, M. R., Dewan, A., Gupta, A., Rangelova, E., & Hassan, Q. (2022). Urban warming of the two most populated cities in the Canadian province of Alberta, and its influencing factors. *Sensors*, *22*(8). Available from https://doi.org/10.3390/s22082894.

English Partnership The Housing Corporation. (2000). Design 1 London, United Kindom Urban Design Compendium 1.

Estoque, R., Murayama, Y., & Myint, S. (2017). Effects of landscape composition and pattern on land surface temperature: An urban heat island study in the megacities of Southeast Asia. *Science of The Total Environment*, *577*. Available from https://doi.org/10.1016/j.scitotenv.2016.10.195.

Fabiani, C., Pisello, A. L., Bou-Zeid, E., Yang, J., & Cotana, F. (2019). Adaptive measures for mitigating urban heat islands: The potential of thermochromic materials to control roofing energy balance. *Applied Energy*, *247*, 155−170. Available from https://doi.org/10.1016/j.apenergy.2019.04.020.

Faedda, S., Plaisant, A., Talu, V., & Tola, G. (2022). The role of urban environment design on health during the COVID-19 pandemic: A scoping review. *Frontiers in Public Health*, *10*. Available from https://doi.org/10.3389/fpubh.2022.791656.

Fan, Y., Wang, Q., Yin, S., & Li, Y. (2019). Effect of city shape on urban wind patterns and convective heat transfer in calm and stable background conditions. *Building and Environment*, *162*. Available from https://doi.org/10.1016/j.buildenv.2019.106288.

FAO. (2000). Land cover classification system: Classification concepts and user manual. United Nations, Rome, Italy.

Founda, D., & Santamouris, M. (2017). Synergies between urban heat island and heat waves in Athens (Greece), during an extremely hot summer (2012). *Scientific Reports*, *7*(1). Available from https://doi.org/10.1038/s41598-017-11407-6.

Foustalieraki, M., Assimakopoulos, M. N., Santamouris, M., & Pangalou, H. (2017). Energy performance of a medium scale green roof system installed on a commercial building using numerical and experimental data recorded during the cold period of the year. *Energy and Buildings*, *135*, 33−38. Available from https://doi.org/10.1016/j.enbuild.2016.10.056.

Fu, J., Wang, Y., Zhou, D., & Cao, S. (2022). Impact of urban park design on microclimate in cold regions using newly developped prediction method. *Sustainable Cities and Society*, *80*. Available from https://doi.org/10.1016/j.scs.2022.103781.

Fu, X., Yao, L., Xu, W., Wang, Y. X., & Sun, S. (2022). Exploring the multitemporal surface urban heat island effect and its driving relation in the Beijing-Tianjin-Hebei urban agglomeration. *Applied Geography*, *144*102714. Available from https://doi.org/10.1016/j.apgeog.2022.102714.

García-Cueto, O. R., Cavazos, M. T., de Grau, P., & Santillán-Soto, N. (2013). Analysis and modeling of extreme temperatures in several cities in northwestern Mexico under climate change conditions. *Theoretical and Applied Climatology*, *116*(1), 211−225. Available from http://doi.org/10.1007/s00704-013-0933-x.

García, D. H., & Díaz, J. A. (2021). Modeling of the Urban Heat Island on local climatic zones of a city using Sentinel 3 images: Urban determining factors. *Urban Climate*, *37*(April). Available from https://doi.org/10.1016/j.uclim.2021.100840.

Georgakis, C., & Santamouris, M. (2017). Determination of the surface and canopy urban heat island in Athens central zone using advanced monitoring. *Climate*, *5*(4), 1−13. Available from https://doi.org/10.3390/cli5040097.

Georgakis, C., Zoras, S., & Santamouris, M. (2014). Studying the effect of "cool" coatings in street urban canyons and its potential as a heat island mitigation technique. *Sustainable Cities and Society*, *13*, 20–31. Available from https://doi.org/10.1016/j.scs.2014.04.002.

Ghiaus, C., Allard, F., Santamouris, M., Georgakis, C., & Nicol, F. (2006). Urban environment influence on natural ventilation potential. *Building and Environment*, *41*(4), 395–406. Available from https://doi.org/10.1016/j.buildenv.2005.02.003.

Giannopoulou, K., Livada, I., Santamouris, M., Saliari, M., Assimakopoulos, M., & Caouris, Y. G. (2011). On the characteristics of the summer urban heat island in Athens, Greece. *Sustainable Cities and Society*, *1*(1), 16–28. Available from https://doi.org/10.1016/j. scs.2010.08.003.

Gobakis, K., Kolokotsa, D., Synnefa, A., Saliari, M., Giannopoulou, K., & Santamouris, M. (2011). Development of a model for urban heat island prediction using neural network techniques. *Sustainable Cities and Society*, *1*(2), 104–115. Available from https://doi.org/ 10.1016/j.scs.2011.05.001.

Gonzalez-Trevizo, M. E., Martinez-Torres, K. E., Armendariz-Lopez, J. F., Santamouris, M., Bojorquez-Morales, G., & Luna-Leon, A. (2021). Research trends on environmental, energy and vulnerability impacts of Urban Heat Islands: An overview. *Energy and Buildings*, *246*. Available from https://doi.org/10.1016/j.enbuild.2021.111051.

Haashemi, S., Weng, Q., Darvishi, A., & Alavipanah, S. (2016). Seasonal variations of the Surface Urban Heat Island in a semi-arid city. *Remote Sensing*, *8*(4). Available from https:// doi.org/10.3390/rs8040352.

Han, D., An, H., Wang, F., Xu, X., Qiao, Z., Wang, M., Sui, X., Liang, S., Hou, X., Cai, H., & Liu, Y. (2022). Understanding seasonal contributions of urban morphology to thermal environment based on boosted regression tree approach. *Building and Environment*, *226*. Available from https://doi.org/10.1016/j.buildenv.2022.109770.

He, B. J. (2022). Green building: A comprehensive solution to urban heat. *Energy and Buildings*, *271*. Available from https://doi.org/10.1016/j.enbuild.2022.112306.

Huang, R., Yang, M., Lin, G., Ma, X., Wang, X., Huang, Q., & Zhang, T. (2022). Cooling effect of green space and water on urban heat island and the perception of residents: A case study of Xi'an city. *International Journal of Environmental Research and Public Health*, *19*(22). Available from https://doi.org/10.3390/ijerph192214880.

Igun, E., Xu, X., Shi, Z., & Jia, G. (2023). Enhanced nighttime heatwaves over African urban clusters. *Environmental Research Letters*, *18*(1), 014001. Available from https://doi.org/ 10.1088/1748-9326/aca920.

Intergovernmental Panel on Climate Change. (2014). IPCC Geneva climate change 2014 synthesis report summary chapter for policymakers.

Jaber, S. M. (2022). On the determination and assessment of the impacts of urban heat islands: A narrative review of literature in the Arab world. *GeoJournal*. Available from https://doi. org/10.1007/s10708-022-10706-4.

Jandaghian, Z., & Akbari, H. (2018a). The effects of increasing surface reflectivity on heat-related mortality in Greater Montreal Area, Canada. *Urban Climate*, *25*, 135–151. Available from https://doi.org/10.1016/j.uclim.2018.06.002.

Jandaghian, Z., & Akbari, H. (2018b). The effect of increasing surface albedo on urban climate and air quality: A detailed study for Sacramento, Houston, and Chicago. *Climate*, *6*(2). Available from https://doi.org/10.3390/cli6020019.

Jato-Espino, D. (2019). Spatiotemporal statistical analysis of the urban heat island effect in a Mediterranean region. *Sustainable Cities and Society*, *46*. Available from https://doi.org/ 10.1016/j.scs.2019.101427.

Kappou, S., Souliotis, M., Papaefthimiou, S., Panaras, G., Paravantis, J., Michalena, E., Hills, J. M., Vouros, A., Ntymenou, A., & Mihalakakou, G. (2022). Cool pavements: State of the art and new technologies. *Sustainability*, *14*(9). Available from https://doi.org/10.3390/su14095159.

Karachaliou, P., Santamouris, M., & Pangalou, H. (2016). Experimental and numerical analysis of the energy performance of a large scale intensive green roof system installed on an office building in Athens. *Energy and Buildings*, *114*, 256–264. Available from https://doi.org/10.1016/j.enbuild.2015.04.055.

Kim, H., Gu, D., & Kim, H. Y. (2018). Effects of Urban Heat Island mitigation in various climate zones in the United States. *Sustainable Cities and Society*, *41*, 841–852. Available from https://doi.org/10.1016/j.scs.2018.06.021.

Kolokotsa, D., Santamouris, M., & Zerefos, S. C. (2013). Green and cool roofs' urban heat island mitigation potential in European climates for office buildings under free floating conditions. *Solar Energy*, *95*, 118–130. Available from https://doi.org/10.1016/j.solener.2013.06.001.

Kousis, I., Manni, M., & Pisello, A. L. (2022). Environmental mobile monitoring of urban microclimates: A review. *Renewable and Sustainable Energy Reviews*, *169*. Available from https://doi.org/10.1016/j.rser.2022.112847.

Ko, J., Schlaerth, H., Bruce, A., Sanders, K., & Ban-Weiss, G. (2022). Measuring the impacts of a real-world neighborhood-scale cool pavement deployment on albedo and temperatures in Los Angeles. *Environmental Research Letters*, *17*(4). Available from https://doi.org/10.1088/1748-9326/ac58a8.

Kwofie, S., Nyamekye, C., Appiah Boamah, L., Owusu Adjei, F., Arthur, R., & Agyapong, E. (2022). Urban growth nexus to land surface temperature in Ghana. *Cogent Engineering*, *9* (1). Available from https://doi.org/10.1080/23311916.2022.2143045.

Kyprianou, I., Artopoulos, G., Bonomolo, A., Brownlee, T., Ávila-Cachado, R., Camaioni, C., Đokić, V., D'Onofrio, R., Đukanović, Z., Fasola, S., Di Giovanni, C. F., Cocci Grifoni, R., Hadjinicolaou, P., Ilardo, G., Jovanović, P., La Grutta, S., Malizia, V., Marchesani, G. E., Ottone, M. F., ... Carlucci, S. (2023). Mitigation and adaptation strategies to offset the impacts of climate change on urban health: A European perspective. *Building and Environment*, *238*. Available from https://doi.org/10.1016/j.buildenv.2023.110226.

Kyriakodis, G. E., & Santamouris, M. (2018). Using reflective pavements to mitigate urban heat island in warm climates - Results from a large scale urban mitigation project. *Urban Climate*, *24*, 326–339. Available from https://doi.org/10.1016/j.uclim.2017.02.002.

Litmaps tool. (2023). Litmaps tool. 2 5. https://www.litmaps.com/.

Li, D., Liao, W., Rigden, A. J., Liu, X., Wang, D., Malyshev, S., & Shevliakova, E. (2019). Urban heat island: Aerodynamics or imperviousness? *Science Advances*, *5*(4). Available from https://doi.org/10.1126/sciadv.aau4299.

Liu, T., Ouyang, S., Gou, M., Tang, H., Liu, Y., Chen, L., Lei, P., Zhao, Z., Xu, C., & Xiang, W. (2023). Detecting the tipping point between heat source and sink landscapes to mitigate urban heat island effects. *Urban Ecosystems*, *26*(1), 89–100. Available from https://doi.org/10.1007/s11252-022-01294-9.

Liu, H., & Weng, Q. (2018). Scaling effect of fused ASTER-MODIS land surface temperature in an urban environment. *Sensors*, *18*(11), 4058. Available from https://doi.org/10.3390/s18114058.

Lontorfos, V., Efthymiou, C., & Santamouris, M. (2018). On the time varying mitigation performance of reflective geoengineering technologies in cities. *Renewable Energy*, *115*, 926–930. Available from https://doi.org/10.1016/j.renene.2017.09.033.

López-Silva, D. V., Méndez-Alonzo, R., Sauceda-Carvajal, D., Sigala-Meza, E., & Zavala-Guillén, I. (2022). Experimental comparison of two extensive green roof designs in Northwest Mexico. *Building and Environment*, *226*. Available from https://doi.org/10.1016/j.buildenv.2022.109722.

Manni, M., Bonamente, E., Lobaccaro, G., Goia, F., Nicolini, A., Bozonnet, E., & Rossi, F. (2020). Development and validation of a Monte Carlo-based numerical model for solar analyses in urban canyon configurations. *Building and Environment*, *170*. Available from https://doi.org/10.1016/j.buildenv.2019.106638.

Manni, M., Lobaccaro, G., Goia, F., Nicolini, A., & Rossi, F. (2019). Exploiting selective angular properties of retro-reflective coatings to mitigate solar irradiation within the urban canyon. *Solar Energy*, *189*, 74−85. Available from https://doi.org/10.1016/j.solener.2019.07.045.

Martinez, S., Machard, A., Pellegrino, A., Touili, K., Servant, L., & Bozonnet, E. (2021). A practical approach to the evaluation of local urban overheating− A coastal city case-study. *Energy and Buildings*, *253*. Available from https://doi.org/10.1016/j.enbuild.2021.111522.

Ma, X., & Peng, S. (2022). Research on the spatiotemporal coupling relationships between land use/land cover compositions or patterns and the surface urban heat island effect. *Environmental Science and Pollution Research*, *29*(26), 39723−39742. Available from https://doi.org/10.1007/s11356-022-18838-3.

Mirabi, E., & Davies, P. J. (2022). A systematic review investigating linear infrastructure effects on Urban Heat Island (UHIULI) and its interaction with UHI typologies. *Urban Climate*, *45*. Available from https://doi.org/10.1016/j.uclim.2022.101261.

Morini, E., Castellani, B., Presciutti, A., Anderini, E., Filipponi, M., Nicolini, A., & Rossi, F. (2017). Experimental analysis of the effect of geometry and façade materials on urban district's equivalent albedo. *Sustainability*, *9*(7), 1245. Available from https://doi.org/10.3390/su9071245.

Morini, E., Touchaei, A. G., Beatrice, C., Federico, R., & Franco, C. (2016). The impact of albedo increase to mitigate the urban heat island in Terni (Italy) using the WRF model. *Sustainability*, *8*(10), 1−14. Available from https://doi.org/10.3390/su8100999.

Morini, E., Touchaei, A. G., Rossi, F., Cotana, F., & Akbari, H. (2018). Evaluation of albedo enhancement to mitigate impacts of urban heat island in Rome (Italy) using WRF meteorological model. *Urban Climate*, *24*, 551−566. Available from https://doi.org/10.1016/j.uclim.2017.08.001.

Nazarian, N., Krayenhoff, E. S., Bechtel, B., Hondula, D. M., Paolini, R., Vanos, J., Cheung, T., Chow, W. T. L., de Dear, R., Jay, O., Lee, J. K. W., Martilli, A., Middel, A., Norford, L. K., Sadeghi, M., Schiavon, S., & Santamouris, M. (2022). Integrated assessment of urban overheating impacts on human life. *Earth's Future*, *10*(8). Available from https://doi.org/10.1029/2022ef002682.

Núñez-Peiró, M., Sánchez-Guevara Sánchez, C., & Neila González, F. J. (2021). Hourly evolution of intra-urban temperature variability across the local climate zones. The case of Madrid. *Urban Climate*, *39*. Available from https://doi.org/10.1016/j.uclim.2021.100921.

Nwakaire, C. M., Onn, C. C., Yap, S. P., Yuen, C. W., & Onodagu, P. D (2020). Urban Heat Island Studies with emphasis on urban pavements: A review. *Sustainable Cities and Society*, *63*. Available from https://doi.org/10.1016/j.scs.2020.102476.

Pan, W., Wang, S., Wang, Y., Yu, Y., Luo, Y., & Yang, J. (2022). Dynamical changes of land use/land cover and their impacts on ecological quality during China's reform periods: A case study of Quanzhou city, China. *PLoS One*, *17*(12), e0278667. Available from https://doi.org/10.1371/journal.pone.0278667.

Paravantis, J., Santamouris, M., Cartalis, C., Efthymiou, C., & Kontoulis, N. (2017). Mortality associated with high ambient temperatures, heatwaves, and the urban heat island in Athens, Greece. *Sustainability*, *9*(4), 1−22. Available from https://doi.org/10.3390/su9040606.

Park, K., Jin, H. G., & Baik, J. J. (2023). Contrasting interactions between urban heat islands and heat waves in Seoul, South Korea, and their associations with synoptic patterns. *Urban Climate*, *49*. Available from https://doi.org/10.1016/j.uclim.2023.101524.

Patel, S., Verma, P., & Shankar Singh, G. (2019). Agricultural growth and land use land cover change in peri-urban India. *Environmental Monitoring and Assessment, 191*. Available from https://doi.org/10.1007/s10661-019-7736-1.

Pezzuto, C., Alchapar, N., & Correa, E. (2022). Urban cooling technologies potential in high and low buildings densities. *Solar Energy Advances, 2*. Available from https://doi.org/10.1016/j.seja.2022.100022.

Pigliautile, I., & Pisello, A. L. (2020). Environmental data clustering analysis through wearable sensing techniques: New bottom-up process aimed to identify intra-urban granular morphologies from pedestrian transects. *Building and Environment, 171*. Available from https://doi.org/10.1016/j.buildenv.2019.106641.

Pirrit, P., Tongue, R., Fontein, P., Sinclair, J., McDonald, C., Zollner, E., Leighton, D., Goodall, K., Fox, D., Whitely, S., Tocker, J., & Dalziel, A. (2005). New Zealand urban design protocol. I.

Pyrgou, A., Hadjinicolaou, P., & Santamouris, M. (2020). Urban-rural moisture contrast: Regulator of the urban heat island and heatwaves' synergy over a mediterranean city. *Environmental Research, 182*(December 2019), 109102. Available from https://doi.org/10.1016/j.envres.2019.109102.

Pyrgou, A., & Santamouris, M. (2018). Increasing probability of heat-related mortality in a mediterranean city Due to urban warming. *International Journal of Environmental Research and Public Health, 15*(8), 1−14. Available from https://doi.org/10.3390/ijerph15081571.

Qian, Y., Chakraborty, T. C., Li, J., Li, D., He, C., Sarangi, C., Chen, F., Yang, X., & Leung, L. R. (2022). Urbanization impact on regional climate and extreme weather: Current understanding, uncertainties, and future research directions. *Advances in Atmospheric Sciences, 39*(6), 819−860. Available from https://doi.org/10.1007/s00376-021-1371-9.

Rajagopal, P., Priya, R. S., & Senthil, R. (2023). A review of recent developments in the impact of environmental measures on urban heat island. *Sustainable Cities and Society, 88*. Available from https://doi.org/10.1016/j.scs.2022.104279.

Reyers, B., Folke, C., Moore, M. L., Biggs, R., & Galaz, V. (2018). Social-ecological systems insights for navigating the dynamics of the anthropocene. *Annual Review of Environment and Resources, 43*, 267−289. Available from https://doi.org/10.1146/annurev-environ-110615-085349.

Roberts, D., Pidcock, R., Chen, Y., Connors, S., & Tignor, M. (2019). Nairobi 2018: Global warming of 1.5°C. An IPCC Special Report on the impacts of global warming of 1.5°C above pre-industrial levels and related global greenhouse gas emission pathways, in the context of strengthening the global response to the threat of climate c.

Rodríguez, L. R., Ramos, J. S., Del Carmen Guerrero Delgado, M., & Álvarez Domínguez, S. (2022). Implications of the Urban Heat Island on the selection of optimal retrofitting strategies: A case study in a Mediterranean climate. *Urban Climate, 44*(April). Available from https://doi.org/10.1016/j.uclim.2022.101234.

Rossi, F., Bonamente, E., Nicolini, A., Anderini, E., & Cotana, F. (2016). A carbon footprint and energy consumption assessment methodology for UHI-affected lighting systems in built areas. *Energy and Buildings, 114*, 96−103. Available from https://doi.org/10.1016/j.enbuild.2015.04.054.

Rossi, F., Castellani, B., Presciutti, A., Morini, E., Anderini, E., Filipponi, M., & Nicolini, A. (2016). Experimental evaluation of urban heat island mitigation potential of retro-reflective pavement in urban canyons. *Energy and Buildings, 126*, 340−352. Available from https://doi.org/10.1016/j.enbuild.2016.05.036.

Rossi, F., Castellani, B., Presciutti, A., Morini, E., Filipponi, M., Nicolini, A., & Santamouris, M. (2015). Retroreflective façades for urban heat island mitigation: Experimental investigation and energy evaluations. *Applied Energy*, *145*, 8−20. Available from https://doi.org/10.1016/j.apenergy.2015.01.129.

Rosso, F., Golasi, I., Castaldo, V. L., Piselli, C., Pisello, A. L., Salata, F., Ferrero, M., Cotana, F., & de Lieto Vollaro, A. (2018). On the impact of innovative materials on outdoor thermal comfort of pedestrians in historical urban canyons. *Renewable Energy*, *118*, 825−839. Available from https://doi.org/10.1016/j.renene.2017.11.074.

Rosso, F., Pisello, A. L., Cotana, F., & Ferrero, M. (2016). On the thermal and visual pedestrians' perception about cool natural stones for urban paving: A field survey in summer conditions. *Building and Environment*. Available from https://doi.org/10.1016/J.BUILDENV.2016.07.028.

Rousta, I., Sarif., Gupta, R. D., Ólafsson, H., Ranagalage, M., Murayama, Y., Zhang, H., & Mushore, T. D. (2018). Spatiotemporal analysis of land use/land cover and its effects on surface urban heat island using Landsat data: A case study of Metropolitan city Tehran (1988−2018). *Sustainability*. Available from https://doi.org/10.3390/SU10124433.

Saha, M., Kafy, A. A., Bakshi, A., Faisal, A. A., Almulhim, A. I., Rahaman, Z. A, Al Rakib, A., Fattah, Md. A., Akter, K. S, Rahman, M. T., Zhang, M., & Rathi, R. (2022). Modelling microscale impacts assessment of urban expansion on seasonal surface urban heat island intensity using neural network algorithms. *Energy and Buildings*, *275*. Available from https://doi.org/10.1016/j.enbuild.2022.112452.

Santamouris, M., Cartalis, C., & Synnefa, A. (2015). Local urban warming, possible impacts and a resilience plan to climate change for the historical center of Athens, Greece. *Sustainable Cities and Society*, *19*, 281−291. Available from https://doi.org/10.1016/j.scs.2015.02.001.

Santamouris, M., Gaitani, N., Spanou, A., Saliari, M., Giannopoulou, K., Vasilakopoulou, K., & Kardomateas, T. (2012). Using cool paving materials to improve microclimate of urban areas - Design realization and results of the flisvos project. *Building and Environment*, *53*, 128−136. Available from https://doi.org/10.1016/j.buildenv.2012.01.022.

Santamouris, M., Papanikolaou, N., Livada, I., Koronakis, I., Georgakis, C., Argiriou, A. A., & Assimakopoulos, D. N. (2001). On the impact of urban climate on the energy consumption of buildings. *Solar Energy*. Available from https://doi.org/10.1016/S0038-092X(00)00095-5.

Santamouris, M., Synnefa, A., & Karlessi, T. (2011). Using advanced cool materials in the urban built environment to mitigate heat islands and improve thermal comfort conditions. *Solar Energy*, *85*(12), 3085−3102. Available from https://doi.org/10.1016/j.solener.2010.12.023.

Santamouris, M. (2023). *Urban overheating—Energy, environmental, and heat-health implications. Urban climate change and heat islands* (1st, pp. 165−225). Amsterdam: Elsevier. Available from 10.1016/B978-0-12-818977-1.00007-7.

Sarif, M. O., Ranagalage, M., Gupta, R. D., & Murayama, Y. (2022). Monitoring urbanization induced surface urban cool island formation in a South Asian megacity: A case study of Bengaluru, India (1989−2019). *Frontiers in Ecology and Evolution*, *10*. Available from https://doi.org/10.3389/fevo.2022.901156.

Schau-Noppel, H., Kossmann, M., & Buchholz, S. (2020). Meteorological information for climate-proof urban planning - The example of KLIMPRAX. *Urban Climate*, *32*. Available from https://doi.org/10.1016/j.uclim.2020.100614.

Shahmohammad, M., Hosseinzadeh, M., Dvorak, B., Bordbar, F., Shahmohammadmirab, H., & Aghamohammadi, N. (2022). Sustainable green roofs: A comprehensive review of influential factors. *Environmental Science and Pollution Research*, *29*(52), 78228−78254. Available from https://doi.org/10.1007/s11356-022-23405-x.

Shao, H., & Kim, G. (2022). A comprehensive review of different types of green infrastructure to mitigate urban heat islands: Progress, functions, and benefits. *Land*, *11*(10). Available from https://doi.org/10.3390/land11101792.

Shirani-bidabadi, N., Nasrabadi, T., Faryadi, S., Larijani, A., & Roodposhti, M. S. (2019). Evaluating the spatial distribution and the intensity of urban heat island using remote sensing, case study of Isfahan city in Iran. *Sustainable Cities and Society*, *45*(May 2018), 686−692. Available from https://doi.org/10.1016/j.scs.2018.12.005.

Shi, L., Luo, Z., Matthews, W., Wang, Z., Li, Y., & Liu, J. (2019). Impacts of urban microclimate on summertime sensible and latent energy demand for cooling in residential buildings of Hong Kong. *Energy*, *189*. Available from https://doi.org/10.1016/j.energy.2019.116208.

Simwanda, M., Ranagalage, M., Estoque, R. C., & Yuji, M. (2019). Spatial analysis of surface urban heat islands in four rapidly growing African cities. *Remote Sensing*, *11*(14). Available from https://doi.org/10.3390/rs11141645.

Skoulika, F., Santamouris, M., Kolokotsa, D., & Boemi, N. (2014). On the thermal characteristics and the mitigation potential of a medium size urban park in Athens, Greece. *Landscape and Urban Planning*, *123*, 73−86. Available from https://doi.org/10.1016/j.landurbplan.2013.11.002.

Soltanifard, H., Kashki, A., & Karami, M. (2022). Analysis of spatially varying relationships between urban environment factors and land surface temperature in Mashhad city, Iran. *The Egyptian Journal of Remote Sensing and Space Science*, *25*(4), 987−999. Available from https://doi.org/10.1016/j.ejrs.2022.10.003.

SSCIPO-LUCC. (1999). The international geosphere-biosphere programme: A study of global change (IGBP) scientific steering committee and international project office of LUCC unpublished content Land-use and land-cover change (LUCC) implementation strategy.

Sultana, S., & Satyanarayana, A. N. V. (2022). Urban heat island: land cover changes, management, and mitigation strategies. In A. Khan, et al. (Eds.), *Global Urban Heat Island Mitigation* (pp. 71−93). Elsevier. Available from https://doi.org/10.1016/B978-0-323-85539-6.00009-3.

Thanvisitthapon, N., Nakburee, A., Khamchiangta, D., & Saguansap, V. (2023). Climate change-induced urban heat Island trend projection and land surface temperature: A case study of Thailand's Bangkok metropolitan. *Urban Climate*, *49*, 101484. Available from https://doi.org/10.1016/J.UCLIM.2023.101484.

U.S. Department of Housing and Human Services. (2016). Washington, D.C. HUD Environmental Justice Strategy 2016−2020.

U.S. Environmental Protection Agency. (2023). Green infrastructure reduce urban heat island effect | https://www.epa.gov/green-infrastructure/reduce-urban-heat-island-effect.

UN Department of Economic and social affairs. (2019). Bern World population prospects 2019.

UN Environment Programme. (2022). Climate action As heatwaves blanket Europe, cities turn to nature for solutions https://www.unep.org/news-and-stories/story/heatwaves-blanket-europe-cities-turn-nature-solutions.

UN-HABITAT. (2020). United Nations Nairobi Unpublished content Global State of Metropolis 2020 − Population Data Booklet.

UNDESA. (2018). The World's Cities in 2018. World Urbanization Prospects: The 2018 Revision, 34.

van Eck, N. J., & Waltman, L. (2017). Citation-based clustering of publications using CitNetExplorer and VOSviewer. *Scientometrics*, 1053−1070. Available from https://doi.org/10.1007/s11192-017-2300-7.

Wang, Y., Berardi, U., & Akbari, H. (2016). Comparing the effects of urban heat island mitigation strategies for Toronto, Canada. *Energy and Buildings*, *114*, 2−19. Available from https://doi.org/10.1016/j.enbuild.2015.06.046.

Wang, Y., Liang, Z., Ding, J., Shen, J., Wei, F., & Li, S. (2022). Prediction of urban thermal environment based on multi-dimensional nature and urban form factors. *Atmosphere*, *13*(9), 1493. Available from https://doi.org/10.3390/atmos13091493.

Wang, Y., Li, Y., Xue, Y., Martilli, A., Shen, J., & Chan, P. W. (2020). City-scale morphological influence on diurnal urban air temperature. *Building and Environment*, *169*(September 2019), 106527. Available from https://doi.org/10.1016/j.buildenv.2019.106527.

Wang, X., Zhang, Y., & Yu, D. (2023). Exploring the relationships between land surface temperature and its influencing factors using multisource spatial big data: A case study in Beijing, China. *Remote Sensing*, *15*(7), 1783. Available from https://doi.org/10.3390/rs15071783.

World Health Organization. (2021). WHO Geneva unpublished content COP26 special report on climate change and health: The health argument for climate action. Geneva.

Wu, H., Huang, B., Zheng, Z., Ma, Z., & Zeng, Y. (2022). Spatial heterogeneity and temporal variation in urban surface albedo detected by high-resolution satellite data. *Remote Sensing*, *14*(23). Available from https://doi.org/10.3390/rs14236166.

Xia, H., Chen, Y., Song, C., Li, J., Quan, J., & Zhou, G. (2022). Analysis of surface urban heat islands based on local climate zones via spatiotemporally enhanced land surface temperature. *Remote Sensing of Environment*, *273*, 112972. Available from https://doi.org/10.1016/j.rse.2022.112972.

Yang, J., & Bou-Zeid, E. (2019). Scale dependence of the benefits and efficiency of green and cool roofs. *Landscape and Urban Planning*, *185*(March 2018), 127–140. Available from https://doi.org/10.1016/j.landurbplan.2019.02.004.

Yang, L., Yu, K., Ai, J., Liu, Y., Yang, W., & Liu, J. (2022). Dominant factors and spatial heterogeneity of land surface temperatures in urban areas: A case study in Fuzhou, China. *Remote Sensing*, *14*(5), 1266. Available from https://doi.org/10.3390/rs14051266.

Yola, L., Adekunle, T. O., & Ayegbusi, O. G. (2022). The impacts of urban configurations on outdoor thermal perceptions: Case studies of Flat Bandar Tasik Selatan and Surya Magna in Kuala Lumpur. *Buildings*, *12*(10). Available from https://doi.org/10.3390/buildings12101684.

Zhang, X., Wang, Y., Zhou, D., Yang, C., An, H., & Teng, T. (2022). Comparison of summer outdoor thermal environment optimization strategies in different residential districts in Xi'an, China. *Buildings*, *12*(9). Available from https://doi.org/10.3390/buildings12091332.

Zheng, X., Chen, L., & Yang, J. (2023). Simulation framework for early design guidance of urban streets to improve outdoor thermal comfort and building energy efficiency in summer. *Building and Environment*, *228*, 109815. Available from https://doi.org/10.1016/j.buildenv.2022.109815.

Zheng, S., Liu, L., Dong, X., Hu, Y., & Niu, P. (2022). Dominance of influencing factors on cooling effect of urban parks in different climatic regions. *International Journal of Environmental Research and Public Health*. Available from https://doi.org/10.3390/IJERPH192315496.

Zheng, Y., & Weng, Q. (2018). High spatial- and temporal-resolution anthropogenic heat discharge estimation in Los Angeles County, California. *Journal of Environmental Management*, *206*, 1274–1286. Available from https://doi.org/10.1016/j.jenvman.2017.07.047.

Zhou, S., Liu, D., Zhu, M., Tang, W., Chi, Q., Ye, S., Xu, S., & Cui, Y. (2022). Temporal and spatial variation of land surface temperature and its driving factors in Zhengzhou City in China from 2005 to 2020. *Remote Sensing*, *14*(17), 4281. Available from https://doi.org/10.3390/rs14174281.

Chapter 8

The impact of heat adaptation on low-income population

Sofia Natalia Boemi[1,2]

[1]*Department of Mechanical Engineering, Process Equipment Design Laboratory (PEDL), Aristotle University of Thessaloniki, Thessaloniki, Greece,* [2]*Cluster of Bioeconomy and Environment of Western Macedonia (CluBE), Kozani, Greece*

8.1 Introduction

The current energy crisis that strongly impacts energy prices worldwide has burdened many consumers in Europe and worldwide, with energy expenditure now taking up a substantial share of their income (OECD, 2023). Low-income households are the vulnerable households most affected by the rising prices of basic utilities (Metropolis, 2023).

In fact, according to Hamadeh et al. (2022), data from high-impact countries, the lower-middle-income population continued its growth in the 2022 analysis, increasing by 108.8 million to 2.39 billion people. Conversely, the middle-income population declined substantially by approximately 75.7 million from 1.38 billion in 2021 to 1.3 billion in the 2022 analysis. In fact, half of the global population lives on less than US$6.85 per person per day. In Europe, energy poverty still affects more than 50 million people (Energy Performance of Buildings [EPBD Recast], 2023).

The primary cause of the energy crisis was the massive increase in European gas prices due to the reduction of the Russian supply. Specifically, supplies of Russian gas have been cut by more than 80% in 2023. Wholesale electricity and gas prices have surged 15-fold since early 2021, severely affecting households and businesses. The primary replacement option is liquefied natural gas. However, its cost has doubled since Russia's February invasion of Ukraine. The problem could well worsen. Europe may experience its first winter without Russian gas, risking higher prices, gas shortages, and a major recession.

That causes a critical situation for households because the energy cost has skyrocketed. According to Metropolis (2023), electricity cost around €60/MWh in February 2012, and 10 years later, it has exceeded €300/MWh.

Mitigation and Adaptation of Urban Overheating. DOI: https://doi.org/10.1016/B978-0-443-13502-6.00008-7
245

Concerning natural gas consumption, it dropped by 20.1% in the period August—November 2022, compared with the average gas consumption for the same months between 2017 and 2021 (Eurostat, 2022). That might be justified by the relatively mild winter in Europe, which helped moderate demand.

In order to protect consumers and especially households from the direct impact of rising prices, national governments adopted measures such as VAT reductions on energy bills or bigger reductions for beneficiaries of subsidized electricity. According to a recent analysis by Bruegel (2023), between September 2021 and February 2023, financial allocation for these compensatory measures amounted to €657 billion across the EU and the UK. Many of these measures are nontargeted, which means that they benefit poor and rich households alike. While these compensatory measures (like energy vouchers and reduction of taxes on gas and electricity) can provide the relief that many households urgently need, they are diverting public finance from structural solutions such as energy efficiency and renewable energy solutions, which reduce households' energy bills and Europe's dependence on (imported) fossil fuels both in the short term and in the long term.

Subsidies on energy bills do not incentivize energy savings or the installation of renewable energy solutions such as rooftop solar, making investments in these measures less attractive. Subsidizing gas and oil prices thus negatively impacts the deployment of these technologies.

The crisis calls for additional behavioral changes to accompany long-term structural and technical solutions to lower gas demand and improve energy efficiency (OECD, 2023). For example, the Greek government started a behavioral change campaign to encourage consumers to reduce their energy consumption. Reducing energy use can not only help curb the current crisis, but if the household's reduction is sustained over time, it can support the transition to net zero.

Recognizing the need for a combination of compensation and structural solutions to the energy crisis, such as energy efficiency and renewable energy, the objective of this paper is to provide an understanding of how national governments can better help their citizens go through by targeting public resources at short-term, low-cost energy efficiency and renewable energy measures, which nonetheless are not in contradiction with long-term measures leading to net zero emissions and can be rolled out to a large number of households.

Diversifying energy sources and reducing energy demand are critical when discussing structural changes to tackle energy poverty and comfort vulnerability. These changes will take time. Improving buildings' energy efficiency, investing in clean-tech solutions, and leading policies are the immediate response to the current energy crisis.

In this chapter, the impact of heat adaptation on the low-income population will be discussed based on how low income, indoor air quality, proper living conditions, and property ownership affect health and vulnerability. Moreover, a discussion of how partial approaches to complex problems like energy poverty, which is linked directly with vulnerability, calls for the need to adopt multiactor approaches and especially behavior changes.

8.2   Energy poverty and vulnerability

The European Commission addressed the concept of energy poverty for the first time in 2009 with the publication of Directives 2009/72/EC and 2009/73/EC, which instructed member states to develop national action plans or other appropriate frameworks to tackle energy poverty (Fig. 8.1).

To support EU countries' efforts to tackle energy poverty in 2020, the commission published the Recommendation on Energy Poverty (Recommendation on Energy Poverty, 14C.E.). The recommendation is a part of the renovation wave strategy and provides guidance on adequate indicators to measure energy poverty. Also, it identifies measures targeting vulnerable groups via access to EU funding programs, including the first proposed energy poverty definition at the EU level.

Building on this recommendation, the Fit for 55 package proposed specific measures to identify key drivers of energy poverty risks for consumers, such as high energy prices, low household income, and poor energy-efficient buildings and appliances, considering structural solutions to vulnerabilities and underlying inequalities. The Fit for 55 package includes:

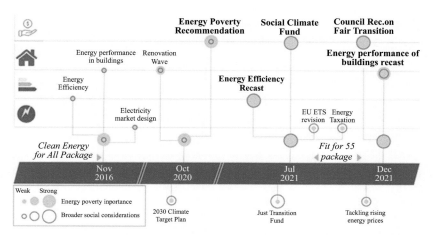

FIGURE 8.1   Presentation of the recent EU climate and energy policy initiatives and their focus on energy poverty and social aspects. *From Vandyck, T., Della Valle, N., Temursho, U., & Weitzel, M. (2023). EU climate action through an energy poverty lens.* Scientific Reports, *13(1). https://doi.org/10.1038/s41598-023-32705-2.*

- The Energy Efficiency Directive (EED) (recast) ("Energy Efficiency Directive [Recast]," n.d.) focuses on alleviating energy poverty and empowering consumers through strengthened requirements on awareness raising and information provision, including creating one-stop shops, technical and financial advice or assistance and consumer protection via out-of-court mechanisms. Furthermore, the proposal for a revised EED introduces an obligation for EU countries to prioritize energy efficiency improvement measures among vulnerable customers, people living in social housing, and, of course, people affected by energy poverty. Also, EED connects climatic conditions with the building's situation and the needed technical measures that should be adopted to address EP. Finally, in article 2 (48), the first official definition is mentioned "*Energy poverty means a household's inability, linked to non-affordability, to meet its basic energy supply needs and a lack of access to essential energy services to guarantee basic levels of comfort and health, a decent standard of living and health, including adequate heating, hot water, cooling, lighting, and energy to power appliances, in the relevant national context, existing social policy and other relevant policies, caused by one or a combination of the following factors: insufficient disposable income, high energy expenditures and poor energy efficiency of homes.*"
- The recast of the Energy Performance of Buildings (EPBD recast) (Energy Performance of Buildings [EPBD Recast], 2023) connects a building's inefficiency with energy poverty and social problems. The EPBD recast represents a comprehensive action plan to renovate the EU's buildings, bring down energy consumption, reduce building-related emissions, and reach a climate-neutral European building stock by 2050. Specifically, it aims to reduce GHG emissions and energy consumption, accelerate building renovation rates, and promote the uptake of renewable energy in buildings.
- RED III ("RED III [RED II Recast]," 2021), in which the "*local and regional authorities are key actors when it comes to bringing Europe closer to achieving its energy and climate objectives. Energy production at the local level is crucial to foster renewable energy production, reduce external energy dependence and decrease energy poverty rates.*"

All these policies address energy poverty and promote vulnerable consumers in an organized manner, both from a social perspective and through the energy sector. So even though energy poverty finally gained the main political momentum, the main concern now is how these efforts will be translated across different MS to define and protect vulnerable consumers.

However, even though according to the EU legislation, energy poverty is caused due to vulnerability, there was a recognition in the literature all these years that typical measurements of energy poverty may not be enough, as they do not sufficiently capture the impact or scope of energy vulnerability.

Vulnerability describes social, economic, physical, environmental, and institutional processes and structures to determine a system's or object's susceptibility and adaptation capacity regarding the way it reacts to dangers, such as the effects of climate change (Birkmann et al., 2012; Leal Filho et al., 2018). Concerning energy poverty and vulnerability, Bouzarovski and Petrova (2015) identified six main factors of energy vulnerability: access, affordability, flexibility, energy efficiency, needs, and practices. These are further expounded upon by Thomson, Bouzarovski, et al. (2017) to include issues such as the inability to invest in new energy infrastructures. Day et al. (2016) and Hearn et al. (2021) stressed the usefulness of defining energy vulnerability as being connected to capability deprivation. Therefore, vulnerability can also be introduced as a major indicator of energy poverty (Boemi & Papadopoulos, 2019a).

The selection of appropriate indicators and approaches for application or development are factors which surfaced in many debates (Boemi & Papadopoulos, 2019b; Gouveia et al., 2019). Therefore, measuring energy poverty is important in understanding the extent and depth of the problem (at a national and regional level) and assessing the impact of dedicated policies. In fact, indicators can be considered key instruments to measure energy poverty. The data collected by their measurement effectively captures different faces of energy poverty situations to build a comprehensive diagnosis. Moreover, the availability of data and selection of adequate indicators is one of the most important aspects of energy poverty analysis. The most commonly used energy poverty indicators are drawn by the EU Statistics on Income and Living Conditions (Energy Poverty Advisory Hub (EPAH), 2022), with two prevailing indicators used to capture EP being self-reported with their inherent limitations and most of them using old data.

The most recent set of indicators is suggested by the Energy Poverty Advisory Hub (Energy Poverty National Indicators: Insights for a More Effective Measuring, 28 C.E.). Its scope is to better describe the energy poverty problem and set the scene for planning and implementing energy poverty mitigation measures, mainly at a local level. That new set of indicators includes critical questions for a better understanding of vulnerability. They can be divided into two categories: the one gathered with in situ audits and help characterize the circumstances that lead to a situation of vulnerability and the second one that derives from the national census and directly depicts energy poverty.

Now the problem goes beyond simply experiencing energy poverty. Living in an adequately heated and cooled home should constitute a civil right, corroborated and enshrined by three United Nations Sustainable Development Goals (UN: Take Action for the Sustainable Development Goals, 2023). UN-SDG 1 and 7 set the background for energy equity and EP eradication analysis across the EU. In the EU, 6.9% of households are struggling to attain adequate warmth (2021 data), 19.2% of households are not comfortable during the summer (2012 data), and 6.4% of households cannot pay their utility bills on time (2021 data), resulting in other difficulties for low-income households and negatively affecting people's health and well-being (Energy Poverty Advisory Hub (EPAH), 2022).

From all these, it is obvious that energy poverty is a growing concern in the European Union, with visible detrimental consequences for the health and well-being of the population, even going beyond the private domain of the home to larger economic and political problems.

8.3 The quality of living conditions

As mentioned before, the key instrument to define energy poverty is indicators. Energy poverty is correlated with low household income, high energy costs, and energy-inefficient buildings, mainly in Europe (Boemi & Papadopoulos, 2019b). Nevertheless, in the developing world, energy poverty is associated with three pillars: household energy access toward cleaner and safer cooking fuels, improving electricity access and efficiency in electricity use, and greater use of renewable energy (Banerjee et al., 2021). All these are associated with the continually increasing energy costs affecting decreasing household incomes of vulnerable social groups in a mixture that can vary according to the prevailing conditions and the financial, regulatory, and fiscal policies. Furthermore, energy consumption by the residential sector, together with the commercial sector, positively influences health outcomes. Productive use of energy by residential households leads to an increased standard of living through higher demand for home lighting, cooking, television, refrigeration, drinking water, hot water, and other home appliances. In the following paragraphs, the impact of energy poverty will be discussed based on indoor air pollution, inefficient buildings, and their energy consumption and health.

8.3.1 Low-income and low indoor conditions

The first pillar of energy poverty and vulnerability is how low income can lead to energy poverty. Income is at the core of health inequalities since income is the most direct resource determining the quality of life and well-being. Adequate income helps people afford necessities, such as food, and live in a better living environment (Park et al., 2022).

A large body of empirical research supports the fact that higher energy poverty is associated with inferior health outcomes (Banerjee et al., 2021). The inability to access energy affects 733 million people globally (World Bank, 2022). Moreover, an increasing literature with empirical economic evidence shows a relationship between poor general and mental health and energy poverty (Awaworyi Churchill & Smyth, 2021; Brown & Vera-

Toscano, 2021; Kahouli, 2020; Rodriguez-Alvarez et al., 2019; Thomson, Snell et al., 2017).

Moreover, people living in inefficient housing conditions with low-quality employment are more likely to suffer from poor health (Brown & Vera-Toscano, 2021). Other studies (Basu, 2009; Kouis et al., 2021) have demonstrated a positive correlation between health aspects like cardiovascular and respiratory mortality and morbidity and high ambient temperature and all-cause.

Evidence for that is given in Fig. 8.2, which shows that the inability to keep the home adequately warm during winter increases the excess to winter mortality.

The most serious problem appears in the Mediterranean countries like Malta, Cyprus, and Spain. According to the WHO (2018), people with certain health conditions such as cardiovascular, pulmonary, or respiratory diseases and arthritis can become greatly sensitive to temperature extremes. Whether a house is too cold or hot, this may exacerbate their symptoms (Liddell et al., 2012). People with poor health conditions may also need

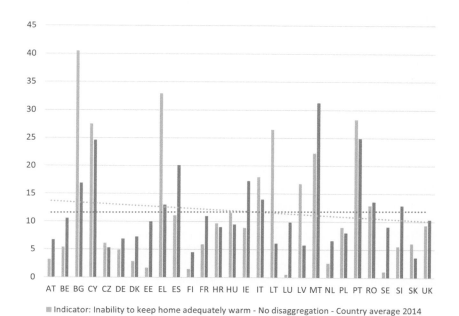

■ Indicator: Inability to keep home adequately warm - No disaggregation - Country average 2014

■ Indicator: Excess winter mortality/deaths 2014

FIGURE 8.2 A comparison between the inability to keep home adequately warm and excess to winter mortality and morbidity for 2014. *Data extracted by Energy Poverty Advisory Hub (EPAH). (2022). Inability to keep home adequately warm. https://indicator.energypoverty.eu/.*

more time at home, which means increased energy demand, especially for electricity (Büchs et al., 2018).

In the last few decades, several Health Impact Assessment studies like the ones mentioned by Kouis et al. (2021) and Baccini et al. (2008) have quantified the health burden as a result of higher temperatures in the Mediterranean regions and other regions by taking into consideration the current temperature-mortality exposure-response functions and future projections of regional temperatures under different climate change scenarios. The results showed that poor health might cause energy poverty, which may cause poor health (Brown & Vera-Toscano, 2021).

In the same direction, Osberghaus and Abeling (2022) showed a significant positive correlation between household income and technical adaptation. Specifically, if an equivalized household income increases by €1000 per month, the estimated probability of heat adaptation increases by two percentage points. Furthermore, the covariates show broad correlations with higher adaptation activities for households in warm regions, middle-aged and male household heads, homeowners, residents of large homes and top floors, people with a high body-mass index, lower levels of education, and those who report the experience of heat-related health problems. These results align with other studies on mitigation and adaptation responses to heatwaves (Demski et al., 2017) and on implementing heat adaptation (De Cian et al., 2019; Kussel, 2018; Murtagh et al., 2019).

Additionally, there is strong epidemiological evidence of human adaptation to increased temperatures. Recent observational studies of health records and historical temperatures showed a decreasing trend in the heat-related impact on mortality and morbidity (Chung et al., 2017; Kouis et al., 2021).

On the other hand, the increasing temperature leads to increasing heatwaves (Kouis et al., 2021). As warned by climate change experts, the world has experienced increasing heatwave intensity, frequency, and duration, and this trend is projected to increase with climate change (He et al., 2023; Marcotullio et al., 2022). Heatwaves, or prolonged high temperatures, are important because they negatively affect human well-being. These projections are alarming for the Mediterranean region since the global warming effects via extreme weather conditions are expressed in a more intensive way there than in other regions (Diffenbaugh & Giorgi, 2012).

Unfortunately, during heatwaves, excess deaths increase with age, and the excess mortality of women tends to be higher than that of men (Brücker, 2005; He et al., 2023) in vulnerable populations, including people with respiratory diseases and children (D'Ippoliti et al., 2010; Patel et al., 2019), and in low-income populations with poor-quality housing, lack of air conditioning, and lack of access to health and social services (Michelozzi et al., 2005). For example, they increase hospital admissions for renal and respiratory diseases (Kovats et al., 2004). Specifically, during the 2003 heatwave in Europe, the peak temperature reached 101.3°F (38.5°C) in the United

Kingdom, and there were more than 50,000 excess deaths estimated in Europe in August 2003 (He et al., 2023); most of the deceased were elderly persons (Garcia-Herrera et al., 2010). During the 2010 heatwave in Russia, the daytime temperature in Moscow reached 100.8°F (38.2°C), and the excess deaths were close to 11,000 between July 6 and August 18, 2010.

8.3.2 Energy-inefficient buildings and energy poverty

Housing characteristics are one of the main factors connected with energy poverty and consumer vulnerability. Specifically, as housing and energy are closely linked, the housing factor becomes an important perspective in examining energy poverty. Decent, warm, comfortable homes require basic energy services, including space heating and cooling, lighting, water heating, cooking, and electricity (Practical Action, 2019). On the one hand, the necessary energy needs of a household are highly correlated with the housing characteristics, such as energy efficiency, building envelope, and thermal insulation (Berger & Höltl, 2019; Burholt & Windle, 2006; Charlier, 2015; Morrison & Shortt, 2008; Seebauer et al., 2019). In general, inefficient energy systems and insulation can increase a household's energy costs to some extent (Papada & Kaliampakos, 2020).

When households cannot afford high energy costs, they choose to reduce their energy consumption. Energy poverty occurs when household energy consumption falls below the energy required to maintain basic decent living conditions. On the other hand, when housing costs account for a high share of household income, this affects the ability of households to pay their energy bills, and households are forced to reduce their energy expenditure, resulting in a situation of underconsumption of energy (Eisfeld & Seebauer, 2022). Furthermore, conversely, unfavorable housing conditions and poor housing quality are often associated with energy poverty; for example, energy-poor households may often experience damp houses and mold on the walls and floors (Healy & Clinch, 2004).

Based on the EPAH data (Fig. 8.3), people suffering from the consequences of poor construction practices of dwellings and humid indoor conditions may not be directly connected to situations of energy poverty. However, it can also be a consequence of an inability to keep the house adequately warm.

As Fig. 8.3 shows, there is a common trend between inefficient living conditions and those that cannot afford to pay their energy bills. In this context, it is known that many people living in energy-inefficient and cold properties struggle to meet their energy needs for comfort and warmth, and they risk developing cold-related health illnesses. Similarly, climate change is increasing the duration of hot weather, leading to a higher risk of energy poverty, overheating, and health problems in most vulnerable households living in energy-inefficient houses (Evola et al., 2014; Kapsalaki et al., 2012).

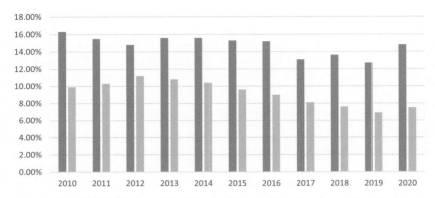

FIGURE 8.3 A comparison between the percentage of the population that live in dwellings with the presence of a leak, damp, and rot and the inability to keep home adequately warm for EU-27. *From Energy Poverty Advisory Hub (EPAH). (2022).* Inability to keep home adequately warm. *https://indicator.energypoverty.eu/.*

Energy poverty literature has shown that higher winter mortality rates are observed in properties built before 1850 having lower energy efficiency ratings. Specifically, 28.2% of winter mortality can appear to be 15% in properties built after 1980 (Semprini et al., 2017). In addition, a poor level of maintenance of indoor housing is associated with damp homes with mold, which leads to a 45% increase in respiratory problems, as well as wheezing, colds, and viral diseases (Santamouris et al., 2007).

In that direction, the recently introduced EPBD mentions that improving the energy efficiency of buildings and appliances has huge potential to ease the energy poverty problem. However, affordability is usually an aspect that is not considered to the needed degree and thus remains unavailable for energy-poor people or their most vulnerable segments. Most vulnerable households do not have the financial capacity to carry out any building improvement or equipment replacement.

Deep renovations are highly costly interventions requiring complex technical and organizational support. Combining that with the fact that approximately 70% of the existing building stock in Mediterranean countries was built during 1960–80 (Semprini et al., 2017), 50 million people in the EU live in energy poverty, unable to adequately heat, cool, or light their homes, and over 20% of poor households live in a dwelling that has mold, damp, or rot, building renovations constitute a long-term solution to address the issue of energy poverty and provide affordable houses for most people vulnerable.

An even more serious challenge is the case of owner-occupied multifamily buildings with inhabitants from mixed social backgrounds. The renovation of such a building, particularly with a significant share of energy-poor homeowners, requires various kinds of support measures. Specifically, for a rental multifamily dwelling, the landlord—tenant problem can be a barrier to energy retrofitting and renovation (Charlier, 2015; Mangold et al., 2018). The resident-owned multifamily dwellings also have internal barriers to larger investments in energy retrofitting (Bardhan et al., 2014). No consensus exists on which data or parameters should be included in optimizing renovation projects. For instance, renovations financed by increased rents in socioeconomically disadvantaged areas risk aggravating societal inequities (Mangold et al., 2018). Countries differ in the policy support they provide for homeownership and the percentage of households in different forms of housing tenure (Stamsø, 2010).

Moreover, housing asset is significantly associated with self-rated health (Costa-Font, 2008) and mortality (Connolly et al., 2010). One of the important issues is the combined effects of housing and income on health.

All in all, to achieve a highly energy-efficient and decarbonized building stock, access to sufficient grants and funding is crucial. It is the only way to meet the 2030 and 2050 energy efficiency targets and reduce the number of people living in energy poverty. Moreover, long-term renovation strategies should become even stronger. These policies should have a strong focus on financing and ensuring that appropriately skilled workers are available for carrying out building renovations, as well as on tackling energy poverty, ensuring electrical and fire safety and improving the energy performance of worst-performing buildings. Finally, the benefits of a building retrofit should include reducing energy consumption, CO_2 emissions, and resources and positively affecting socioeconomic aspects. That would contribute to the identification and location of the most vulnerable social sectors and those suffering from energy poverty, thus helping to fight against social inequalities that may arise from applying the different climate action measures.

8.4   Addressing heat vulnerability: recommendations and solutions

Low-income areas in the developing world are expected to grow in population, contributing to the disparities in heat wave exposure, and due to less income, they will not be able to adopt any heat-adapting technologies and healthcare for those affected. Literature evidence shows that populations contributing the least to anthropogenic climate change will continue to bear climate change impacts, even though, historically, higher-income countries contribute to most greenhouse gases (Open Access Government, 2023).

On the other hand, heat vulnerability currently creates severe challenges for urban areas, especially in large cities with extremely high population densities and rapidly growing transport infrastructures. Moreover, urban overheating caused by the UHI phenomenon is one of the driving factors of increased cooling energy consumption, contributing to global greenhouse gas emissions and further intensifying the urban heat island effect, creating the feedback loop. Therefore, it is essential to adopt heat mitigation technologies and strategies to reduce the buildings' cooling loads (Adilkhanova et al., 2023). So, although higher-income areas generally have better access to climate change adaptations, they will still face rolling blackouts or brownouts as electricity demand continues to rise worldwide.

To address any heat vulnerability issues, the measures that should be adopted can be divided into four main categories based on their impact.

8.4.1 Political measures

The Renovation Wave and the Recovery and Resilience Plans are part of the answer. Even though the ambitious climate targets that the EU has set will require the transition to speed up in a way that may increase inequalities with regard to access to energy, any investments, including the Social Climate Fund, should address and deliver a well-being economy based on the EU's fundamental principles of democratic values and energy democracy. Furthermore, any financial instrument should support large-scale EU investments in building renovations and affordable energy-efficient social housing.

8.4.2 Social engagement and behavioral changes

The role of civil society should be strengthened by providing direct assistance to the most vulnerable. That will allow energy-poor households to overcome barriers linked to information searching to change the status quo and make enrollment in an energy efficiency program easier, including access to energy renovation schemes or switching to a better contract or energy supplier. Moreover, reminders are also efficient tools for encouraging households to consume less energy. There are other behavioral principles that can be followed in the short and long term (Fig. 8.4), such as successful informational campaigns, which tend to provide a set of clear and actionable guidelines and social norms which can influence information campaigns' effectiveness (OECD, 2023).

- *Ways of energy production.* The decarbonization of energy systems and fast penetration of renewables allow for solutions for the problem of energy poverty during low-carbon energy transitions. Moving toward a low-carbon future from the point of view of the cost burden for low-income, energy-poor and vulnerable households will not cause any problems but allow for the use of cheaper and more effective clean energy sources in the future to reduce energy poverty. As renewable energy sources have almost zero marginal operation costs and lower market clearing prices, these processes can positively impact vulnerable

Behavioral principles affecting short -term action	Possible behavioral interventions
Social norms – tendency to conform to the behavior and beliefs of others in the social network. For example, individuals are more likely to conserve energy if they think others in their network are doing it.	Social norms interventions (e.g. energy savings motivated by shared sense of solidarity for Ukraine).
Status quo bias – aversion to changing one's habits, such as habitual heating temperatures at home.	Providing feedback on consumption; changing defaults.
Intention – action gap – discrepancy between one's intentions and actions, such as an awareness of climate risks which does not translate in specific energy conservation action.	Goal setting
Forgetfulness – attention is limited and easily distracted, such as forgetting lights on at home.	Reminders.
Costly information acquisition – knowledge of which energy behaviors to adopt might not be easily available to consumers and they will not pro-actively search for it.	Communication and awareness campaigns
Bevahioral principles affecting long-term action	Possible behavioral interventions
Optimism bias – predisposition to systematically overestimate the probability of positive events and underestimate the probability of negative events. For example, tendency to attribute the highest level of severity to environmental risks that are geographically far.	Communication and awareness campaigns
Prevent bias – tendency to underestimate risks perceived to be far in the future as climate change.	Commitment devices and goals
Risk misperceptions – tendency to misinterpret the likelihood of complex events, such as climate disasters or energy shocks.	Changes in defaults; communication and awareness campaigns

FIGURE 8.4 Examples of behavioral barriers that can affect energy consumption in the short and long term and possible responses. *From OECD. (2023).* Confronting the energy crisis: Changing behaviours to reduce energy consumption. *https://oecdecoscope.blog/2023/04/17/confronting-the-energy-crisis-changing-behaviours-to-reduce-energy-consumption/.*

consumers if the costs of low-carbon interventions are not fully and equally covered by end consumers through energy bills. Numerous scholars proved that renewable energy sources have already reached cost parity with conventional energy sources. Moreover, they will even overcome them soon since their capacities will continue to develop by providing low-cost advanced solar photovoltaic or windmill air turbine designs.

- *Energy upgrade of buildings.* It is clear that household energy efficiency improvements allow for addressing energy poverty issues. Though there are many energy subsidies to help vulnerable households deal with energy poverty, these measures allow for short-term results in combating energy poverty. However, policies aiming to increase energy efficiency in homes provide long-term results and increase the quality of housing and reduce the burden of energy expenditures for low-income people. However, for any measure to be successful, it is necessary to highlight that energy efficiency improvements positively impact a fiscal situation in the country by dropping expenditures on energy-related subsidies, allowing more spending for needed public investments. In addition, diminished energy subsidy expenditures can allow accumulating financial resources to support energy renovation, education, and public health and help households not make a trade-off between heating and eating.

All in all, all these measures should be accompanied by new educational opportunities. Building upon well-established human capital is important. Findings for the developing world (Apergis et al., 2022) provided insights into the education-energy poverty nexus, supporting a negative and statistically significant relationship between education and energy poverty. Specifically, results indicated that households with a low level of education had restricted access to clean energy forms, such as electricity, and mostly use black or gray energy that generates high levels of carbon emissions. This study also uncovered positive and statistically significant relationships between energy poverty with other covariates, such as economic growth, energy intensity, and electricity consumption. Moreover, it unraveled environmental pollution's negative and statistically significant effect on energy poverty. While for developed countries, findings (Boemi et al., 2017) showed a correlation between educational level and energy poverty and financial crisis, meaning that the extent of education could affect the reaction of individuals against energy poverty during the financial crisis.

8.5 Conclusion

Efforts to tackle energy poverty and mitigate vulnerability have far proved insufficient. The reduction of the energy use of a household can help curb the current energy and climate crisis. However, if sustained over time, it can support the transition to net zero and carbon neutrality. What is needed is the identification of the psychological factors that influence energy conservation behavior because changing behavior results in a change of habits, energy consumption, and health. Therefore, tackling the problem of energy poverty is one of the most urgent challenges that policymakers face. So far, several measures aimed at addressing the structural roots of energy poverty have been proposed and adopted. Nevertheless, one should bear in mind that any vulnerability mitigation technologies and strategies that support the change of the energy model or mitigate the energy and climate crisis could be effective without the right policy and find tools. After all, energy poverty is a problem of how energy markets operate and are regulated.

References

Adilkhanova, I., Santamouris, M., & Yun, G. Y. (2023). Coupling urban climate modeling and city-scale building energy simulations with the statistical analysis: Climate and energy implications of high albedo materials in Seoul. *Energy and Buildings*, *290*, 113092. Available from https://doi.org/10.1016/j.enbuild.2023.113092.

Apergis, N., Polemis, M., & Soursou, S.-E. (2022). Energy poverty and education: Fresh evidence from a panel of developing countries. *Energy Economics*, *106*, 105430. Available from https://doi.org/10.1016/j.eneco.2021.105430.

Awaworyi Churchill, S., & Smyth, R. (2021). Energy poverty and health: Panel data evidence from Australia. *Energy Economics*, *97*. Available from https://doi.org/10.1016/j.eneco.2021.105219, http://www.elsevier.com/inca/publications/store/3/0/4/1/3/.

Baccini, M., Biggeri, A., Accetta, G., Kosatsky, T., Katsouyanni, K., Analitis, A., Anderson, H. R., Bisanti, L., D'Iippoliti, D., Danova, J., Forsberg, B., Medina, S., Paldy, A., Rabczenko, D., Schindler, C., & Michelozzi, P. (2008). Heat effects on mortality in 15 European cities. *Epidemiology*, *19*(5), 711–719. Available from https://doi.org/10.1097/EDE.0b013e318176bfcd.

Banerjee, R., Mishra, V., & Maruta, A. A. (2021). Energy poverty, health and education outcomes: Evidence from the developing world. *Energy Economics*, *101*, 105447. Available from https://doi.org/10.1016/j.eneco.2021.105447.

Bardhan, A., Jaffee, D., Kroll, C., & Wallace, N. (2014). Energy efficiency retrofits for U.S. housing: Removing the bottlenecks. *Regional Science and Urban Economics*, *47*(1), 45–60. Available from https://doi.org/10.1016/j.regsciurbeco.2013.09.001, http://www.elsevier.com/inca/publications/store/5/0/5/5/7/0/.

Basu, R. (2009). High ambient temperature and mortality: A review of epidemiologic studies from 2001 to 2008. *Environmental Health: A Global Access Science Source*, *8*(1). Available from https://doi.org/10.1186/1476-069X-8-40.

Berger, T., & Höltl, A. (2019). Thermal insulation of rental residential housing: Do energy poor households benefit? A case study in Krems, Austria. *Energy Policy*, *127*, 341–349. Available from https://doi.org/10.1016/j.enpol.2018.12.018, http://www.journals.elsevier.com/energy-policy/.

Birkmann, J., Bach, C., & Vollmer, M. (2012). Tools for resilience building and adaptive spatial governance. *Raumforschung und Raumordnung | Spatial Research and Planning*, *70*(4). Available from https://doi.org/10.1007/s13147-012-0172-0.

Boemi, S. N., & Papadopoulos, A. M. (2019a). Energy poverty and energy efficiency improvements: A longitudinal approach of the Hellenic households. *Energy and Buildings*, *197*, 242–250. Available from https://doi.org/10.1016/j.enbuild.2019.05.027.

Boemi, S. N., & Papadopoulos, A. M. (2019b). Monitoring energy poverty in Northern Greece: The energy poverty phenomenon. *International Journal of Sustainable Energy*, *38*(1), 74–88. Available from https://doi.org/10.1080/14786451.2017.1304939, http://www.tandfonline.com/toc/gsol20/current.

Boemi, S. N., Avdimiotis, S., & Papadopoulos, A. M. (2017). Domestic energy deprivation in Greece: A field study. *Energy and Buildings*, *144*, 167–174. Available from https://doi.org/10.1016/j.enbuild.2017.03.009.

Bouzarovski, S., & Petrova, S. (2015). A global perspective on domestic energy deprivation: Overcoming the energy poverty-fuel poverty binary. *Energy Research and Social Science*, *10*, 31–40. Available from https://doi.org/10.1016/j.erss.2015.06.007, http://www.journals.elsevier.com/energy-research-and-social-science/.

Brown, H., & Vera-Toscano, E. (2021). Energy poverty and its relationship with health: Empirical evidence on the dynamics of energy poverty and poor health in Australia. *SN Business & Economics, 1*(10). Available from https://doi.org/10.1007/s43546-021-00149-3.

Brücker, G. (2005). Vulnerable populations: Lessons learnt from the summer 2003 heat waves in Europe. *Eurosurveillance, 10*(7), 1−2. Available from https://doi.org/10.2807/esm.10.07.00551-en.

Bruegel. (2023). *The fiscal side of Europe's energy crisis: The facts, problems and prospects.*

Büchs, M., Bahaj, A. B., Blunden, L., Bourikas, L., Falkingham, J., James, P., Kamanda, M., & Wu, Y. (2018). Sick and stuck at home − How poor health increases electricity consumption and reduces opportunities for environmentally-friendly travel in the United Kingdom. *Energy Research and Social Science, 44*, 250−259. Available from https://doi.org/10.1016/j.erss.2018.04.041, http://www.journals.elsevier.com/energy-research-and-social-science/.

Burholt, V., & Windle, G. (2006). Keeping warm? Self-reported housing and home energy efficiency factors impacting on older people heating homes in North Wales. *Energy Policy, 34*(10), 1198−1208. Available from https://doi.org/10.1016/j.enpol.2004.09.009.

Charlier, D. (2015). Energy efficiency investments in the context of split incentives among French households. *Energy Policy, 87*, 465−479. Available from https://doi.org/10.1016/j.enpol.2015.09.005, http://www.journals.elsevier.com/energy-policy/.

Chung, Y., Noh, H., Honda, Y., Hashizume, M., Bell, M. L., Guo, Y. L. L., & Kim, H. (2017). Temporal changes in mortality related to extreme temperatures for 15 cities in Northeast Asia: Adaptation to heat and maladaptation to cold. *American Journal of Epidemiology, 185*(10), 907−913. Available from https://doi.org/10.1093/aje/kww199, http://aje.oxfordjournals.org/.

Connolly, S., O'Reilly, D., & Rosato, M. (2010). House value as an indicator of cumulative wealth is strongly related to morbidity and mortality risk in older people: A census-based cross-sectional and longitudinal study. *International Journal of Epidemiology, 39*(2), 383−391. Available from https://doi.org/10.1093/ije/dyp356.

Costa-Font, J. (2008). Housing assets and the socio-economic determinants of health and disability in old age. *Health & Place, 14*(3), 478−491. Available from https://doi.org/10.1016/j.healthplace.2007.09.005.

Day, R., Walker, G., & Simcock, N. (2016). Conceptualising energy use and energy poverty using a capabilities framework. *Energy Policy, 93*, 255−264. Available from https://doi.org/10.1016/j.enpol.2016.03.019, http://www.journals.elsevier.com/energy-policy/.

De Cian, E., Pavanello, F., Randazzo, T., Mistry, M. N., & Davide, M. (2019). Households' adaptation in a warming climate. *Air conditioning and thermal insulation choices. Environmental Science and Policy, 100*, 136−157. Available from https://doi.org/10.1016/j.envsci.2019.06.015, http://www.elsevier.com/wps/find/journaldescription.cws_home/601264/description#description.

Demski, C., Capstick, S., Pidgeon, N., Sposato, R. G., & Spence, A. (2017). Experience of extreme weather affects climate change mitigation and adaptation responses. *Climatic Change, 140*(2), 149−164. Available from https://doi.org/10.1007/s10584-016-1837-4, http://www.wkap.nl/journalhome.htm/0165-0009.

Diffenbaugh, N. S., & Giorgi, F. (2012). Climate change hotspots in the CMIP5 global climate model ensemble. *Climatic Change, 114*(3−4), 813−822. Available from https://doi.org/10.1007/s10584-012-0570-x.

Directorate-General for Energy. *Energy Poverty National Indicators: Insights for a more effective measuring.* Directorate-General for Energy unpublished content. https://energy-poverty.

ec.europa.eu/discover/publications/publications/energy-poverty-national-indicators-insights-more-effective-measuring_en.

D'Ippoliti, D., Michelozzi, P., Marino, C., de'Donato, F., Menne, B., Katsouyanni, K., Kirchmayer, U., Analitis, A., Medina-Ramón, M., Paldy, A., Atkinson, R., Kovats, S., Bisanti, L., Schneider, A., Lefranc, A., Iñiguez, C., & Perucci, C. A. (2010). The impact of heat waves on mortality in 9 European cities: Results from the EuroHEAT project. *Environmental Health*, *9*(1). Available from https://doi.org/10.1186/1476-069x-9-37.

Eisfeld, K., & Seebauer, S. (2022). The energy austerity pitfall: Linking hidden energy poverty with self-restriction in household use in Austria. *Energy Research & Social Science*, *84*, 102427. Available from https://doi.org/10.1016/j.erss.2021.102427.

Energy Efficiency Directive (recast). (2021). *European Parliament and of the Council. Energy Efficiency Directive (EED) (recast), 2021 Proposal for a Directive of the European Parliament and of the Council on Energy Efficiency (recast)*. https://www.europarl.europa.eu/doceo/document/TA-9-2022-0315_EN.html.

Energy Poverty Advisory Hub (EPAH). (2022). 2023 5 16 Inability to keep home adequately warm. https://indicator.energypoverty.eu/.

EU Commission. Recommendation on energy poverty. EU Commission unpublished content. https://eur-lex.europa.eu/legal-content/EN/TXT/?uri = CELEX:32020H1563&qid = 1606124119302.

European Parliament. (2023). *Energy performance of buildings*. https://www.europarl.europa.eu/doceo/document/TA-9-2023-0068_EN.html.

Eurostat. (2022). *EU gas consumption down by 20.1%*. https://ec.europa.eu/eurostat/web/products-eurostat-news/w/DDN-20221220-3.

Evola, G., Margani, G., & Marletta, L. (2014). Cost-effective design solutions for low-rise residential Net ZEBs in Mediterranean climate. *Energy and Buildings*, *68*, 7−18. Available from https://doi.org/10.1016/j.enbuild.2013.09.026.

Garcia-Herrera, R., Díaz, J., Trigo, R. M., Luterbacher, J., & Fischer, E. M. (2010). A review of the European summer heat wave of 2003. *Critical Reviews in Environmental Science and Technology*, *40*(4), 267−306. Available from https://doi.org/10.1080/10643380802238137, http://www.tandf.co.uk/journals/titles/10643389.asp.

Gouveia, J. P., Palma, P., & Simoes, S. G. (2019). Energy poverty vulnerability index: A multi-dimensional tool to identify hotspots for local action. *Energy Reports*, *5*, 187−201. Available from https://doi.org/10.1016/j.egyr.2018.12.004, http://www.journals.elsevier.com/energy-reports/.

Hamadeh, N., Rompaey, C., Metreau, E., & Eapen, S.G. (2022). *New World Bank country classifications by income level: 2022−2023*. World Bank Blogs.

He, R., Tsoulou, I., Thirumurugesan, S., Morgan, B., Gonzalez, S., Plotnik, D., Senick, J., Andrews, C., & Mainelis, G. (2023). Effect of heatwaves on PM2.5 levels in apartments of low-income elderly population. A case study using low-cost air quality monitors. *Atmospheric Environment*, *301*, 119697. Available from https://doi.org/10.1016/j.atmosenv.2023.119697.

Healy, J. D., & Clinch, J. P. (2004). Quantifying the severity of fuel poverty, its relationship with poor housing and reasons for non-investment in energy-saving measures in Ireland. *Energy Policy*, *32*(2), 207−220. Available from https://doi.org/10.1016/S0301-4215(02)00265-3.

Hearn, A. X., Sohre, A., & Burger, P. (2021). Innovative but unjust? Analysing the opportunities and justice issues within positive energy districts in Europe. *Energy Research & Social Science*, *78*, 102127. Available from https://doi.org/10.1016/j.erss.2021.102127.

Kahouli, S. (2020). An economic approach to the study of the relationship between housing hazards and health: The case of residential fuel poverty in France. *Energy Economics*, *85*, 104592. Available from https://doi.org/10.1016/j.eneco.2019.104592.

Kapsalaki, M., Leal, V., & Santamouris, M. (2012). A methodology for economic efficient design of Net Zero Energy Buildings. *Energy and Buildings*, *55*, 765−778. Available from https://doi.org/10.1016/j.enbuild.2012.10.022.

Kouis, P., Psistaki, K., Giallouros, G., Michanikou, A., Kakkoura, M. G., Stylianou, K. S., Papatheodorou, S. I., & Paschalidou, A. (2021). Heat-related mortality under climate change and the impact of adaptation through air conditioning: A case study from Thessaloniki, Greece. *Environmental Research*, *199*. Available from https://doi.org/10.1016/j. envres.2021.111285, http://www.elsevier.com/inca/publications/store/6/2/2/8/2/1/index.htt.

Kovats, R. S., Hajat, S., & Wilkinson, P. (2004). Contrasting patterns of mortality and hospital admissions during hot weather and heat waves in Greater London, UK. *Occupational and Environmental Medicine*, *61*(11), 893−898. Available from https://doi.org/10.1136/ oem.2003.012047.

Kussel, G. (2018). Adaptation to climate variability: Evidence for German households. *Ecological Economics*, *143*, 1−9. Available from https://doi.org/10.1016/j.ecole-con.2017.06.039, http://www.elsevier.com/inca/publications/store/5/0/3/3/0/5.

Leal Filho, W., Echevarria Icaza, L., Neht, A., Klavins, M., & Morgan, E. A. (2018). Coping with the impacts of urban heat islands. A literature based study on understanding urban heat vulnerability and the need for resilience in cities in a global climate change context. *Journal of Cleaner Production*, *171*, 1140−1149. Available from https://doi.org/10.1016/j. jclepro.2017.10.086.

Liddell, C., Morris, C., McKenzie, S. J. P., & Rae, G. (2012). Measuring and monitoring fuel poverty in the UK: National and regional perspectives. *Energy Policy*, *49*, 27−32. Available from https://doi.org/10.1016/j.enpol.2012.02.029.

Mangold, M., Österbring, M., Overland, C., Johansson, T., & Wallbaum, H. (2018). Building ownership, renovation investments, and energy performance-A study of multi-family dwellings in gothenburg. *Sustainability*, *10*(5), 1684. Available from https://doi.org/10.3390/ su10051684.

Marcotullio, P. J., Keßler, C., & Fekete, B. M. (2022). Global urban exposure projections to extreme heatwaves. *Frontiers in Built Environment*, *8*. Available from https://doi.org/ 10.3389/fbuil.2022.947496, http://journal.frontiersin.org/journal/built-environment.

Metropolis, B. (2023). *Energy crisis and vulnerability*. https://www.barcelona.cat/metropolis/en/ contents/energy-crisis-and-vulnerability.

Michelozzi, P., de Donato, F., Bisanti, L., Russo, A., Cadum, E., DeMaria, M., D'Ovidio, M., Costa, G., & Perucci, C. A. (2005). The impact of the summer 2003 heat waves on mortality in four Italian cities. *Eurosurveillance*, *10*(7), 11−12. Available from https://doi.org/ 10.2807/esm.10.07.00556-en.

Morrison, C., & Shortt, N. (2008). Fuel poverty in Scotland: Refining spatial resolution in the Scottish Fuel Poverty Indicator using a GIS-based multiple risk index. *Health & Place*, *14* (4), 702−717. Available from https://doi.org/10.1016/j.healthplace.2007.11.003.

Murtagh, N., Gatersleben, B., & Fife-Schaw, C. (2019). Occupants' motivation to protect residential building stock from climate-related overheating: A study in southern England. *Journal of Cleaner Production*, *226*, 186−194. Available from https://doi.org/10.1016/j.jcle-pro.2019.04.080, https://www.journals.elsevier.com/journal-of-cleaner-production.

OECD. (2023). *Confronting the energy crisis: Changing behaviours to reduce energy consumption.* https://oecdecoscope.blog/2023/04/17/confronting-the-energy-crisis-changing-behaviours-to-reduce-energy-consumption/.

Open Access Government. (2023). *Lowest-income populations face 40% more exposure to heat waves.* https://www.openaccessgovernment.org/heat-waves-lowest-income/129609/.

Osberghaus, D., & Abeling, T. (2022). Heat vulnerability and adaptation of low-income households in Germany. *Global Environmental Change, 72,* 102446. Available from https://doi.org/10.1016/j.gloenvcha.2021.102446.

Papada, L., & Kaliampakos, D. (2020). Being forced to skimp on energy needs: A new look at energy poverty in Greece. *Energy Research & Social Science, 64,* 101450. Available from https://doi.org/10.1016/j.erss.2020.101450.

Park, G.-R., Grignon, M., Young, M., & Dunn, J. R. (2022). How do housing asset and income relate to mortality? A population-based cohort study of 881220 older adults in Canada. *Social Science & Medicine, 314,* 115429. Available from https://doi.org/10.1016/j.socscimed.2022.115429.

Patel, D., Jian, L., Xiao, J., Jansz, J., Yun, G., Lin, T., & Robertson, A. (2019). Joint effects of heatwaves and air quality on ambulance services for vulnerable populations in Perth, western Australia. *Environmental Pollution, 252,* 532−542. Available from https://doi.org/10.1016/j.envpol.2019.05.125, https://www.journals.elsevier.com/environmental-pollution.

Practical Action. (2019). *Poor People's Energy Outlook.* https://practicalaction.org/poor-peoples-energy-outlook/.

RED III (RED II recast). (2021). *European Parliament and of the Council, 2021. Recommendation on energy poverty, 2020. Commission Recommendation (EU) 2020/1563 of 14 October 2020 on energy poverty\nRED III (recast RED2 directive).* https://eur-lex.europa.eu/legal-content/EN/TXT/?uri = CELEX%3A52021PC0557.

Rodriguez-Alvarez, A., Orea, L., & Jamas, T. (2019). Fuel poverty and well-being: A consumer theory and stochastic frontier approach. *Energy Policy, 131,* 22−32. Available from https://doi.org/10.1016/j.enpol.2019.04.031, https://www.sciencedirect.com/science/article/pii/S0301421519302800?via%3Dihub.

Santamouris, M., Kapsis, K., Korres, D., Livada, I., Pavlou, C., & Assimakopoulos, M. N. (2007). On the relation between the energy and social characteristics of the residential sector. *Energy and Buildings, 39*(8), 893−905. Available from https://doi.org/10.1016/j.enbuild.2006.11.001.

Seebauer, S., Friesenecker, M., & Eisfeld, K. (2019). Integrating climate and social housing policy to alleviate energy poverty: An analysis of targets and instruments in Austria. *Energy Sources, Part B: Economics, Planning and Policy, 14*(7−9), 304−326. Available from https://doi.org/10.1080/15567249.2019.1693665, http://www.tandf.co.uk/journals/titles/15567249.asp.

Semprini, G., Gulli, R., & Ferrante, A. (2017). Deep regeneration vs shallow renovation to achieve nearly Zero Energy in existing buildings: Energy saving and economic impact of design solutions in the housing stock of Bologna. *Energy and Buildings, 156,* 327−342. Available from https://doi.org/10.1016/j.enbuild.2017.09.044.

Stamsø, M. A. (2010). Housing and welfare policy - Changing relations? A cross-national comparison. *Housing, Theory and Society, 27*(1), 64−75. Available from https://doi.org/10.1080/14036090902764216.

Thomson, H., Snell, C., & Bouzarovski, S. (2017). Health, well-being and energy poverty in Europe: A comparative study of 32 European countries. *International Journal of Environmental Research and Public Health, 14*(6), 584. Available from https://doi.org/10.3390/ijerph14060584.

Thomson, H., Bouzarovski, S., & Snell, C. (2017). Rethinking the measurement of energy poverty in Europe: A critical analysis of indicators and data. *Indoor and Built Environment, 26* (7), 879−901. Available from https://doi.org/10.1177/1420326X17699260.

Unitd Nations. (2023). *Take action for the Sustainable Development Goals.* Available from https://www.un.org/sustainabledevelopment/sustainable-development-goals/.

World Bank. (2022). *COVID-19 slows progress toward universal energy access.* Available from https://www.worldbank.org/en/news/press-release/2022/06/01/report-covid-19-slows-progress-towards-universal-energy-access.

World Health Organisation (WHO). (2018). *Housing and health guidelines.* Available from https://www.ncbi.nlm.nih.gov/books/NBK535294/.

Chapter 9

Regional climatic change and aged population. Adaptive measures to support current and future requirements

Konstantina Vasilakopoulou[1,2]
[1]School of the Built Environment, Faculty of Arts, Design and Architecture, University of New South Wales, Sydney, NSW, Australia, [2]UNSW Ageing Futures Institute, UNSW Sydney, NSW, Australia

Highlights

- People's resilience to global climate change depends on their gender, ethnicity, financial and educational status, and their age.
- Older age is often associated with more physical and mental health issues, reduced mobility, disabilities, and comorbidities, making older adults more susceptible to suffering serious consequences from climate change.
- Some of the consequences of climate change affecting mostly the older population are the increased heat, the rise in the occurrence of infectious diseases, famine, and nature degradation.
- Measures and strategies that can minimize the impacts of climate change on older adults range from policies that reduce energy poverty to designing better physical environments.

9.1 Introduction

Climate change is a global phenomenon caused mainly by the extensive emissions of greenhouse gases. The greatest emitters include power generation, manufacturing, transportation, and the building sector, while the destruction of natural greenhouse gases' absorbers, that is, the forests, also contributes significantly (United Nations, 2023). The main result of global climate change is the temperature increase, leading to the rising seawater levels and extreme weather phenomena all around the world. From place to place, these major changes can be translated into different impacts of varying

Mitigation and Adaptation of Urban Overheating. DOI: https://doi.org/10.1016/B978-0-443-13502-6.00009-9

severity (I.P.C.C., 1998). Regional climatic change refers to the alteration of the local climate compared to historical averages and the consequences on the local population, human or other.

Regional vulnerability to climate change depends on many factors, including the local environmental conditions, the environmental, social, and financial policies and practices, the resilience of the infrastructure, and so forth. Coastal areas, for example, already suffer from the rise of water levels and changing shorelines, causing extreme floods and impacting infrastructure, the local ecosystems, and humans. According to NASA (2023), the areas that will be most impacted by the rising levels of water in the near future are tropical and subtropical river deltas at the U.S. East Coast and Gulf Coast, Asia, and islands, the populations of which might be in need of relocation by the end of this century.

On the other hand, there is scientific evidence that recent mass movements in mountains can be attributed to global warming and increased precipitation. Other issues in mountainous communities are speculated to include increased rock instability and rockfall, floods of glacier lakes, ice falls and avalanches, and other chain reactions (Stoffel & Huggel, 2012).

Similarly, the environmental results of climate change have a significant impact on human mental and physical health, with different population groups being affected in varying ways. People's resilience depends on their gender, ethnicity, financial and educational status, and of course their age. The following paragraphs briefly present the effects of climate change on the health of the aged population and attempt to identify ways to alleviate the consequences of global climatic change on this vulnerable population.

9.2   Effects of climate change on the health of older adults

The improvement of public health practices as well as the widespread availability of clean water and nutritious food has led to an important reduction in the number of deaths during early and middle life and an increase in the average life of the human population (Kirkwood, 2017). Older age is often associated with more physical and mental health issues, reduced mobility, disabilities, and comorbidities (Karlamangla et al., 2007), and older adults are considered more susceptible to suffering serious consequences from climate change (Phifer & Norris, 1989). It is thus very interesting to examine the effects of climate change on this vulnerable population group, especially since older adults represent a constantly increasing portion of the world population (World Health Organisation, 2018).

9.2.1   Vulnerability to heat

One of the most important consequences of climate change is the increase in temperature, especially in the cities. The urban heat island effect, a phenomenon caused by both the climate change-driven temperature increase and the changes in the urban fabric increases the need for mechanical cooling, leading to more anthropogenic heat being released to the urban environment and thus contributing to the problem. The cities are currently the habitat for 55% of the world's population, while the urban population will reach 68% of the total by 2050 (United Nations, 2023). More older adults live in cities than in the past, and the urban heat has a significant impact on their lives.

Older adults are more sensitive to extreme temperatures that exceed their physiological adaptive capacity due to impairments in their thermoregulatory system, preventing them from promptly adjusting to very high or low environmental temperatures (Hughes & Natarajan, 2019). Evidence also shows that older adults have lower thirst levels and receive less water when dehydrated than younger adults, while it takes longer for older people to recover from dehydration, making the risks from high environmental temperatures even more severe (Kenny et al., 2010).

Preexisting health issues increase their risk for complications as well. The individuals at greater risk from heat are those with obesity, cardiovascular disease, respiratory disease, and diabetes mellitus (Kenny et al., 2010). The mechanisms that are impaired in the bodies of older adults with these health conditions include the heat-sensing and/or heat-dissipating abilities (Vroman et al., 1983), adequate ability to increase cardiac output, the control of blood flow in the skin, and thus the ability of the skin to dissipate heat, sweating response, and so forth (Kenny et al., 2010).

Medication often consumed by older people, such as diuretics, sedatives, tranquilizers, and certain heart and blood pressure drugs, can contribute to heat intolerance and reduce sweating or cause other issues that increase vulnerability to heat (Calvin, 2018). There is also evidence that during heat waves, not only the number of hospital admissions due to mental disorders (organic illnesses, including symptomatic mental disorders; dementia; mood disorders; neurotic, stress-related, and somatoform disorders; disorders of psychological development; and senility) is increased, but mortalities attributed to mental and behavioral disorders in people between 65 and 74 years also tend to rise (Hansen et al., 2008).

9.2.2 Infectious diseases

Changes in ecosystems and land cover, as well as the increase in temperature and extreme weather effects, can have impacts on infectious diseases. Even though it is hard to associate climate change characteristics with infectious diseases, various scientific teams have investigated the occurrence and geographic range of diseases, such as malaria, dengue, yellow fever, various types of viral encephalitis, schistosomiasis, leishmaniasis, Lyme disease, and onchocerciasis (McMichael et al., 2006). The vulnerability of people with comorbidities and dysfunctional and/or aged immune systems, such as older adults, to these conditions is increased (Yoshikawa, 2000).

9.2.3 Famine

The strains put on the environment by climate change, especially floods, and droughts have a direct impact on food crops and the food industry (I.P.C.C., 1998). The agricultural sector is facing changing levels of productivity; wood product industries need to adapt to the changing wood supplies; fruit and vegetable growing areas are shifting affecting food production, and so forth. Studies show that even though global crop production will probably remain stable in the coming years, regional crop production changes might lead to hunger, especially in poor nations (Parry et al., 2004). Mortality during periods of famine is highest for vulnerable individuals, such as children and older adults (Young & Jaspars, 1995).

The occurrence of famine due to climate change is not uniform among the population or the geographic location. People of low income are much more vulnerable to climate change and its consequences. In countries where the older population is poorer than other age groups, such as the OECD, Africa, and America, older people are very vulnerable to undernutrition and agricultural prices impacted by climate change (Balasubramanian, 2018).

9.3 Emotional resilience and mental health

Climate change and its expressions, that is, the increasing ambient temperatures, the extreme weather events, changes to the ecosystems, and so forth, can lead to multiple social challenges. Properties and businesses can suffer serious damage, loss of human lives or injuries can occur, inequities might be strengthened, and the need for displacement of communities can arise. All these issues can incite posttraumatic stress disorder, major depressive disorder, anxiety, depression, complicated grief, survivor guilt, vicarious trauma, recovery fatigue, substance abuse, and suicidal ideation in the people affected (Hayes et al., 2018).

Even though older adults seem to have better emotional regulation than younger people (Charles & Luong, 2013), they can get very much attached to their home or the environment, where they have lived for a long time. It is characteristic that older adults present greater resistance to evacuating during an extreme phenomenon or disaster. A study investigating the behaviors of older people during evacuation due to a natural disaster showed that the main reasons older people would not evacuate would be to protect their property (13.7%), they did not find a reason to leave (12.2%), they would have no transportation (14.4%), or they would have nowhere to go (9.4%) (Rosenkoetter et al., 2007). Even when older people decide to leave their homes, they are more likely to have reduced mobility, vision, and hearing impairments and depend on carers, making it more difficult to evacuate their homes or other facilities, go to safer locations, and remain safe during the evacuation.

It can also be assumed that the emotional consequences of nature degradation on older indigenous populations with a strong spiritual connection to the land can be significant (Drissi, 2020). The studies that have investigated the outcomes of climate change on mental health suggest that the degradation of the land can be directly or indirectly linked to mental health issues of indigenous populations (Speldewinde et al., 2009).

9.4 Adaptive measures to support current and future requirements

Older adults belong to a vulnerable category of the global population due to their lower resilience, preexisting health conditions, mental and emotional fragility, and social status, making them the first victims of the effects of climate change. The statistics of recent climate change-related events prove this fact. The heatwave that hit Western Europe in 2003 killed thousands of people. Among the 14,000 heat-related deaths in France, 80% were people aged over 75. 75% of the victims of Hurricane Katrina that struck New Orleans in 2005 were over the age of 60, even though this group comprised only 16% of the population. The floods in Argentina in 2013 resulted in the death of approximately 200 people, 70% of whom were over the age of 60 (High Commissioner for Human Rights, 2022). Epidemiological studies predict an even darker future for the older population. Hajat et al. (2014) concluded that the risk of heat- or cold-related deaths will increase in the United Kingdom by the end of this century, with the greatest risks occurring in people older than 85 years.

The vulnerability of older people to climate change has been recognized by the United Nations Human Rights Council (Resolution 44/7 -A/HRC/RES/44/7), which has called for adaptation measures against the effects of climate change that will recognize the unique challenges the older population is facing as well as the use of this group's knowledge and experience to effectively respond to climate change and protect the future generations. The following paragraphs attempt to provide an overview of the most important adaptive measures that can support the aging population to be more resilient to the effects of climate change.

9.4.1 Built environment mitigation strategies

To improve the quality of life of the elderly and reduce the health risks for this vulnerable population group, heat mitigation and adaptation technologies may be used in the built environment to decrease the intensity of the heat sources that cause overheating and increase the resilience to extreme climatic conditions (Akbari et al., 2016). Heat mitigation technologies involve the use of advanced materials for the envelope of the buildings and urban infrastructure, the additional use of greenery, the use of evaporative systems, the use of solar control devices, and the use of low-temperature environmental heat sinks (Santamouris, 2016).

Recent studies have shown that the implementation of heat mitigation technologies in cities can significantly decrease the levels of heat-related mortality. More specifically, it has been found that an increase in the urban albedo can reduce the magnitude of heat-related mortality in the elderly and global population in a city between 0.1 and 4 deaths per day on average, corresponding to a 19.8% decrease in the deaths per degree of temperature drop or 1.8% reduction of the mortality per 0.1 increase of the urban albedo (Santamouris & Fiorito, 2021). Moreover, it has been found that an increase in the green infrastructure in a city can decrease the daytime peak temperature up to 1°C, on average, while a temperature decrease by 0.1% caused by the rise of the urban green infrastructure can decrease the magnitude of heat-related mortality of the elderly and global population by 3% on average (Santamouris & Osmond, 2020).

The most recent development in mitigation technologies is the innovative reflective and fluorescent materials for urban environments and buildings, which, combined with well-designed greenery systems, can contribute to the decreasing peak temperature of cities up to 3 °C, reduce energy consumption by 40%, on average, improve indoor and outdoor thermal comfort, and at the same time decrease heat-related mortality by at least 30% (Santamouris et al., 2017).

For the mitigation systems to have an effect on the lives of the vulnerable population groups, they need to be applied to neighborhoods, buildings, and structures where older people live and frequent and to systems that they use. For example, hospitals, care homes, public transport, parks, and so forth need to have excellent thermal comfort conditions and be energy-autonomous and easily accessible, and the staff needs to be trained to efficiently respond to older people's needs during a heatwave or an extreme weather event.

9.4.2 Health monitoring and environmental control

The inability of some older people to sense the increases in their body temperature can lead to serious health risks. Building automations and wearable sensors are technologies that enable environmental control, health tracking, and heat exposure sensing to improve resilience to extreme weather events. These technologies have proven to be reliable and can easily be implemented in cases where older people live alone, especially in remote areas or in care settings (Cheong et al., 2020; Pham et al., 2020). They can also offer knowledge about which mitigation and adaptation strategies are most relevant and can be combined with information provision channels to alert communities, policymakers, and individuals to take adaptive actions (Hass & Ellis, 2019).

9.4.3 Policies responses

Older people with comorbidities who are more likely to be impacted by the effects of climate change often have low socioeconomic status and/or might live in housing with higher energy needs or areas of poor environmental conditions (Evans, 2013). Vulnerability should inform environmental and social policies, which ought to provide incentives for older people to retrofit their energy-wasting houses and provide reliable energy. Without measures like this, the energy demand during extreme phenomena will continue to rise, leading to service interruptions and the inability to use electric devices for establishing thermal comfort, running life support equipment, contacting emergency services or carers, preserving food, and so forth (Klinger et al., 2014).

Governments also have the duty to alleviate energy poverty and provide financial assistance to people that really need to consume energy during extreme phenomena. For example, many older people might avoid turning on their air conditioning to save energy. Due to their compromised thermo-regulatory system, they might easily get dehydrated or overheated with often serious health consequences.

Other policies required to ensure the safety of the older population include mechanisms to enable food security, to recover homes and livelihoods after events that have caused displacement, increase public health capacity, invest in climate change adaptation and mitigation, train emergency service providers on how to support the elderly and frail, and so forth (Watts et al., 2015).

9.4.4 Appropriate information provision

Cultural and educational barriers or special communication needs might limit older people's ability to get information and warnings relevant to climate change and extreme weather phenomena. Even though the older population is getting better at using technology, there still is a significant part of the population that does not have mobile phones or cannot access the internet and relies on the television or friends and family to provide vital information. The "gray digital divide" (Mubarak & Suomi, 2022) is a reality that needs to be tackled by training older people on how to use technology, but also by providing more inclusive information to the population, using diverse media, forms, and pathways. Communicating warnings is a scientific field in itself, and messages designed for older people need to be designed based on their limitations and capacities (Mayhorn, 2005).

9.5 Conclusions

Climate change impacts the frequency and intensity of extreme weather events, such as droughts, extreme heat, extreme precipitation, hurricanes, tornadoes, and wildfires (Centre for Climate and Energy Solutions, 2023). Extreme weather phenomena can lead to the distraction of habitats and communities, loss of property, and compromise of infrastructure. These changes to the physical environment have financial, social, and health implications on the lives of the local populations. The local population might be forced to evacuate their homes and neighborhoods, mourn the loss of the natural environment as they knew it, and lose properties, people, and animals. Apart from the emotional cost, climatic changes are shown to affect human health, including mortality and morbidity from extreme heat, cold, drought, or storms; changes in air and water quality; and changes in the ecology of infectious diseases (Patz et al., 2005).

The consequences of climate change on the health of the population depend on cultural and social parameters, with specific groups, such as women, children, older people, and people with disabilities, being the most impacted. The responsibility of policymakers and governments to provide mitigation and adaptation systems and solutions to vulnerable populations is significant.

Older people, their carers, and families must be included in all climate action initiatives to ensure that their specific needs are addressed. Moreover, older people, especially indigenous populations, can hugely contribute to the tackling of climate change, by participating in movements promoting climate action and sharing their diverse experiences and knowledge about the environment. Today more than ever, it is vital to ensure that the health of this growing part of the global population is protected from the effects of climate change to ensure a better quality of life and their participation in the lives of the younger generations.

References

Akbari, H., Cartalis, C., Kolokotsa, D., Muscio, A., Pisello, A. L., Rossi, F., Santamouris, M., Synnefa, A., Wong, N. H., & Zinzi, M. (2016). Local climate change and urban heat island mitigation techniques - The state of the art. *Journal of Civil Engineering and Management*, *22*(1), 1–16. Available from https://doi.org/10.3846/13923730.2015.1111934, http://www.tandfonline.com/loi/tcem20.

Balasubramanian, M. (2018). Climate change, famine, and low-income communities challenge Sustainable Development Goals. *The Lancet Planetary Health*, *2*(10), e421–e422. Available from https://doi.org/10.1016/s2542-5196(18)30212-2.

Calvin, K. (2018). National Institute on Aging. *National Institute on aging heat-related health dangers for older adults soar during the summer*. Available from https://www.nih.gov/news-events/news-releases/heat-related-health-dangers-older-adults-soar-during-summer.

Centre for Climate and Energy Solutions. (2023). *Extreme weather and climate change*, 7 26. 2023. Available from https://www.c2es.org/content/extreme-weather-and-climate-change/.

Charles, S. T., & Luong, G. (2013). Emotional experience across adulthood: The theoretical model of strength and vulnerability integration. *Current Directions in Psychological Science, 22*(6), 443−448. Available from https://doi.org/10.1177/0963721413497013, http://cdp.sagepub.com/content/by/year.

Cheong, S. M., Bautista, C., & Ortiz, L. (2020). Sensing physiological change and mental stress in older adults from hot weather. *IEEE Access, 8*, 70171−70181. Available from https://doi.org/10.1109/ACCESS.2020.2982153, http://ieeexplore.ieee.org/xpl/RecentIssue.jsp?punumber = 6287639.

Drissi, S. (2020). *United Nations Environment Programme Indigenous peoples and the nature they protect.* Available from https://www.unep.org/news-and-stories/story/indigenous-peoples-and-nature-they-protect.

Evans, G. (2013). Ageing and climate change: A society-technology-design discourse. *Design Journal, 16*(2), 239−258. Available from https://doi.org/10.2752/175630613X13584367985027, http://docserver.ingentaconnect.com/deliver/connect/bloomsbury/14606925/v16n2/s7.pdf?, United Kingdom.

Hajat, S., Vardoulakis, S., Heaviside, C., & Eggen, B. (2014). Climate change effects on human health: Projections of temperature-related mortality for the UK during the 2020s, 2050s and 2080s. *Journal of Epidemiology and Community Health, 68*, 641−648.

Hansen, A., Bi, P., Nitschke, M., Ryan, P., Pisaniello, D., & Tucker, G. (2008). The effect of heat waves on mental health in a temperate Australian city. *Environmental Health Perspectives, 116*(10), 1369−1375. Available from https://doi.org/10.1289/ehp.11339.

Hass, A. L., & Ellis, K. N. (2019). Using wearable sensors to assess how a heatwave affects individual heat exposure, perceptions, and adaption methods. *International Journal of Biometeorology, 63*(12), 1585−1595. Available from https://doi.org/10.1007/s00484-019-01770-6, http://www.link.springer.de/link/service/journals/00484/index.htm.

Hayes, K., Blashki, G., Wiseman, J., Burke, S., & Reifels, L. (2018). Climate change and mental health: Risks, impacts and priority actions. *International Journal of Mental Health Systems, 12*(1). Available from https://doi.org/10.1186/s13033-018-0210-6.

High Commissioner for Human Rights. (2022). *Summary of the panel discussion on the human rights of older persons in the context of climate change.* United Nations. Available from https://doi.org/10.1017/S0020818300012881.

Hughes, C., & Natarajan, S. (2019). Summer thermal comfort and overheating in the elderly. *Building Services Engineering Research and Technology, 40*(4), 426−445. Available from https://doi.org/10.1177/0143624419844518, http://bse.sagepub.com/archive/.

I.P.C.C.. (1998). The regional impacts. In *An assessment of vulnerability*, UK: Cambridge University Press. Available from https://www.ipcc.ch/report/the-regional-impacts-of-climate-change-an-assessment-of-vulnerability/.

Karlamangla, A., Tinetti, M., Guralnik, J., Studenski, S., Wetle, T., & Reuben, D. (2007). Comorbidity in older adults: Nosology of impairment, diseases, and conditions. *Journal of Gerontology: Medical Sciences, 62A*(3), 296−300.

Kenny, G. P., Yardley, J., Brown, C., Sigal, R. J., & Jay, O. (2010). Heat stress in older individuals and patients with common chronic diseases. *CMAJ. Canadian Medical Association Journal, 182*(10), 1053−1060. Available from https://doi.org/10.1503/cmaj.081050, http://www.cmaj.ca/cgi/reprint/182/10/1053.

Kirkwood, T. B. L. (2017). Why and how are we living longer? *Experimental Physiology, 102*(9), 1067−1074. Available from https://doi.org/10.1113/EP086205, http://onlinelibrary.wiley.com/journal/10.1111/(ISSN)1469-445X.

Klinger, C., Landeg, O., & Murray, V. (2014). Power outages, extreme events and health: A systematic review of the literature from 2011−2012. *PLoS Currents*. Available from https://doi.org/10.1371/currents.dis.04eb1dc5e73dd1377e05a10e9edde673, http://www.plos.org/publications/currents/.

Mayhorn, C. B. (2005). Cognitive aging and the processing of hazard information and disaster warnings. *Natural Hazards Review*, 6(4), 165−170. Available from https://doi.org/10.1061/(ASCE)1527-6988(2005)6:4(165).

McMichael, A. J., Woodruff, R. E., & Hales, S. (2006). Climate change and human health: Present and future risks. *Lancet*, 367(9513), 859−869. Available from https://doi.org/10.1016/S0140-6736(06)68079-3, http://www.journals.elsevier.com/the-lancet/.

Mubarak, F., & Suomi, R. (2022). Elderly forgotten? Digital exclusion in the information age and the rising grey digital divide. *Inquiry (United States)*, 59. Available from https://doi.org/10.1177/00469580221096272, https://journals.sagepub.com/home/INQ.

NASA. (2023). *NASA Which areas of the world will be most affected by sea-level rise over the next century*. Available from https://sealevel.nasa.gov/faq/17/which-areas-of-the-world-will-be-most-affected-by-sea-level-rise-over-the-next-century-and-after-that/#:∼:text = Hot%20spots%20include%20the%20U.S.,Coast%2C%20Asia%2C%20and%20islands.

Parry, M. L., Rosenzweig, C., Iglesias, A., Livermore, M., & Fischer, G. (2004). Effects of climate change on global food production under SRES emissions and socio-economic scenarios. *Global Environmental Change*, 14(1), 53−67. Available from https://doi.org/10.1016/j.gloenvcha.2003.10.008.

Patz, J. A., Campbell-Lendrum, D., Holloway, T., & Foley, J. A. (2005). Impact of regional climate change on human health. *Nature*, 438(7066), 310−317. Available from https://doi.org/10.1038/nature04188, http://www.nature.com/nature/index.html.

Pham, S., Yeap, D., Escalera, G., Basu, R., Wu, X., Kenyon, N. J., Hertz-Picciotto, I., Ko, M. J., & Davis, C. E. (2020). Wearable sensor system to monitor physical activity and the physiological effects of heat exposure. *Sensors (Switzerland)*, 20(3). Available from https://doi.org/10.3390/s20030855, https://www.mdpi.com/1424-8220/20/3/855/pdf.

Phifer, J. F., & Norris, F. H. (1989). Psychological symptoms in older adults following natural disaster: Nature, timing, duration, and course. *Journal of Gerontology*, 44(6), S207−S212. Available from https://doi.org/10.1093/geronj/44.6.S207.

Rosenkoetter, M. M., Covan, E. K., Cobb, B. K., Bunting, S., & Weinrich, M. (2007). Perceptions of older adults regarding evacuation in the event of a natural disaster: Populations at risk across the lifespan: Empirical studies. *Public Health Nursing*, 24(2), 160−168. Available from https://doi.org/10.1111/j.1525-1446.2007.00620.x.

Santamouris, M., Ding, L., Fiorito, F., Oldfield, P., Osmond, P., Paolini, R., Prasad, D., & Synnefa, A. (2017). Passive and active cooling for the outdoor built environment − Analysis and assessment of the cooling potential of mitigation technologies using performance data from 220 large scale projects. *Solar Energy*, 154, 14−33. Available from https://doi.org/10.1016/j.solener.2016.12.006.

Santamouris, M., & Fiorito, F. (2021). On the impact of modified urban albedo on ambient temperature and heat related mortality. *Solar Energy*, 216, 493−507. Available from https://doi.org/10.1016/j.solener.2021.01.031.

Santamouris, M., & Osmond, P. (2020). Increasing green infrastructure in cities: Impact on ambient temperature, air quality and heat-related mortality and morbidity. *Buildings*, 10(12). Available from https://doi.org/10.3390/buildings10120233.

Santamouris, M. (2016). Innovating to zero the building sector in Europe: Minimising the energy consumption, eradication of the energy poverty and mitigating the local climate change. *Solar Energy*, 128, 61−94. Available from https://doi.org/10.1016/j.solener.2016.01.021.

Speldewinde, P. C., Cook, A., Davies, P., & Weinstein, P. (2009). A relationship between environmental degradation and mental health in rural Western Australia. *Health and Place*, *15*(3), 880−887. Available from https://doi.org/10.1016/j.healthplace.2009.02.011.

Stoffel, M., & Huggel, C. (2012). Effects of climate change on mass movements in mountain environments. *Progress in Physical Geography*, *36*(3), 421−439. Available from https://doi.org/10.1177/0309133312441010, https://journals.sagepub.com/home/PPG.

United Nations. (2023). *Causes and effects of climate change*. Available from https://www.un.org/en/climatechange/science/causes-effects-climate-change.

Vroman, N. B., Buskirk, E. R., & Hodgson, J. L. (1983). Cardiac output and skin blood flow in lean and obese individuals during exercise in the heat. *Journal of Applied Physiology*, *55*(1), 69−74. Available from https://doi.org/10.1152/jappl.1983.55.1.69.

Watts, N., Adger, W. N., Agnolucci, P., Blackstock, J., Byass, P., Cai, W., Chaytor, S., Colbourn, T., Collins, M., Cooper, A., Cox, P. M., Depledge, J., Drummond, P., Ekins, P., Galaz, V., Grace, D., Graham, H., Grubb, M., Haines, A., ... Costello, A. (2015). Health and climate change: Policy responses to protect public health. *The Lancet*, *386*(10006), 1861−1914. Available from https://doi.org/10.1016/s0140-6736(15)60854-6.

World Health Organisation. (2018). *World Health Organisation Ageing*. Available from https://www.who.int/news-room/fact-sheets/detail/ageing-and-health#:~:text = Every%20country%20in%20the%20world,in%202020%20to%201.4%20billion.

Yoshikawa, T. T. (2000). Epidemiology and unique aspects of aging and infectious diseases. *Clinical Infectious Diseases*, *30*(6), 931−933. Available from https://doi.org/10.1086/313792.

Young, H., & Jaspars, S. (1995). Nutrition, disease and death in times of famine. *Disasters*, *19*(2), 94−109. Available from https://doi.org/10.1111/j.1467-7717.1995.tb00361.x.

Chapter 10

The impact of heat adaptation on socioeconomically vulnerable populations

Lauren Ferguson and Mavrogianni Anna

Institute for Environmental Design and Engineering (IEDE), Bartlett School of Environment, Energy and Resources (BSEER), Bartlett Faculty of the Built Environment, University College London (UCL), London, United Kingdom

10.1 Introduction

According to the Intergovernmental Panel on Climate Change (IPCC) *Synthesis Report for the* 6th Assessment Report, finalized in March 2023 (Lee et al., 2023), weather and climate extremes globally are affected by anthropogenic climate change, and there is *high confidence* that this will have a wide range of adverse consequences for the natural and human environment. There is also *high confidence* that climate change impacts will be disproportionately felt by socioeconomically vulnerable communities, which historically contributed less to greenhouse gas emissions causing global warming. The risks for low-income, urban populations are also expected to be magnified due to the urban heat island effect (Santamouris, 2020; Santamouris & Kolokotsa, 2015). The IPCC AR6 report highlights that despite some progress having been made across several sectors to adapt to a warming climate, significant gaps still exist; this indicates that the current levels of implementation of climate change adaptation measures are insufficient, in particular in developing countries and low-income settings.

It is fundamental that climate change adaptation strategies are designed and implemented through an equity, inclusion, and just transition lens (Lee et al., 2023), through a framework that recognizes compounding inequalities. The COVID-19 pandemic intersected with climate change to further magnify inequalities across regions. Post-pandemic, it has been argued (Mattar et al., 2021) that, although climate adaptation and development policies have been

Mitigation and Adaptation of Urban Overheating. DOI: https://doi.org/10.1016/B978-0-443-13502-6.00010-5

dominated thus far by economic priorities, climate and social justice and equity should be prioritized; this should be achieved through increased support to people and communities that are likely to be the most affected and least able to adapt due to historical injustices and current vulnerabilities. This will be necessary to avoid maladaptation, defined as *the allocation of exclusionary visions of what and for whom adaptation is for* (Forsyth & McDermott, 2022).

This chapter examines the impact of heat exposure on socioeconomically vulnerable and low-income populations, in particular, in the context of a warming climate, and investigates the multifaceted manner in which climate change may exacerbate or forge new socioeconomic and health inequalities.

10.2 Heat exposure through a climate justice lens

There has been an emergent understanding that issues of justice are entwined with climate change impacts across different scales, regions, and populations. In the first instance, as discussed above, it has been widely argued that the nations least responsible for contributing to greenhouse gas emissions historically are and will be the ones hardest hit by its adverse effects (Meyer & Roser, 2010; Newell et al., 2021). It has been estimated (Althor et al., 2016) that 20 of the 36 countries with the highest historical greenhouse gas emissions are among the least vulnerable to adverse climate change effects, whereas 11 of the 17 countries with low or moderate historical emissions are likely to be worst affected by a warming climate. The global inequity in the responsibility for climate change and the burden of its impacts is demonstrated in Fig. 10.1.

A plethora of studies to date (Cevik & Jalles, 2023; Ebi & Hess, 2020; Watts et al., 2021) have highlighted the imperative to identify and quantify the ways in which ongoing and future human climate change will magnify existing global inequalities. The role of the built environment, as a "modifier" of environmental exposures, in shaping these health inequalities and the ability of a community to respond to climate hazards, is also increasingly recognized (Gelormino et al., 2015; Klinsky & Mavrogianni, 2020; Renalds et al., 2010; Smith et al., 2022).

A multivalent approach to climate justice has been proposed (Klinsky & Mavrogianni, 2020), in relation to the built environment, which includes, in the context of climate change adaptation and mitigation, the three pillars of climate justice:

- *Distributive justice* refers to the way in which benefits and burdens of climate change are divided across different stakeholders.
- *Procedural justice* relates to the fairness and inclusiveness of decision-making.
- *Recognition justice* is linked to the identification of stakeholders, needs, and priorities in a policy framing context.

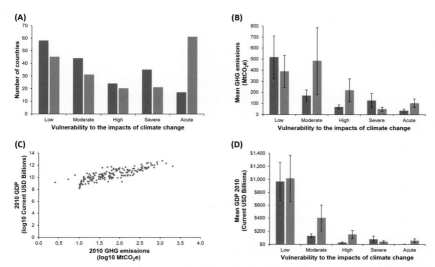

FIGURE 10.1 (A) Number of countries in each climate change vulnerability category, derived from development assistance research associates vulnerability data, for 2010 (*blue bars*) and 2030 (*green bars*). (B) Mean greenhouse gas emissions for 2010, derived from climate analysis indicators tool greenhouse gas emissions data, shown in CO_2 equivalent units and climate vulnerability categories for 2010 (blue bars, with standard error) and 2030 (*green bars*, with standard error). (C) Gross domestic product (GDP) shown in current US$ (in billions), derived from the World Bank GDP 2010 data and 2010 greenhouse gas emissions. (D) Mean GDP for 2010 shown in current US$ (in billions) and climate change vulnerability for 2010 (*blue bars*, with standard error) and 2030 (*green bars*, with standard error). *From Althor, G., Watson, J. E. M., & Fuller, R. A. (2016).* Global mismatch between greenhouse gas emissions and the burden of climate change. Scientific Reports, 6. *https://doi.org/10.1038/srep20281. This work is licensed under a Creative Commons Attribution 4.0 International License. The images or other third party material in this article are included in the article's Creative Commons license, unless indicated otherwise in the credit line. To view a copy of this license, visit http://creativecommons. org/licenses/by/4.0/.*

It is, thus, of fundamental importance that heat exposure, and its exacerbation due to climate change, is viewed through a climate justice lens. This has been reflected in increased public awareness and media interest, following recent extreme heat episodes, such as the 2018 heatwave: Heat was described by The Guardian in August 2018 as *the next big inequality issue* (Fleming et al., 2018), that will divide the global population into *the cool haves and the hot have-nots*, whereas The New York Times in 2020 described heat-related inequity being *at the boiling point* emphasizing that *a hotter planet does not hurt equally* (Sengupta, 2020).

With regard to *distributive* justice, it is likely that low-income and, more broadly, socioeconomically vulnerable populations will be disproportionately affected by excess heat exposure. Globally, areas that are most susceptible to temperature increases, are likely to be occupied by low-income communities.

Notably, this is observed across and within countries, as well as across and within cities (Hsu et al., 2021). It is worth bearing in mind that such inequalities do not only relate to heat-related risks but compound risks; for example, low-income communities may also be more likely to live in flood or drought prone areas (Maskrey et al., 2023). Not only are the burdens of heat exposure felt unequally across socioeconomic groups, low-income groups may also lack the financial resources to adapt, cope, and recover from extreme heat. Declining heat-related mortality throughout the latter half of the twentieth century has been partly attributed to the widespread and continued uptake of air conditioning (Sera et al., 2020), and uptake is projected to rise twofold and fourfold across Europe and India, respectively, by 2050 (Colelli et al., 2023). This significant rise in electricity demand is likely to pose significant challenges for lower-income countries, particularly in the Global South, who are already feeling some of the worst effects of climate change and presently lack the grid capacity to distribute air conditioning units to rapidly growing urban populations (Sherman et al., 2022). Across countries where the use of air conditioning is already pervasive, a lack of access is associated with higher heat-related deaths in ethnic minority and low-income populations (Madrigano et al., 2015; O'Neill et al., 2005).

Crucially, *procedural justice* issues emerge at the planning stage, long before an extreme heat event occurs. It is well documented that disadvantaged groups may be less likely to participate in key decision-making procedures for long-term planning and policies aiming to mitigate the effects of climate change at the national, citywide, or neighborhood level (Holland, 2017). In a *recognition* climate justice context, long-standing historical injustices, such as systemic racism, may result in the needs of many at-risk communities to be neglected (Gutschow et al., 2021).

In summary, heat is not solely an environmental hazard, but a social and economic justice issue that requires a sociotechnical approach that addresses underlying inequalities. The next section describes in more detail how such inequalities manifest at different scales, from the global to the regional, neighborhood, building, and individual level.

10.3   Impacts of heat on inequalities

10.3.1   At the global and regional level

As pointed out above, there is a significant geographical variation of heat exposure and linked adverse health effects across the globe, and populations in countries that have contributed least to global warming over the decades are likely to be located in warmer geographical regions, which are projected to experience excess levels of further warming. Many of these settings are developing countries or countries with an appreciable proportion of low-income populations; thus, they may lack the resources and infrastructure to respond to heatwaves. This includes both public health infrastructure that could facilitate short-term emergency responses during a hot spell, as well as long-term planning that could encourage the adoption of urban and building design strategies that mitigate overheating risks. In some developing countries and low-income communities around the world, heat risks may be magnified due to limited access to clean water, which can increase the risk of dehydration and other heat-health symptoms. This further creates a vicious circle of heat impacts (Cappelli et al., 2021; Islam & Winkel, 2017) as a result of (1) currently high heat exposures, (2) that will worsen in the future, combined with (3) limited adaptation potential, contributing to the intensification of underlying heat-related health and socioeconomic inequalities (Fig. 10.2).

10.3.2   At the urban/neighborhood level

Urban areas are particularly vulnerable to heatwaves due the urban heat island effect, which can exacerbate the effects of heatwaves (Kolokotsa & Santamouris, 2015; Santamouris, 2020). In addition, it is worth examining intracity differences in heat exposure by the socioeconomic group. For example, low-income populations are overrepresented in many cities around the

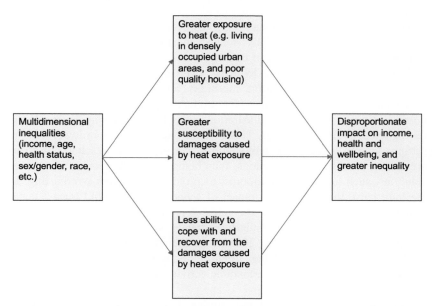

FIGURE 10.2 Vicious circle of heat impacts and inequalities for disadvantaged groups. *Adapted from Islam, S. N., & Winkel, J. (2017). Climate Change and Social Inequality. Department of Economic & Social Affairs (DESA) Working Paper No. 152 ST/ESA/2017/DWP/ 152, United Nations (UN), New York, USA. Available from https://www.un.org/esa/desa/papers/ 2017/wp152_2017.pdf, ©2017 United Nations. Used with the permission of the United Nations.*

world. Lower-income groups are likely to live in warmer parts of the city heat island "hot spots" (Huang et al., 2011), and these neighborhoods are also likely to be more densely populated and predominantly characterized by human-made surfaces that absorb and re-emit the heat in the urban environment, with lower levels of green areas that could potentially both reduce the heat island effect as well as provide heat protection through shade during a heat event (Hsu et al., 2021). The importance of public cool spaces in the urban environment is particularly important for lower-income groups, as they may have limited heat protection options indoors and they are, thus, more likely to seek refuge in the public space. Such spaces may be acutely needed by the most vulnerable segments of the population, such as rough sleepers, during a heatwave.

Higher population density in cities is combined with housing density (Macintyre et al., 2018), which results in smaller, more compact dwellings, which may increase heat transfer through adjacent properties and could limit cooling potential, for example, through cross ventilation, hence increasing overheating risk. This results in a "triple jeopardy" scenario where location within the urban heat island, poor quality housing with limited cooling potential, and individual vulnerability overlap to create high heat risk areas (Taylor et al., 2015). It is worth noting that, in cities such as London, some heat-vulnerable populations, such as older people, may be concentrated in suburban rather than core urban areas; however, there may still be a significant proportion of older, socially isolated and low-income individuals living in more central, warmer parts of the city.

Disproportionate exposure to urban overheating not only is limited to direct effects on thermal comfort, morbidity, and mortality, it can also result in a wide range of consequences for productivity, education, and socioeconomic activities. By way of illustration, school closures may be inevitable during excessively hot periods. This is more likely to occur in lower-income settings, due to higher local temperatures, but also because schools in these areas may have budget limitations, thus being unable to provide short term solutions, such as fans in classrooms, to protect pupils from overheating. In the long term, the cumulative impact of school closures could lower educational attainment in some of the most deprived areas, further deepening socioeconomic inequalities, similarly to trends observed during the COVID-19 pandemic (Leal Filho et al., 2023). Low-income individuals are also more likely to carry out manual labor, often outdoors, which may be hindered by excess heat, therefore limiting economic activity (Day et al., 2019; Kjellstrom et al., 2018).

Similarly to regional-level inequalities, lower-income groups living in high risk areas for urban overheating are negatively affected not only by the current thermal conditions linked to urbanization and poorly designed outdoor/indoor environments, but also by the potential lack of investment in heat mitigation plans for their neighborhoods (Smith et al., 2022), such as urban interventions to increase vegetation and water-permeable surfaces, green/cool roofs, shading structures, and improved ventilation pathways. However, urban cooling solutions, such as the provision of green or other public cool spaces, should also be carefully thought through to tailor such spaces to the specific needs of heat-vulnerable populations. For instance, in some low-income communities, heat risk may intersect with sociocultural dimensions that may act as barriers to the use of such spaces (Kabisch & Haase, 2014).

10.3.3   At the building level

The physical characteristics of the buildings people live in have been found to be a greater determinant of indoor temperatures, and therefore risk, than the heat island in global cities (Oikonomou et al., 2012; Taylor et al., 2015). Dwelling features identified as proxies for indoor overheating are living in a top floor flat or bungalow, with higher levels of internal wall and floor insulation, smaller floor areas, and a lack of operable windows (Lomas & Porritt, 2017; Taylor et al., 2023).

Flats are consistently found to have a higher risk of overheating than other dwelling types in monitoring and modeling studies (Beizaee et al., 2013; Sharifi et al., 2019; Taylor et al., 2016). High-rise flats will receive less solar shading from neighboring buildings, and top floor flats located in both high-rise and converted buildings will be directly below the roof and, therefore, less protected from high outdoor temperatures. This also applies to bungalows, which occupy only one story and are consistently found to overheat. Flats and bungalows are more likely to be occupied by vulnerable, low-income (Ferguson et al., 2021) and older individuals, compounding the risk of exposure to environmental heat and susceptibility to developing heat-related health impacts.

The extent and type of insulation can also play a role in modifying over-heating risk: Those living in top-floor flats without roof insulation had an increased risk of mortality during the 2003 heatwave in France (Vandentorren et al., 2006). Roof insulation, window glazing, and external wall insulation may reduce overheating, whereas internally retrofitted walls and floor insulation can increase indoor overheating (Lomas & Porritt, 2017; Taylor et al., 2023). Insulation can both increase and decrease dwelling susceptibility to indoor overheating depending on whether the building is poorly or well designed and will depend on a number of other factors such as the outdoor climate, extent of shading the dwelling receives, its orientation, and occupation (Fosas et al., 2017).

The interaction of occupants with features of the building can modify indoor temperatures significantly. A monitoring study in 26 low-energy homes located in Inverness, Scotland, demonstrated that some dwellings experienced overheating, while others with a very similar design did not

(Morgan et al., 2017). Authors attributed most of the observed variance to differences in occupant adaptation to high internal temperatures, such as window opening, and varied indoor temperature preferences. Operation of windows may be constrained by features of the outdoor environment and occupant concerns related to outdoor air pollution, noise, crime, safety, and security. Window operation is, thus, an area of vulnerability to high indoor temperatures at the interplay between features of the outdoor environment and the physical characteristics of the building. An individual's perception of temperature may also influence how they respond to indoor heat, and this was cited as a driver of differences in overheating between vulnerable and nonvulnerable households in Exeter, UK (Vellei et al., 2017). The study also found that the homes of overcrowded households were more likely to overheat. It has been shown that median bedroom indoor temperature increased with the number of occupants in a sample of 823 dwellings broadly representative of the English stock (Petrou et al., 2019). Within England, 3.5% of all households are overcrowded, with this effect bias in low-income, rented, and ethnic minority households (Wilson & Barton, 2021).

Although fuel poverty studies in heating dominated climatic contexts have focused on winter space heating needs and how these correlate with disposable income and dwelling energy efficiency, an increasing body of literature is placing emphasis on inequalities related to being able to meet a household's cooling needs and summer fuel poverty (Fabbri, 2015; Haddad et al., 2022; Kolokotsa & Santamouris, 2015; Sánchez-Guevara Sánchez et al., 2017; Santamouris et al., 2007; Tabata & Tsai, 2020). An extension of current fuel poverty definitions is, thus, proposed (Sánchez-Guevara Sánchez et al., 2017) for cooling dominated countries but also regions that are projected to experience significant increases in overheating risk. Such definitions should quantify the ability of a household to maintain indoor temperatures at safe and comfortable levels during summer without adverse consequences for their disposable income and ability to cover basic household needs.

10.3.4 At the household/individual level

Across the population, individual factors affect one's vulnerability to heat-related morbidity and mortality: Older individuals (typically above 65 years of age), people living with chronic physical (cardiovascular, respiratory) or mental health conditions, and underlying comorbidities, are more vulnerable to heat health effects (Arbuthnott & Hajat, 2017).

While there is a concerted, global effort to climate-proof buildings, especially in higher-income, historically space heating-dominated countries, a number of demographic shifts may be increasing population risk to environmental heat. People are living on average longer, with older age being a risk factor for increased heat vulnerability. Older individuals are also more likely to live alone, and a number of observational studies following global heatwaves have identified higher deaths among socially isolated individuals (Fouillet et al., 2006; Naughton et al., 2002), which remains even after controlling for age (Wan et al., 2022). The underpinning mechanisms behind this vulnerability are not well understood but may be related to the reliance of individuals with poorer underlying health on formal networks of care, such as national health and social care services. These services may face disruption during extreme weather events, where individuals who are socially isolated will not have the informal networks to bring this discontinuity of care to attention (Wistow et al., 2015). The landmark 1995 heatwave in Chicago exposed a number of underlying social conditions which foster inequalities in heat-related mortality, where poorer, socially isolated older citizens living in areas with higher neighborhood crime were most affected (Klinenberg, 2015).

The number of single person households in the European Union has increased by 28.5% between 2009 and 2021 (Eurostat, 2021) and has steadily risen in the United States since the 1960s (United States Census Bureau, 2022). Increasing evidence suggests rapid urbanization and higher density living may be contributing factors to social isolation (Lai et al., 2021). Nowhere is this paradox more evident than in London, UK, the largest metropolis in western Europe, where proportionally more individuals report feeling lonely than other areas of the United Kingdom, linked to higher neighborhood overcrowding, multiunit accommodation and lower levels of greenness (Hammoud et al., 2021; Lai et al., 2021).

Fig. 10.3 illustrates the combination of spatial distribution of heat exposure, income, and one-person households in London.

The evidence for heatwaves and periods of sustained high temperatures exacerbating mental health conditions varies in strength, but continues to build (Liu et al., 2021; Thompson et al., 2018). Mental health conditions are an additional heat-health risk factor rising in many developed economies, overlapping with neighborhood deprivation, social isolation, and urban densification (Curtis et al., 2006).

10.4 Synergies with other environmental hazards

In housing occupied by low-income households, poor air quality and excessive heat can combine to increase health hazards for occupants. High concentrations of outdoor air pollution are often a proxy for deprived neighborhoods in many large cities globally (Clark et al., 2017; Fairburn et al., 2019). There is a noted spatial overlap between urban heat islands and areas of high outdoor air pollution (Ulpiani, 2021). Although both outdoor environmental hazards are associated with increased risk of mortality, cardiovascular and respiratory illness, and mental health conditions, there is growing evidence that the combined exposure from ambient air pollution and heat may have a

(A) Average daily outdoor temperature (°C) in July 2020

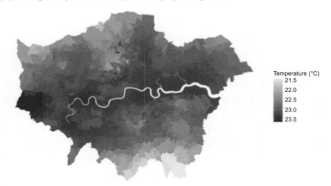

(B) Spatial distribution of low-income individuals across London LSOAs

(C) Spatial distribution of one-person households across London LSOAs

FIGURE 10.3 (A) The average daily outdoor temperature in July 2020 from modeled 1 km UK climate observations (Hollis et al., 2022), a notable heatwave period in the United Kingdom (Thompson et al., 2022). (B) The spatial distribution of low-income individuals across London Lower Super Output Areas. Individuals in receipt of Universal Credit, a means-tested social security payment, is used as a proxy of low income (UK DWP, 2022). (C) The spatial distribution of one-person households, across London Lower Super Output Areas. Data is from the latest 2021 UK Census estimates (ONS, 2022).

synergistic effect on population health (Grigorieva & Lukyanets, 2021; Yitshak-Sade et al., 2018).

Within the indoor environment, concerns about outdoor air pollution could deter window opening or other natural ventilation options. In turn, this could result in the accumulation of not only heat but also indoor air contaminants generated by daily cooking, personal hygiene, and cleaning practices or off-gassing of poor quality furniture and furnishings. Concentrations may be higher in lower-income dwellings, which may be overall smaller (in terms of both floor area as well as ceiling height) and/or overcrowded (Ferguson et al., 2021). Smaller homes, such as flats, are more likely to share party walls with neighboring dwellings, increasing the risk of heat exchange and air pollution infiltration from neighbors. Additionally, population subgroups which have an elevated risk of developing health impacts following exposure to high indoor temperatures and air pollution concentrations may spend an increased amount of time at home, indoors, relative to the wider population (Ferguson et al., 2021; Holgate et al., 2021).

Implementing solutions that address both poor air quality and overheating in low-income dwellings is crucial to resolving these problems. This could include programs that offer financial support for housing renovations that holistically address climate change mitigation and adaptation, such as better insulation and ventilation, as well as aid with initiatives to monitor air quality and reduce pollution in the neighborhood.

10.5 Conclusions

Building on the existing climate justice frameworks, a systems thinking approach is recommended that will map the way in which heat vulnerabilities are exacerbated by systemic socioeconomic inequalities and how this could further perpetuate inequalities. To address the challenge of equitable heat adaptation across different scales and population groups, cocreation of urban overheating solutions with vulnerable and low-income groups is required to identify solutions that best suit their needs. This should ensure that cooling solutions with cobenefits for other socioeconomic and health outcomes are prioritized, for example, green infrastructure, and that heat-vulnerable communities are empowered through the provision of tools and assistance to deal with excessive temperatures, now and in the future.

References

Althor, G., Watson, J. E. M., & Fuller, R. A. (2016). Global mismatch between greenhouse gas emissions and the burden of climate change. *Scientific Reports, 6*. Available from https://doi.org/10.1038/srep20281, http://www.nature.com/srep/index.html.

Arbuthnott, K. G., & Hajat, S. (2017). The health effects of hotter summers and heat waves in the population of the United Kingdom: A review of the evidence. *Environmental Health: A Global*

Access Science Source, 16. Available from https://doi.org/10.1186/s12940-017-0322-5, http://www.ehjournal.net/home/.

Beizaee, A., Lomas, K. J., & Firth, S. K. (2013). National survey of summertime temperatures and overheating risk in English homes. *Building and Environment, 65,* 1−17. Available from https://doi.org/10.1016/j.buildenv.2013.03.011.

Cappelli, F., Costantini, V., & Consoli, D. (2021). The trap of climate change-induced "natural" disasters and inequality. *Global Environmental Change, 70,* 102329. Available from https://doi.org/10.1016/j.gloenvcha.2021.102329.

Cevik, S., & Jalles, J. T. (2023). For whom the bell tolls: Climate change and income inequality. *Energy Policy, 174.* Available from https://doi.org/10.1016/j.enpol.2023.113475, http://www.journals.elsevier.com/energy-policy/.

Clark, L. P., Millet, D. B., & Marshall, J. D. (2017). Changes in transportation-related air pollution exposures by race-ethnicity and socioeconomic status: Outdoor nitrogen dioxide in the United States in 2000 and 2010. *Environmental Health Perspectives, 125*(9). Available from https://doi.org/10.1289/EHP959, https://ehp.niehs.nih.gov/wp-content/uploads/2017/09/EHP959.alt_.pdf.

Colelli, F. P., Wing, I. S., & Cian, E. D. (2023). Air-conditioning adoption and electricity demand highlight climate change mitigation-adaptation tradeoffs. *Scientific reports, 13*(1), 4413. Available from https://doi.org/10.1038/s41598-023-31469-z.

Curtis, S., Copeland, A., Fagg, J., Congdon, P., Almog, M., & Fitzpatrick, J. (2006). The ecological relationship between deprivation, social isolation and rates of hospital admission for acute psychiatric care: A comparison of London and New York City. *Health and Place, 12*(1), 19−37. Available from https://doi.org/10.1016/j.healthplace.2004.07.002, http://www.elsevier.com/locate/healthplace.

Day, E., Fankhauser, S., Kingsmill, N., Costa, H., & Mavrogianni, A. (2019). Upholding labour productivity under climate change: An assessment of adaptation options. *Climate Policy, 19* (3), 367−385. Available from https://doi.org/10.1080/14693062.2018.1517640, http://www.tandfonline.com/loi/tcpo20.

Ebi, K. L., & Hess, J. J. (2020). Health risks due to climate change: Inequity in causes and consequences. *Health Affairs, 39*(12), 2056−2062. Available from https://doi.org/10.1377/hlthaff.2020.01125, http://www.healthaffairs.org/.

Eurostat. (2021). *Household composition statistics.* Eurostat. Available from https://ec.europa.eu/eurostat.

Fabbri, K. (2015). Building and fuel poverty, an index to measure fuel poverty: An Italian case study. *Energy, 89,* 244−258. Available from https://doi.org/10.1016/j.energy.2015.07.073, http://www.elsevier.com/inca/publications/store/4/8/3/.

Fairburn, J., Schüle, S. A., Dreger, S., Hilz, L. K., & Bolte, G. (2019). Social inequalities in exposure to ambient air pollution: A systematic review in the WHO European region. *International Journal of Environmental Research and Public Health, 16*(17). Available from https://doi.org/10.3390/ijerph16173127, https://www.mdpi.com/1660-4601/16/17/3127/pdf.

Ferguson, L., Taylor, J., Zhou, K., Shrubsole, C., Symonds, P., Davies, M., & Dimitroulopoulou, S. (2021). Systemic inequalities in indoor air pollution exposure in London, UK. *Buildings and Cities, 2*(1), 425−448. Available from https://doi.org/10.5334/bc.100, https://journal-buildingscities.org/articles/10.5334/bc.100/galley/133/download/.

Filho, W. L., Balasubramanian, M., Abeldaño Zuñiga, R. A., & Sierra, J. (2023). The effects of climate change on children's education attainment. *Sustainability, 15*(7), 6320. Available from https://doi.org/10.3390/su15076320.

Fleming, A., Michaelson, R., Youssef, A., & Holmes, O. (2018, August 18). Sweltering cities Heat: the next big inequality issue. *The Guardian*. Available from https://www.theguardian. com/cities/2018/aug/13/heat-next-big-inequality-issue-heatwaves-world.

Forsyth, T., & McDermott, C. L. (2022). When climate justice goes wrong: Maladaptation and deep co-production in transformative environmental science and policy. *Political Geography*, *98*. Available from https://doi.org/10.1016/j.polgeo.2022.102691, http://www. elsevier.com/inca/publications/store/3/0/4/6/5/index.htt.

Fosas, D., Coley, D. A., Natarajan, S., Herrera, M., Fosas de Pando, M., & Ramallo-Gonzalez, A. (2017). Mitigation versus adaptation: Does insulating dwellings increase overheating risk. *Building and Environment*, *143*, 740–759. Available from https://doi.org/10.1016/j.buildenv. 2018.07.033.

Fouillet, A., Rey, G., Laurent, F., Pavillon, G., Bellec, S., Guihenneuc-Jouyaux, C., Clavel, J., Jougla, E., & Hémon, D. (2006). Excess mortality related to the August 2003 heat wave in France. *International Archives of Occupational and Environmental Health*, *80*(1), 16–24. Available from https://doi.org/10.1007/s00420-006-0089-4.

Gelormino, E., Melis, G., Marietta, C., & Costa, G. (2015). From built environment to health inequalities: An explanatory framework based on evidence. *Preventive Medicine Reports*, *2*, 737–745. Available from https://doi.org/10.1016/j.pmedr.2015.08.019.

Grigorieva, E., & Lukyanets, A. (2021). Combined effect of hot weather and outdoor air pollution on respiratory health: Literature review. *Atmosphere*, *12*(6). Available from https://doi. org/10.3390/atmos12060790, https://www.mdpi.com/2073-4433/12/6/790/pdf.

Gutschow, B., Gray, B., Ragavan, M. I., Sheffield, P. E., Philipsborn, R. P., & Jee, S. H. (2021). The intersection of pediatrics, climate change, and structural racism: Ensuring health equity through climate justice. *Current Problems in Pediatric and Adolescent Health Care*, *51*(6). Available from https://doi.org/10.1016/j.cppeds.2021.101028, http://www.elsevier.com/inca/ publications/store/6/2/3/2/9/1/index.htt.

Haddad, S., Paolini, R., Synnefa, A., De Torres, L., Prasad, D., & Santamouris, M. (2022). Integrated assessment of the extreme climatic conditions, thermal performance, vulnerability, and well-being in low-income housing in the subtropical climate of Australia. *Energy and Buildings* (272). Available from https://doi.org/10.1016/j.enbuild.2022.112349, https:// www.journals.elsevier.com/energy-and-buildings.

Hammoud, R., Tognin, S., Bakolis, I., Ivanova, D., Fitzpatrick, N., Burgess, L., Smythe, M., Gibbons, J., Davidson, N., & Mechelli, A. (2021). Lonely in a crowd: Investigating the association between overcrowding and loneliness using smartphone technologies. *Scientific Reports*, *11*(1). Available from https://doi.org/10.1038/s41598-021-03398-2, http://www. nature.com/srep/index.html.

Holgate, S., Grigg, J., Arshad, H., Carslaw, N., Cullinan, P., Dimitroulopoulou, S., Greenough, A., Holland, M., Jones, B., Linden, P., Sharpe, T., Short, A., Turner, B., Ucci, M., Vardoulakis, S., Stacey, H., & Hunter, L. (2021). Health effects of indoor air quality on children and young people. *Issues in Environmental Science and Technology*, *50*(2021), 151–188. Available from https://doi.org/10.1039/9781839160431-00151, http://www.rsc.org/.

Hollis, D., McCarthy, M., Kendon, M., & Legg, T. (2022). *HadUK-Grid Gridded Climate Observations on a 1km grid over the UK, v1.1.0.0 (1836-2021)*. Available from https:// www.metoffice.gov.uk/research/climate/maps-and-data/data/haduk-grid/haduk-grid.

Holland, B. (2017). Procedural justice in local climate adaptation: Political capabilities and transformational change. *Environmental Politics*, *26*(3), 391–412. Available from https://doi.org/ 10.1080/09644016.2017.1287625, http://www.tandf.co.uk/journals/titles/09644016.asp.

Hsu, A., Sheriff, G., Chakraborty, T., & Manya, D. (2021). Disproportionate exposure to urban heat island intensity across major US cities. *Nature Communications*, *12*(1). Available from https://doi.org/10.1038/s41467-021-22799-5, http://www.nature.com/ncomms/index.html.

Huang, G., Zhou, W., & Cadenasso, M. L. (2011). Is everyone hot in the city? Spatial pattern of land surface temperatures, land cover and neighborhood socioeconomic characteristics in Baltimore, MD. *Journal of Environmental Management*, *92*(7), 1753−1759. Available from https://doi.org/10.1016/j.jenvman.2011.02.006.

Islam, S.N., & Winkel, J. (2017). *Climate Change and Social Inequality*. Department of Economic & Social Affairs (DESA) Working Paper No. 152 ST/ESA/2017/DWP/152, United Nations (UN), New York, USA. Available from https://www.un.org/esa/desa/papers/2017/wp152_2017.pdf.

Kabisch, N., & Haase, D. (2014). Green justice or just green? Provision of urban green spaces in Berlin, Germany. *Landscape and Urban Planning*, *122*, 129−139. Available from https://doi.org/10.1016/j.landurbplan.2013.11.016, http://www.elsevier.com/inca/publications/store/5/0/3/3/4/7.

Kjellstrom, T., Freyberg, C., Lemke, B., Otto, M., & Briggs, D. (2018). Estimating population heat exposure and impacts on working people in conjunction with climate change. *International Journal of Biometeorology*, *62*(3), 291−306. Available from https://doi.org/10.1007/s00484-017-1407-0, http://www.link.springer.de/link/service/journals/00484/index.htm.

Klinenberg, E. (2015). *Heat wave: A social autopsy of disaster in Chicago*. University of Chicago Press.

Klinsky, S., & Mavrogianni, A. (2020). Climate justice and the built environment. *Buildings and Cities*, *1*(1), 412−428. Available from https://doi.org/10.5334/bc.65, https://journal-buildingscities.org/about/.

Kolokotsa, D., & Santamouris, M. (2015). Review of the indoor environmental quality and energy consumption studies for low income households in Europe. *Science of the Total Environment*, *536*, 316−330. Available from https://doi.org/10.1016/j.scitotenv.2015.07.073, http://www.elsevier.com/locate/scitotenv.

Lai, K. Y., Sarkar, C., Kumari, S., Ni, M. Y., Gallacher, J., & Webster, C. (2021). Calculating a national anomie density ratio: Measuring the patterns of loneliness and social isolation across the UK's residential density gradient using results from the UK Biobank study. *Landscape and Urban Planning*, *215*. Available from https://doi.org/10.1016/j.landurbplan.2021.104194, http://www.elsevier.com/inca/publications/store/5/0/3/3/4/7.

Lee H., Calvin K., Dasgupta D., Krinner G., Mukherji A., Thorne P., Trisos C., Romero J., Aldunce P., Barrett K., Blanco G., Cheung W.W.L., Connors S.L., Denton F., Diongue-Niang A., Dodman D., Garschagen M., Geden O., Hayward B., ... Zommers, Z. (2023).

Climate change 2023: Synthesis report. Contribution of Working Groups I, II and III to the Sixth Assessment Report of the Intergovernmental Panel on Climate Change. Intergovernmental Panel on Climate Change. Available from https://doi.org/10.59327/IPCC/AR6-9789291691647.001.

Liu, J., Varghese, B. M., Hansen, A., Xiang, J., Zhang, Y., Dear, K., Gourley, M., Driscoll, T., Morgan, G., Capon, A., & Bi, P. (2021). Is there an association between hot weather and poor mental health outcomes? A systematic review and meta-analysis. *Environment International* (153). Available from https://doi.org/10.1016/j.envint.2021.106533, http://www.elsevier.com/locate/envint.

Lomas, K. J., & Porritt, S. M. (2017). Overheating in buildings: Lessons from research. *Building Research and Information*, *45*(1−2), 1−18. Available from https://doi.org/10.1080/09613218.2017.1256136, http://www.tandf.co.uk/journals/titles/09613218.asp.

Macintyre, H. L., Heaviside, C., Taylor, J., Picetti, R., Symonds, P., Cai, X. M., & Vardoulakis, S. (2018). Assessing urban population vulnerability and environmental risks across an urban area during heatwaves — Implications for health protection. *Science of the Total Environment*, *611*, 678–690. Available from https://doi.org/10.1016/j.scitotenv.2017.08.062, http://www.elsevier.com/locate/scitotenv.

Madrigano, J., Ito, K., Johnson, S., Kinney, P. L., & Matte, T. (2015). A case-only study of vulnerability to heat wave–related mortality in New York City (2000–2011). *Environmental Health Perspectives*, *123*(7), 672–678. Available from https://doi.org/10.1289/ehp.1408178, http://ehp.niehs.nih.gov/wp-content/uploads/123/7/ehp.1408178.alt.pdf.

Maskrey, A., Jain, G., & Lavell, A. (2023). The social construction of systemic risk: Towards an actionable framework for risk governance. *Disaster Prevention and Management: An International Journal*. Available from https://doi.org/10.1108/DPM-07-2022-0155, http://www.emeraldinsight.com/info/journals/dpm/dpm.jsp.

Mattar, S. D., Jafry, T., Schröder, P., & Ahmad, Z. (2021). Climate justice: Priorities for equitable recovery from the pandemic. *Climate Policy*, *21*(10), 1307–1317. Available from https://doi.org/10.1080/14693062.2021.1976095, http://www.tandfonline.com/loi/tcpo20.

Meyer, L. H., & Roser, D. (2010). Climate justice and historical emissions. *Critical Review of International Social and Political Philosophy*, *13*(1), 229–253. Available from https://doi.org/10.1080/13698230903326349.

Morgan, C., Foster, J. A., Poston, A., & Sharpe, T. R. (2017). Overheating in Scotland: Contributing factors in occupied homes. *Building Research and Information*, *45*(1–2), 143–156. Available from https://doi.org/10.1080/09613218.2017.1241472, http://www.tandf.co.uk/journals/titles/09613218.asp.

Naughton, M. P., Henderson, A., Mirabelli, M. C., Kaiser, R., Wilhelm, J. L., Kieszak, S. M., Rubin, C. H., & McGeehin, M. A. (2002). Heat-related mortality during a 1999 heat wave in Chicago. *American Journal of Preventive Medicine*, *22*(4), 221–227. Available from https://doi.org/10.1016/S0749-3797(02)00421-X.

Newell, P., Srivastava, S., Naess, L. O., Torres Contreras, G. A., & Price, R. (2021). Toward transformative climate justice: An emerging research agenda. *Wiley Interdisciplinary Reviews: Climate Change*, *12*(6). Available from https://doi.org/10.1002/wcc.733, http://onlinelibrary.wiley.com/journal/10.1002/(ISSN)1757-7799.

Oikonomou, E., Davies, M., Mavrogianni, A., Biddulph, P., Wilkinson, P., & Kolokotroni, M. (2012). Modelling the relative importance of the urban heat island and the thermal quality of dwellings for overheating in London. *Building and Environment*, *57*, 223–238. Available from https://doi.org/10.1016/j.buildenv.2012.04.002.

O'Neill, M. S., Zanobetti, A., & Schwartz, J. (2005). Disparities by race in heat-related mortality in four US cities: The role of air conditioning prevalence. *Journal of Urban Health*, *82*(2), 191–197. Available from https://doi.org/10.1093/jurban/jti043.

ONS. (2022). *Families and households in the UK: 2021.* UK Office for National Statistics (ONS).

Petrou, G., Symonds, P., Mavrogianni, A., Mylona, A., & Davies, M. (2019). The summer indoor temperatures of the English housing stock: Exploring the influence of dwelling and household characteristics. *Building Services Engineering Research and Technology*, *40*(4), 492–511. Available from https://doi.org/10.1177/0143624419847621, http://bse.sagepub.com/archive/.

Renalds, A., Smith, T. H., & Hale, P. J. (2010). A systematic review of built environment and health. *Family & Community Health*, *33*(1), 68–78. Available from https://doi.org/10.1097/fch.0b013e3181c4e2e5.

Sánchez-Guevara Sánchez, C., Mavrogianni, A., & González, F. J. Neila (2017). On the minimal thermal habitability conditions in low income dwellings in Spain for a new definition of fuel poverty. *Building and Environment, 114*, 344−356. Available from https://doi.org/10.1016/j.buildenv.2016.12.029, http://www.elsevier.com/inca/publications/store/2/9/6/index.htt.

Santamouris, M., Kapsis, K., Korres, D., Livada, I., Pavlou, C., & Assimakopoulos, M. N. (2007). On the relation between the energy and social characteristics of the residential sector. *Energy and Buildings, 39*(8), 893−905. Available from https://doi.org/10.1016/j.enbuild.2006.11.001.

Santamouris, M., & Kolokotsa, D. (2015). On the impact of urban overheating and extreme climatic conditions on housing, energy, comfort and environmental quality of vulnerable population in Europe. *Energy and Buildings, 98*, 125−133. Available from https://doi.org/10.1016/j.enbuild.2014.08.050.

Santamouris, M. (2020). Recent progress on urban overheating and heat island research. Integrated assessment of the energy, environmental, vulnerability and health impact. *Synergies with the global climate change. Energy and Buildings, 207*. Available from https://doi.org/10.1016/j.enbuild.2019.109482, https://www.journals.elsevier.com/energy-and-buildings.

Sengupta, S. (2020). *This is inequity at the boiling point.* The New York Times.

Sera, F., Hashizume, M., Honda, Y., Lavigne, E., Schwartz, J., Zanobetti, A., Tobias, A., Iñiguez, C., Vicedo-Cabrera, A. M., Blangiardo, M., Armstrong, B., & Gasparrini, A. (2020). Air conditioning and heat-related mortality: A multi-country longitudinal study. *Epidemiology (Cambridge, Mass.)*, 779−787. Available from https://doi.org/10.1097/EDE.0000000000001241, http://journals.lww.com/epidem/pages/default.aspx.

Sharifi, S., Saman, W., & Alemu, A. (2019). Identification of overheating in the top floors of energy-efficient multilevel dwellings. *Energy and Buildings, 204*. Available from https://doi.org/10.1016/j.enbuild.2019.109452, https://www.journals.elsevier.com/energy-and-buildings.

Sherman, P., Lin, H., & McElroy, M. (2022). Projected global demand for air conditioning associated with extreme heat and implications for electricity grids in poorer countries. *Energy and Buildings, 268*. Available from https://doi.org/10.1016/j.enbuild.2022.112198, https://www.journals.elsevier.com/energy-and-buildings.

Smith, G. S., Anjum, E., Francis, C., Deanes, L., & Acey, C. (2022). Climate change, environmental disasters, and health inequities: The underlying role of structural inequalities. *Current Environmental Health Reports, 9*(1), 80−89. Available from https://doi.org/10.1007/s40572-022-00336-w, http://link.springer.com/journal/40572.

Tabata, T., & Tsai, P. (2020). Fuel poverty in Summer: An empirical analysis using microdata for Japan. *Science of the Total Environment, 703*. Available from https://doi.org/10.1016/j.scitotenv.2019.135038, http://www.elsevier.com/locate/scitotenv.

Taylor, J., Davies, M., Mavrogianni, A., Shrubsole, C., Hamilton, I., Das, P., Jones, B., Oikonomou, E., & Biddulph, P. (2016). Mapping indoor overheating and air pollution risk modification across Great Britain: A modelling study. *Building and Environment, 99*, 1−12. Available from https://doi.org/10.1016/j.buildenv.2016.01.010, http://www.elsevier.com/inca/publications/store/2/9/6/index.htt.

Taylor, J., McLeod, R., Petrou, G., Hopfe, C., Mavrogianni, A., Castaño-Rosa, R., Pelsmakers, S., & Lomas, K. (2023). Ten questions concerning residential overheating in Central and Northern Europe. *Building and Environment, 234*. Available from https://doi.org/10.1016/j.buildenv.2023.110154, http://www.elsevier.com/inca/publications/store/2/9/6/index.htt.

Taylor, J., Wilkinson, P., Davies, M., Armstrong, B., Chalabi, Z., Mavrogianni, A., Symonds, P., Oikonomou, E., & Bohnenstengel, S. I. (2015). Mapping the effects of urban heat island, housing, and age on excess heat-related mortality in London. *Urban Climate, 14*, 517−528. Available from https://doi.org/10.1016/j.uclim.2015.08.001, http://www.journals.elsevier.com/urban-climate/.

Thompson, R., Hornigold, R., Page, L., & Waite, T. (2018). Associations between high ambient temperatures and heat waves with mental health outcomes: A systematic review. *Public Health, 161,* 171–191. Available from https://doi.org/10.1016/j.puhe.2018.06.008, http://www.elsevier.com/inca/publications/store/6/4/5/7/2/7/645727.pub.htt.

Thompson, R., Landeg, O., Kar-Purkayastha, I., Hajat, S., Kovats, S., & O'connell, E. (2022). Heatwave mortality in summer 2020 in England: An observational study. *International Journal of Environmental Research and Public Health, 19*(10). Available from https://doi.org/10.3390/ijerph19106123, https://www.mdpi.com/1660-4601/19/10/6123/pdf?version = 1652858943.

UK DWP. (2022). UK Department for Work and Pensions, Stat-Xplore tool.

Ulpiani, G. (2021). On the linkage between urban heat island and urban pollution island: Three-decade literature review towards a conceptual framework. *Science of the Total Environment, 751.* Available from https://doi.org/10.1016/j.scitotenv.2020.141727, http://www.elsevier.com/locate/scitotenv.

United States Census Bureau. (2022). *Annual Social and Economic Supplements.* Available from https://www.census.gov/data/datasets/time-series/demo/cps/cps-asec.html.

Vandentorren, S., Bretin, P., Zeghnoun, A., Mandereau-Bruno, L., Croisier, A., Cochet, C., Ribéron, J., Siberan, I., Declercq, B., & Ledrans, M. (2006). August 2003 heat wave in France: Risk factors for death of elderly people living at home. *European Journal of Public Health, 16*(6), 583–591. Available from https://doi.org/10.1093/eurpub/ckl063.

Vellei, M., Ramallo-González, A. P., Coley, D., Lee, J., Gabe-Thomas, E., Lovett, T., & Natarajan, S. (2017). Overheating in vulnerable and non-vulnerable households. *Building Research and Information, 45*(1–2), 102–118. Available from https://doi.org/10.1080/09613218.2016.1222190, http://www.tandf.co.uk/journals/titles/09613218.asp.

Wan, K., Feng, Z., Hajat, S., & Doherty, R. M. (2022). Temperature-related mortality and associated vulnerabilities: Evidence from Scotland using extended time-series datasets. *Environmental Health: A Global Access Science Source, 21*(1). Available from https://doi.org/10.1186/s12940-022-00912-5, https://ehjournal.biomedcentral.com/.

Watts, N., Amann, M., Arnell, N., Ayeb-Karlsson, S., Beagley, J., Belesova, K., Boykoff, M., Byass, P., Cai, W., Campbell-Lendrum, D., Capstick, S., Chambers, J., Coleman, S., Dalin, C., Daly, M., Dasandi, N., Dasgupta, S., Davies, M., Di Napoli, C., . . . Costello, A. (2021). The 2020 report of the Lancet countdown on health and climate change: Responding to converging crises. *The Lancet, 397*(10269), 129–170. Available from https://doi.org/10.1016/S0140-6736(20)32290-X, http://www.journals.elsevier.com/the-lancet/.

Wilson, W., & Barton, C. (2021). *Overcrowded housing (England).* Research Briefing. House of Commons Library. Available from https://commonslibrary.parliament.uk/research-briefings/sn01013/.

Wistow, J., Dominelli, L., Oven, K., Dunn, C., & Curtis, S. (2015). The role of formal and informal networks in supporting older people's care during extreme weather events. *Policy and Politics, 43*(1), 119–135. Available from https://doi.org/10.1332/030557312X655855, http://docserver.ingentaconnect.com/deliver/connect/tpp/03055736/v43n1/s7.pdf?expires = 1423206497&id = 80715474&titleid = 777&accname = Elsevier + BV&checksum = 3C54B25919242941EF5EF1A49A7E53CF.

Yitshak-Sade, M., Bobb, J. F., Schwartz, J. D., Kloog, I., & Zanobetti, A. (2018). The association between short and long-term exposure to PM2.5 and temperature and hospital admissions in New England and the synergistic effect of the short-term exposures. *Science of the Total Environment, 639,* 868–875. Available from https://doi.org/10.1016/j.scitotenv.2018.05.181, http://www.elsevier.com/locate/scitotenv.

Chapter 11

Urban overheating governance on the mitigation and adaptation of anthropogenic heat emissions

Elmira Jamei[1,2], Majed Abuseif[3,4], Amirhosein Ghaffarianhoseini[5] and Ali Ghaffarianhoseini[5]

[1]*College of Engineering and Science, Victoria University, Melbourne, VIC, Australia,* [2]*Institute of Sustainable Industries and Liveable Cities, Victoria University, Melbourne, VIC, Australia,* [3]*School of Engineering and Built Environment, Griffith University, Gold Coast, QLD, Australia,* [4]*Green Infrastructure Research Labs (GIRLS), Cities Research Institute, Griffith University, Gold Coast, QLD, Australia,* [5]*Department of Built Environment Engineering, School of Future Environments, Auckland University of Technology, Auckland, New Zealand*

11.1 Introduction

By 2050, more than 2.5 billion people are expected to live in urban areas, as cities play an important role in social and economic development at global, regional, and local scales. Urban areas are the nodal points for political and economic hubs, decision making, technology, innovation, and knowledge. Metropolitans in particular are key drivers for global change and significantly impact sociocultural, economic, political, and environmental dimensions of urban dwellers' lives (Hall & Pfeiffer, 2013; Kraas, 2007).

The rapid growth in population densities and urban developments triggers the need to build additional urban infrastructure and remove natural green coverage in urban areas. This replacement leads to the formation of urban heat island (UHI), which is also defined as "the higher air and surface temperature in urban settings compared with surrounding rural areas." This phenomenon is mainly caused by the alterations in the physical changes in the energy balance of surfaces (Oke et al., 2017; Stewart, 2019) and the rejected heat emissions from anthropogenic sources (buildings, transportation, and human metabolism) (Chow et al., 2014; Sailor, 2011).

Mitigation and Adaptation of Urban Overheating. DOI: https://doi.org/10.1016/B978-0-443-13502-6.00016-6
295

UHI increases minimum and maximum air temperatures, and this trend has been observed since the 1950s across all climate zones (Stocker, 2013). Since 1980, urban areas have also significantly increased in the number of heat waves (Mishra et al., 2015). The combined effect of local-scale UHI and increased number of extreme hot days leads to exacerbated overheating in urban areas (Chapman et al., 2019; Kotharkar & Surawar, 2016; Krayenhoff et al., 2018; Wouters, 2017). As a result, the number of urban population who become vulnerable to urban overheating and extreme heat (Pelling & Garschagen, 2019) is increasing.

The primary definition of overheating strongly focuses on energy use in buildings, indoor environmental conditions, and public health of urban dwellers from building and urban design perspectives (Santamouris et al., 2015; Taylor, 2015). A more accurate definition of "urban overheating" is presented as "the exceedance of locally defined thermal thresholds that correspond to negative impacts on people (e.g., health, comfort, and productivity) and associated urban systems."

Urban overheating is driven by global climate change and rapid urban growth, and it is affected by complex interactions among building-, city-, and global-scale climates. Urban overheating has several negative impacts on human life and well-being, overall energy consumption, liveability, and sustainability. The level of vulnerability to urban overheating is high for certain groups, including low socioeconomy people and those with underlying health conditions (e.g., elderly and people with disability and cardiovascular disease). Therefore, researchers have identified methods to save vulnerable people's life and formulate strategies to reduce heat-related mortality and morbidity in such communities. These strategies are multidisciplinary solutions and vary from reducing the sensitivity and heat exposure to increasing adaptive capacities in governance.

The discourse on mitigation of urban climate change has only recently received momentum in the political and scientific arenas. As a result, the overall understanding of urban overheating and mitigation methods has progressed well over the past 2 decades. However, an integrated governance approach on mitigating urban overheating is lacking. In this chapter, we aim to focus on one of the key drivers of urban overheating, that is, anthropogenic heat. We firstly define the "anthropogenic heat" concept and major factors that lead to its formation. Then, we discuss the role that governance can play in mitigating anthropogenic heat through a case study approach.

11.2 Anthropogenic heat and its definition and sources

The anthropogenic heat flux term can be considered a source term in the volume that accounts for emission of heat from vehicles, building HVAC systems, and other sources, including human metabolism of individuals within the street canyons. In fact, the anthropogenic heat resale is the accumulation of the heat from human-induced activities in addition to the urban long-wave radiation budget. Various sources lead to the formation of anthropogenic heat. Anthropogenic heat is related to not only the energy consumption but also to the sectors that generate the highest level of energy (transportation, buildings, and industry) and human metabolism. The most important sources that contribute to the formation of anthropogenic heat are presented in the following sections.

11.2.1   Metabolism

The metabolism rate can vary from 100 to 300 depending on the intensity of the activity. The metabolism rate varies during the course of the day and increases as the level of activity rises. According to Baxter et al. (1987) and Fanger (1972), the metabolism rate for a 70 kg sleeping man is 75 W.

The magnitude of anthropogenic heat from human metabolism in a city depends on its population density. Given that cities accommodate larger population densities in urban cores and most people reside in buildings, the anthropogenic heat in an urban area is mainly a result of building sector occupancy load, heating—cooling load, and human metabolism. Therefore, most studies consider human metabolism as part of their building energy model (e.g., Heiple & Sailor, 2008; Hsieh et al., 2007) or totally remove it (e.g., Grimmond, 1992; Sailor & Lu, 2004). These studies assume that a major part of the metabolic heat occurs in buildings. Moreover, outdoor human metabolism is generally less than 1% of the total anthropogenic heat in a city. Therefore, it can be removed from the calculations.

11.2.2   Industry

The major part of the energy use in the industrial area can be directly converted into sensible heat. However, some cases utilize evaporative cooling towers to remove heat. Therefore, estimating the magnitude of sensible anthropogenic heat is difficult. Reasonable estimates should be in place with regard to equipment features and usage within the predominant industries in the study area. After the spatial distribution and magnitude of energy consumption are determined, the overall energy use during the year can be calculated (e.g., Sailor & Fan, 2004; Sailor & Lu, 2004; Torrance & Shun, 1976). The monthly or annual energy use of the power utility and other governmental energy agency can be achieved based on the study area/city. Then, land use data can be used to apportion the industrial sector energy use into the study area within the city where the industry is prevalent.

11.2.3   Buildings

Energy use in buildings can be classified into three main categories: lighting, plug, and appliance and HVAC loads. This load largely depends on the occupancy timetable, day type (working or nonworking), and building type (residential or commercial).

The energy use from the HVAC is a complex function of occupancy, internal, and environmental loads. After the external heat enters the building (E), the internal heat is generated by lighting (L), plug loads (P), and human metabolism (M). Therefore, the cooling system works extra and uses a higher level of energy (AC) to transfer the heat to the outdoor. The rejected heat R is represented as follows:

$$\text{As } R = E + P + M + L + AC \tag{11.1}$$

The magnitude of the rejected heat (including sensible and latent heat) largely depends on the mechanical equipment used to cool the building.

Environmental loads are mainly associated with environmental heating or cooling of building interiors (e.g., loss of heat to outside via conduction, leakage, natural ventilation, infiltration, and solar radiation penetration through windows). The rejected heat by air conditioning in summer is larger than the thermal load caused by the energy use of the building. For example, the rejected heat in summer from an office building in Texas can be 70% higher than the total energy use of the building (Sailor & Brooks, 2009). Some of the air conditioning systems use evaporative cooling to remove the heat. As a result, the majority of the heat can be removed as evaporated water.

A major complexity in quantifying anthropogenic emissions from the building sector is the calculation of vertical location of such emissions. The reason is that some emissions occur via conduction from the building envelop and others through air exchange via facades and windows and doors. The highest proportion of anthropogenic heat emission from a building sector is associated with heat rejections from the HVAC and ventilation equipment. Therefore, identifying where these systems are situated (e.g., ground and roof) is critical. The reason is that the height in which the anthropogenic emissions is removed from the building is an important factor in its accurate calculation.

11.2.4 Vehicles

The anthropogenic heat from vehicles is mainly generated with the combustion of the gas or diesel fuel. A research study showed that the heat generated by a vehicle can be up to 45 MJ/kg (Annamalai & Puri, 2006). A chemical process also initiates where the water evaporates when the fuel is combusted. For each liter of gas or diesel fuel combusted, 1 kg of water vapor will be produced. For 1 km of driving, nearly 100 g of water will be generated.

Identifying the temporal and spatial distribution of vehicles on roads within a city is a key challenge in estimating the anthropogenic emissions from vehicles. The distribution changes substantially over the course of a day. The actual fuel economy of vehicle also largely depends on the type and distribution of vehicles on roads. It is also subject to change over the course of the day. The fuel economy depends on the city and the country as well, but it generally ranges from 8 to 16 km/L (20−40 miles per gallon).

The transport departments in some countries track the vehicle use in different forms. In some cases, the data and statistics related to the average distance traveled by a vehicle in a city and hourly traffic count data are unavailable for all road types. As a result, reaching a certain level of disaggregation and assignment of diurnal profiles is critical. For example, a diurnal profile with the typical morning and evening traffic peak times was developed by Hallenbeck et al. (1997) for US cities. The study showed that nearly 16% of the daily traffics occur between 16:00 and 18:00, and 13% of them occur between 7:00 and 9:00 in the peak morning hour. A research by Sailor & Lu (2004) showed that the anthropogenic heat emission caused by vehicles in US cities is as high as 300 W/m^2 at peak hours.

11.3 Governance and mitigation of urban overheating

The adverse impacts of urban overheating on human health and liveability have encouraged researchers to pay additional attention toward identifying the responsible institutions in the governance of urban overheating, particularly anthropogenic heat. The governance of urban overheating is defined as "an extension of—or perhaps even an explicit component of—climate change governance more broadly defined (Fröhlich & Knieling, 2013)." Urban overheating governance is also explained as "an aspect of climate adaptation, for which a rich suite of definitions, conceptual models, and theories have been proposed" (Keith & Meerow, 2022; Moser & Ekstrom, 2010).

Over the last 2 decades, an incentive discourse on governance has been developed, which has significantly impacted political and academic arenas (Elander, 2002; Stren & Cameron, 2005). Most of these discourses and approached are identified under the label of the governance model, and each encompasses various normativisms and ontologies (Alcántara et al., 2004; Grindle, 2007; Peters & Pierre, 2000). However, some common principles exist in nearly all governance approaches.

Urban overheating occurs at different scales (e.g., local and regional scales) (Georgescu, 2015; Jay, 2021). In the context of climate change, the key challenge for adaptation and mitigation is to divide the responsibility among authorities and resources and agreements on mitigation methods and their implementation (Moser & Ekstrom, 2010). The lack of clarity with respect to the responsibility and accountability of each institution and accessibility to financial resources and overall ignorance toward the urban overheating issue are some of the key governance barriers in mitigating urban overheating in urban areas (Keith et al., 2019).

Most of the contemporary governance models are well established in the way they deal with natural hazards such as air or noise pollution. Several strategies can deal with such hazards (e.g., provision of incentives, national to local regulatory structures, regulatory structures, and workplace protections) (Keith et al., 2021). In the urban overheating context, the contemporary governance models strongly focus on the impacts (e.g., health and energy) rather than prevention. For example, the World Health Organisation and the World Meteorological Organisation have released guidance on implementing heat-health warning systems to raise awareness among public and communities on the adverse impacts of heat waves (McGregor, 2015).

In certain cases, structural planning documents provide strategies to address the urban overheating issue. These documents are often endorsed by a local council with varying degrees of regulatory authority. These regulations are often integrated into broader plans (e.g., general plans or sustainability/resilience plans) (Gabbe et al., 2021).

The challenges to govern urban overheating are reportedly consistent with those identified in the climate change governance literature. For example, the lack of clarity with regard to the role and responsibility of practitioners with respect to the urban climate governance is identified as a key barrier in tackling the governance of urban overheating issue (Guyer, 2019; Mees et al., 2015). The other key challenge in the governance arena is associated with the responsible authority for urban growth and developments in the context of urban overheating and the governance that impacts the processes (Mahlkow et al., 2016). These challenges are particularly more intense in developing countries where the rapid population growth and the need to provide infrastructure are inevitable (Birkmann et al., 2010).

In the planning arena, informal urban developments cannot be adjusted or modified to mitigate urban overheating. Therefore, the mitigation measures have to be transparent and additional governance models should be considered to go beyond state-centered formal mechanisms. In planning to conduct appropriate governance for tackling urban overheating in formal settlements, urban dwellers should be consulted and included in the decision-making process; they should contribute to the more effective implementation of mitigation strategies by using their lived experience (Guardaro et al., 2020; Marschütz, 2020). Therefore, citizen engagement and community consultations have been promoted in the contemporary governance models.

Generation of digital tweens for cities and scenario planning workshops have demonstrated promising results in engagement with communities and better shaping governance strategies for urban overheating (Iwaniec et al., 2020). Notably, participation of the private sector in the implementation of mitigation strategies is essential due to the limited land owned by governmental agencies in urban areas. Measures to increase such participation and collaborations may include public–private partnerships, incentives, and financing. These factors can all contribute to speeding up the process of identifying solutions for urban overheating.

Technology in the form of IOT can act as a mediator, but its utilization in governance arena should be controlled and balanced. IOT can play a critical role in enabling access of governance agencies to accurate data on urban climate and allowing them to understand the impact of urban climate on urban systems (Hamstead, 2020; Hondula et al., 2018; Yin & Zhu, 2020). However, widespread sensing raises potential social and legal issues with respect to privacy and data security.

Adaptive governance is a promising model to tackle urban overheating, and it is also an effective model in the context of urban ecology (Bettini et al., 2015; Green, 2016; Larson et al., 2015). This model provides the highest level of flexibility to adapt during paradigm shift. The adaptive model highlights the need for a paradigm shift toward an integrative perspective in planning and governance structures, as well as actions and agency of self-regulating agents. This model also indicates that the mitigation of urban overheating cannot be addressed adequately via the sole implementation of large-scale structural measures in modifying the built environment. Planning and multilevel governance structures should be considered to ensure that all relevant aspects are captured and relevant stakeholders are included. Therefore, governance in this model tends to broaden the scope.

A recent approach suggested in Meerow and Newell (2019) is practical and inclusive for decision makers and communities and has five Ws for urban resilience (for whom, what, when, where, and why). According to this approach, addressing the governance challenge with respect to urban overheating cannot be conducted without considering the sociopolitical aspect and process that form urban areas. Therefore, all parties should participate to better address the urban overheating issue. Thus, we must consider for whom, what, when, where, and why these efforts are being directed.

11.4   Melbourne (case study)

Cities worldwide have implemented various strategies and initiatives to mitigate urban overheating. These efforts promote sustainability and resilience in the face of climate change to addressing the immediate and long-term impacts of urban overheating.

According to *Victoria's Climate Science Report 2019*, the climate in Victoria has already changed over the last few decades and is increasingly becoming hotter and drier with more frequent hot days and more droughts (Krayenhoff et al., 2018). The Victorian government has agreed to commit to the Paris Climate Agreement and has developed the Climate Change Act in 2017. It enables Melbourne City to implement a long-term plan to cut its emissions for contributing to the drop in human-induced heat.

Melbourne Municipality is 37.7 km^2 and covers the central business district and inner suburbs. The key contributing factors in Melbourne's anthropogenic heat are the heat generated through the energy use in buildings and transport. Therefore, the governance strategies aimed to develop ways to reduce the emissions and heat from these sources. Melbourne has set its key strategies to become 100% reliant on renewable energy, which realize zero emission from the building and transportation sectors. The following sections discuss some of the governance strategies developed by Melbourne City to reduce human-induced heat in the urban area.

11.4.1 Strategic priority 1: Melbourne renewable energy project

One of the initiatives proposed by Melbourne City is to reduce the magnitude of human-induced heat under the "Melbourne Renewable Energy Project." This project aims to unite Melbourne's leading businesses, higher educations, and stakeholders in developing a power purchase agreement. This plan enables the generation of 39 turbines with 80 MW capacity windfarm at Crolands in regional Victoria. This project aims to build up on the current 18% adopted renewable energy and increase it by 3% (Baxter et al., 1987). This project also will work as a proof of evidence for the other stakeholders in the state and will accelerate renewable energy uptake as a power source.

A key challenge in implementing renewable energy is to determine the responsibility and role of state government in rolling out the governance incentives. The Victorian government committed to achieve 40% renewable energy by 2025. Achieving this target would be impossible without thoughtful planning and governance at the federal and state level and science-based emission reduction plans. Therefore, further investments on renewable energy are needed to reduce the human-induced heat caused by energy usage.

11.4.2 Strategic priority 2: zero emissions from buildings and precincts

Melbourne City has collaborated with residential and commercial building owners and tenants to incentivize the improvements in environmental performance of their buildings. One of the initiatives that was developed to address this objective is the "1200 Buildings Programme." It aims to raise awareness among building owners on energy efficiency literacy, retrofitting strategies, and overall perception toward environmental sustainability.

However, addressing this target in Melbourne is challenging because most of the buildings in Melbourne are not as efficient as they could be. As a result, the benchmark for evaluating the energy performance of building portfolio is not fully developed (particularly for apartments). Therefore, residents are not aware of the ways in which they can improve the building energy performance, which can lead to increased human-induced heat in the urbanized area.

The other challenge is that renters for both residential and commercial buildings cannot retrofit the buildings without permission from landlords and their landlords are also not incentivized to improve the energy performance of their buildings. Thus, the landlords will put the pressure of high costs to the corporate tenants.

These buildings also often retain the heat, and they become uninhabitable during heat waves and power cut. This situation becomes a critical issue for public health, given that the number and intensity of heat waves are increasing rapidly.

One of the strategies to reduce the human-induced heat is to ensure that both new and existing residential and commercial buildings become carbon-neutral by 2050. This aim is also well aligned with Australia's Carbon Positive Road Map.

One of Melbourne's first 6-star building is a 10-story office called "Council House 2." It has a novel water and energy-saving technology, sustainable building material, green roof, and bike storage to promote active transportation. As a result of Melbourne's renewable energy project, this building has been running with 100% renewable energy since 2019. The amount of energy use per employee in this building is less than half of the consumption in Council House 1, which was built in 1970. After the successful implementation of this project, Melbourne City has progressively continued building 6-star buildings across the municipality (e.g., East Melbourne Library, Art Play, and Library at the Dock).

The City of Melbourne, as a major urban center, has recognized the crucial role that buildings play in contributing to anthropological heat. Thus, it has prioritized initiatives that aim to increase energy efficiency in buildings, as shown in Table 11.1.

11.4.3 Strategic priority 3: zero emissions from transportation

The third key contributor to Melbourne's human-induced heat is the public transport sector. Most of the heat generated by transportation is associated with the use of private cars and freights. A large number of daytime population live and work in Melbourne. Around 900,000 workers, tourists and residents who may use public transport, cycle, or drive to surrounding suburbs contribute to the generation of human-induced heat. This trend is estimated to increase with the rapid growth in rail infrastructure in Melbourne. Thus, the heat generated by the public transport sector needs support and mitigation. This problem can be addressed via planning for more public open space, parks, and more pedestrian spaces. Introduction of solar powered trams is another strategy in addressing this challenge. The Victorian

TABLE 11.1 Melbourne City initiatives in reducing the energy consumption in buildings.

1200 Buildings Programme	This program is a collaboration among Melbourne City, the Victorian Government, and industry partners to retrofit 1,200 buildings in the city for improving energy efficiency and reducing greenhouse gas (GHG) emissions. The program includes building assessments, financial incentives, and technical support (City of Melbourne, 2023a).
Smart Blocks Programme	This program provides resources and support for apartment buildings to improve energy efficiency and reduce energy costs. The program includes building assessments, energy-saving advice, and support for implementing energy-efficient upgrades (City of Melbourne, 2023b).
Environmental Upgrade Finance	This program provides access to finance for building owners to implement energy-efficient upgrades. The finance is repaid through council rates over a period of up to 10 years (Melbourne, 2023a).
Sustainable Melbourne Fund	This initiative provides access to finance for building owners to implement sustainability upgrades, including energy-efficient measures. The finance is repaid through council rates for up to 20 years (Melbourne, 2023b).
Residential Efficiency Scorecard	This program is a collaboration between Melbourne City and the Victorian Government. It provides a rating system for the energy efficiency of residential buildings. The program includes assessments of building energy performance, recommendations for energy-efficient upgrades, and information on available incentives and funding (State Government of Victoria, 2023)

government ran a tender in 2017 to assist with building solar farms that can accommodate Melbourne tram's network. This initiative alone could cut the emission by 80,000 tons of GHG emissions every year to cover 493 trams. It can consequently contribute to reduced anthropogenic heat caused by transportation.

The method of promotion and incentivization of electric vehicles until the availability of adequate supply of renewable energy to support the electricity grid is also proposed to decrease the anthropogenic heat in this sector. A longer-term strategy should explore avenues on the way to enable cars and buses to be powered by 100% renewable energy.

The governance regarding transportation in Melbourne is influenced by the Victorian government, specifically the Integrated Transport Planning; it is a holistic approach to transportation planning that considers the various modes of transport, including road, rail, bus, cycling, and walking, as well as land use, social and economic considerations, and environmental sustainability (Victorian Auditor-General's Office, 2023). The city's government has recognized the need to reduce reliance on private cars, which are a major contributor to air pollution and GHG emissions (Krayenhoff et al., 2018). Therefore, the city's transportation strategy includes various initiatives to encourage sustainable modes such as cycling, walking, and public transportation.

One of the main components of Melbourne's transportation strategy is to construct dedicated bike lanes and pedestrian walkways. The city has invested heavily in developing an extensive network of bike lanes that connect various parts of the city, which makes cycling safer and easier for people. The bike lanes are often separated from vehicular traffic to provide a safe and comfortable cycling experience. Similarly, the city has also developed a network of pedestrian walkways that are designed to encourage walking and provide safe and accessible routes for pedestrians (City of Melbourne, 2023c, 2023d).

Expanding Melbourne's public transportation network is another key transportation strategy component of the city. The city has a well-connected public transportation system that includes trains, trams, and buses. In recent years, the city has invested heavily in expanding and upgrading its public transportation infrastructure to make it more reliable and efficient. This initiative includes constructing new train and tram lines, introducing high-capacity buses, and implementing smart transportation systems (Victoria State Government, 2023a, 2023b).

Melbourne's transportation strategy is also supported by various policy measures that aim to encourage sustainable transportation modes. For instance, the city offers incentives for employers to encourage their employees to use sustainable transportation modes such as cycling and public transportation. The incentives may include subsidies for public transportation passes or the provision of on-site facilities such as bike racks and showers. This policy measure helps promote sustainable transportation options and reduce reliance on private cars (Melbourne City Greens, 2023; Tourism & Transport Forum, 2011).

11.5 Auckland (case study)

On par with other cities around the globe with initiatives to mitigate urban overheating, key cities in New Zealand, including Auckland, are also affected by the negative consequences of urban overheating (Jalali, 2022). As a result, there are strategies in place for this matter, primarily under the larger umbrella of emission reduction plans at national and city levels. These endeavors are deemed to contribute significantly to tackling the impacts of climate change in New Zealand.

In 2022, the New Zealand government established a comprehensive Emissions Reduction Plan for addressing climate issues in the coming 15 years (Barnett, 2022), which will directly result in reducing the urban overheating issue, most specifically in the case of Auckland, New Zealand's most populous city. The plan outlines specific goals and measures to achieve them, encompassing all areas of the government and economy, including transportation, energy, construction, waste management, forestry, and agriculture. The objective is to impact all sectors in reducing emissions.

This plan, aimed at reducing emissions, directs New Zealand, and in particular Auckland, toward a trajectory that contributes to global endeavors that restrict the increase in temperature to 1.5°C above the pre-industrial era. The strategy necessitates an additional 11.5 megatons of carbon dioxide equivalent (Mt CO2-e) reduction between 2022 and 2025. This is equal to approximately 4.3 to 5.5 million gasoline-powered vehicles being driven annually for 10,000 kilometers (Barnett, 2022).

To effectively account for and reduce these emissions, a consumption approach is employed, as recommended by the Climate Change Commission (2021). This approach focuses on two main types of building-related emissions: operational carbon emissions, which arise from the energy and resources used in operating the building, and embodied carbon emissions, which are associated with the manufacture, use, and disposal of building materials and products throughout their lifespan. Embodied carbon emissions encompass emissions from the production, transportation, and disposal of these materials, emphasizing the need for a comprehensive strategy to tackle building-related emissions. To mitigate urban overheating and reduce GHG emissions, almost 9.4% of which were attributed to the building industry in 2018, the government provided a list of recommendations (Barnett, 2022). Decreasing the operational and embodied carbon in construction materials and promoting low-emission building designs were highlighted as key concepts, coupled with promises to revise the Building Code and promote the significance of building energy performance monitoring and assessment.

By 2023, Auckland needs to halve its emissions, and by 2050, the goal is to reach net-zero emissions. Te Tāruke-ā-Tāwhiri, Auckland's climate plan, has been adopted by Auckland Council, and the implementation of strategies and changes is underway. Auckland's emissions are primarily driven by several factors, which Tāruke-ā-Tāwhiri prioritizes in the following eight categories: Transport, Natural Environment, Food, Built Environment, Communities and Coast, Te Puāwaitanga ō te Tātai (prosperity and resilience of Māori communities), Economy, and Energy and Industry (Auckland, 2020). Similarly, to combat urban overheating in Auckland, it is crucial to refer to Auckland's Urban Ngahere (Forest) Strategy, which aims to understand, grow, and protect urban forests in Auckland with core principles such as preferences for native species, diversity, healthy trees, ecological connections, and more (Auckland, 2019). In a recent report by the Climate Change

Commission, it is highlighted that New Zealand should accelerate the shift to a low-emission and climate-resilient future by adopting efficient initiatives and strategies, enabling the country to meet its emission reduction targets. In this regard, the highest priorities are set, including EV charging infrastructure, new renewable electricity generation, healthier and more resilient buildings, user-friendly public transport, Iwi/Māori emissions reduction, forestry, and more (Climate Change Commission, 2021).

In line with this strategy, not only are advanced sustainable development spectrums considered as a means to reduce carbon emissions in current and future building and urban developments in New Zealand, but specific focus has also been directed toward the progression and upkeep of intelligent infrastructure, particularly from a transportation perspective. Currently, various research endeavors concerning the application of advanced and emerging technologies such as artificial intelligence (AI) to support this sector are being developed. Recent advancements in AI enable otherwise disparate and inaccessible information about a subject area to be made available to decision-making models, resulting in a step change in their performance (Allam & Dhunny, 2019). Natural language processing techniques make the interpretation and presentation of the results easily accessible to a wide range of users, from small individual operators to government regulators (Khurana et al., 2023). When applied to New Zealand's transportation system, this step change in capability should, in principle, significantly improve and accelerate operational, strategic, and regulatory decision-making, and when applied to respective emissions, allow for significantly faster and lower-cost pathways to emission reductions.

Regarding the transportation domain, from a broader perspective, the plan outlines recommendations for reducing reliance on cars by enhancing public transport options for low-income citizens, increasing support for walking and cycling initiatives, and promoting e-bike usage (Barnett, 2022). The government is encouraging the adoption of low-emission vehicles through incentives such as the Clean Car Discount scheme, expanding access to these vehicles for low-income households, and improving the EV charging infrastructure nationwide. Efforts to decarbonize heavy transport and freight are underway, with funding allocated to the freight sector for purchasing zero- and low-emission trucks, a mandate requiring zero-emission public transport buses by 2025, and support for the uptake of low-carbon liquid fuels through sustainable aviation fuel mandates and biofuel obligations. In addition to forming plans to decarbonize heavy freight transportation and developing the baseline for an interconnected network of EV charging stations, the New Zealand government is making crucial decisions to improve the reach, frequency, and quality of public transportation as a whole, while making it available to a wider range of users with varied backgrounds (Barnett, 2022). It is essential to ensure that such ground-breaking efforts are done following a sustainable approach; otherwise, it will defeat the purpose. Furthermore, the Vision Mātauranga policy of the plan can actualize the scientific potential of the indigenous Māori knowledge and holistic outlook, emphasizing the importance of nature in the development of advanced technologies.

The government's 2050 vision for energy and industry, which aims to create a highly renewable, sustainable, and efficient energy system, is considered another key initiative contributing to the mitigation of urban overheating. By transitioning away from fossil fuels and toward increased renewable electricity generation and low-emission fuels, the energy sector will produce fewer GHG emissions, subsequently reducing the UHI effect. Moreover, the adoption of clean technologies and enhanced energy efficiency in buildings can lead to decreased energy consumption for cooling and heating, which in turn helps to mitigate urban overheating. Additionally, promoting economic development through sectors like hydrogen, bioenergy, and electrification can encourage the growth of greener, more sustainable urban spaces, further alleviating the issue of urban overheating.

11.6 Conclusion

There is no doubt that the increased urban air temperature affects the health and well-being of urban dwellers, particularly vulnerable groups (e.g., elderlies, children, and those with underlying disease). The governance decisions and short- and long-term urban planning strategies significantly impact heat mitigation and health of future generations. They can also pave the way for future environmental, social, and economic roadmaps.

Human-induced or anthropogenic heat is a major contributor to the increase in urban air temperature. The human-induced heat can be reduced if not only the government organizations but also public institutions, investors, and businesses take steps in decreasing emissions.

This chapter presents the concept of anthropogenic heat, discusses the key factors that contribute to its formation, and highlights the significance of mitigating anthropogenic heat for the health of urban dwellers. This chapter is then continued by providing an insight on various types of governance. Governance strategies of Melbourne and Auckland are presented as case studies to demonstrate the way governance decisions can reduce human-induced heat.

This chapter concludes that there is no one-size-fits-all solution to mitigate urban overheating. Effective governance requires a multifaceted approach that considers the local context and involvement of various stakeholders. Learning from best practices in this domain provides valuable insights for effective governance of urban overheating.

This chapter also concludes that effective governance strategy requires a high level of literacy on the cause and effects of urban overheating and on human-induced activities that can accelerate the heat. Therefore, educating communities and stakeholders and raising awareness on the impact of resident's behavior and stakeholder's decision making on mitigating the increased urban air temperature can greatly benefit the governments. For instance, preplanning and providing information on incorporating the mitigation of anthropogenic heat in each stage of planning can be considered in the council's agenda.

The other important step toward effective governance in tackling this challenge is to ensure that all stakeholders, including government agencies, private sector organizations, and local communities, are working together in a participatory way. This will result in a decision making that is evidence-based and that captures vulnerable areas and groups that are more prone to human-induced heat.

The involvement of the community in all aspects of governance, from policy development to implementation, will contribute to a better understanding of the issue at the local scale and establish a place-based approach to tackle the issue. Community involvement can take different forms, including public consultations, awareness campaigns, and education programs. In the meantime, public consultations provide an opportunity for the community to provide feedback and input on governance strategies. Awareness campaigns also help educate the community on the impacts of urban overheating and the need for sustainable development practices. Education programs can also provide knowledge and skills to community members and enable them to take effective actions in mitigating urban overheating.

Engagement of community with stakeholders will also assist in better understanding the need of the community with respect to the issue, and the priority areas under risk will be identified to inform better decision making. The partnerships and collaboration among the community, stakeholders, and the government will help mobilize resources, increase knowledge sharing, and promote the adoption of sustainable development practices. Partnerships and collaborations can involve different stakeholders, including government agencies, private sector organizations, civil society, and academic institutions. In the meantime, academic institutions can provide research and technical expertise to the table, and the government agencies can develop policy and regulatory frameworks. Notably, private sector organizations can play an important role in providing financial resources and expertise. The coordination and cooperation among these disciplines can enable the integration of different approaches and expertise, which leads to comprehensive and effective governance and planning interventions.

Finally, it is necessary to measure, report, and evaluate the progress and reflect on the past performances. Progress can be reviewed each year to develop thoughtful short- and long-term implementation. Developing monitoring and evaluation mechanisms is crucial to implement effective governance of urban overheating, particularly the heat generated by human activities. These mechanisms allow authorities to track progress, identify areas that need improvement, and evaluate the effectiveness of interventions. Developing monitoring and evaluation mechanisms at different levels of governance, including citywide, neighborhood, and project levels, is important. Monitoring and evaluation should be ongoing, with data collected and analyzed regularly. This allows authorities to respond quickly to emerging issues and make adjustments to interventions as needed. It is also important to communicate monitoring and evaluation results to the stakeholders and the public to build transparency and accountability.

11.7   Challenges and future directions

A major challenge in addressing anthropogenic heat is the lack of data and understanding of its sources and distribution in urban environments. The complex interactions among urban form, land use, building materials, and energy use also cause difficulty in accurately quantifying the amount of anthropogenic heat generated in different areas of the city.

Another challenge is the limited understanding of the interactions between anthropogenic heat and other urban environmental stressors, such as air pollution and UHI effects. These stressors can exacerbate the impacts of anthropogenic heat, and strategies to address anthropogenic heat must also consider the broader urban environmental context.

In addition, the potential trade-offs among different mitigation strategies, such as UHI strategies, can challenge the attainment of sustainable development goals. For example, some strategies may have unintended consequences, such as increased energy use for air conditioning in buildings with cool roofs. Therefore, careful consideration and integrated planning are necessary to ensure that mitigation strategies are effective and sustainable.

Feedback loops and nonlinear effects should also be considered. For example, increased use of air conditioning may lead to increased energy consumption and GHG emissions, which in turn may exacerbate climate change impacts. Similarly, UHIs may affect patterns of precipitation and cloud cover, which can further impact urban temperatures. These complex feedback loops require careful consideration and interdisciplinary research to understand and address them fully.

Future directions must also focus on the need for greater policy coordination and integration. This initiative may involve the development of new policies and regulations to promote sustainable energy use and reduce the overall demand for energy in urban environments. It may also involve integrating urban heat mitigation strategies into broader urban planning and development policies.

References

Alcántara, J. I., Moore, B. C., & Marriage, J. (2004). Comparison of three procedures for initial fitting of compression hearing aids. II. Experienced users, fitted unilaterally. *International Journal of Audiology*, *43*(1), 3−14.

Allam, Z., & Dhunny, Z. A. (2019). On big data, artificial intelligence and smart cities. *Cities (London, England)*, *89*, 80−91.

Annamalai, K., & Puri, I. K. (2006). *Combustion science and engineering*. CRC press.

Auckland Council. (2019). Auckland's urban Ngahere (forest) strategy, Accessed 08.03.23.

Auckland Council. (2020). *Te Tāruke-ā-Tāwhiri, Auckland's climate plan*, Accessed 08.03.23.

Barnett, M., (2022). Towards a productive, sustainable and inclusive low emissions economy. Te Hau mārohi ki anamata.

Baxter, L. R., Jr, Mazziotta, J. C., Phelps, M. E., Selin, C. E., Guze, B. H., & Fairbanks, L. (1987). Cerebral glucose metabolic rates in normal human females versus normal males. *Psychiatry Research*, *21*(3), 237−245.

Bettini, Y., Brown, R. R., & de Haan, F. J. (2015). Exploring institutional adaptive capacity in practice: Examining water governance adaptation in Australia. *Ecology and Society*, *20*(1).

Birkmann, J., Garschagen, M., Kraas, F., & Quang, N. (2010). Adaptive urban governance: New challenges for the second generation of urban adaptation strategies to climate change. *Sustainability Science*, *5*, 185−206.

Chapman, S., Watkins, N. W., & Stainforth, D. A. (2019). Warming trends in summer heatwaves. *Geophysical Research Letters*, *46*(3), 1634−1640.

Chow, W. T., Volo, T. J., Vivoni, E. R., Jenerette, G. D., & Ruddell, B. L. (2014). Seasonal dynamics of a suburban energy balance in Phoenix, Arizona. *International Journal of Climatology*, *34*(15), 3863−3880.

City of Melbourne. (2023a). 1200 Buildings [Online]. Available from <https://www.melbourne.vic.gov.au/SiteCollectionDocuments/1200-buildings-advice.PDF> Accessed 12.03.23.

City of Melbourne. (2023b). Competition to run innovative Smart Blocks apartment program. City of Melbourne. <https://www.melbourne.vic.gov.au/news-and-media/Pages/competition-to-run-innovative-smart-blocks-apartment-program.aspx> Accessed 12.03.23.

City of Melbourne. (2023c). Cycling lanes and routes. <https://www.melbourne.vic.gov.au/parking-and-transport/cycling/Pages/cycling-lanes-and-routes.aspx#:~:text = Major%20on%2Droad%20cycling%20routes,Footscray%20Road%20and%20Harbour%20Esplanade> Accessed 08.03.23.

City of Melbourne. (2023d). Walking. City of Melbourne. <https://participate.melbourne.vic.gov.au/transportstrategy/walking> Accessed 08.03.23.

Climate Change Commission. (2021). Ināia tonu nei: a low emissions future for Aotearoa.

Elander, I. (2002). Partnerships and urban governance. *International Social Science Journal*, *54*(172), 191−204.

Fanger, P. O. (1972). *Thermal comfort, analysis and application in environmental engineering* (ed). New York: McGraw Hill.

Fröhlich, J., & Knieling, J. (2013). Conceptualising climate change governance. *Climate change governance*, 9−26.

Gabbe, C., Pierce, G., Petermann, E., & Marecek, A. (2021). Why and how do cities plan for extreme heat? *Journal of Planning Education and Research*, 0739456X211053654.

Georgescu, M., et al. (2015). Prioritizing urban sustainability solutions: Coordinated approaches must incorporate scale-dependent built environment induced effects. *Environmental Research Letters*, *10*(6), 061001

Green, O. O., et al. (2016). Adaptive governance to promote ecosystem services in urban green spaces. *Urban Ecosystems*, *19*, 77−93.

Grimmond, C. (1992). The suburban energy balance: Methodological considerations and results for a mid-latitude west coast city under winter and spring conditions. *International Journal of Climatology*, *12*(5), 481−497.

Grindle, M. S. (2007). Good enough governance revisited. *Development Policy Review*, *25*(5), 533–574.

Guardaro, M., Messerschmidt, M., Hondula, D. M., Grimm, N. B., & Redman, C. L. (2020). Building community heat action plans story by story: A three neighborhood case study. *Cities (London, England)*, *107*, 102886

Guyer, J. L. (2019). *African niche economy: Farming to feed ibadan*. Edinburgh University Press.

Hall, P., & Pfeiffer, U. (2013). *Urban future 21: A global agenda for twenty-first century cities*. Routledge.

Hallenbeck, M. E., Rice, M., Smith, B., Cornell-Martinez, C., & Wilkinson, J. (1997). *Vehicle volume distributions by classifications*. United States: Federal Highway Administration.

Hamstead, Z., et al. (2020). Thermally resilient communities: Creating a socio-technical collaborative response to extreme temperatures. *Buildings and Cities*, *1*(1).

Heiple, S., & Sailor, D. J. (2008). Using building energy simulation and geospatial modeling techniques to determine high resolution building sector energy consumption profiles. *Energy and Buildings*, *40*(8), 1426–1436.

Hondula, D. M., Kuras, E. R., Longo, J., & Johnston, E. W. (2018). Toward precision governance: Infusing data into public management of environmental hazards. *Public Management Review*, *20*(5), 746–765.

Hsieh, C.-M., Aramaki, T., & Hanaki, K. (2007). Estimation of heat rejection based on the air conditioner use time and its mitigation from buildings in Taipei City. *Building and Environment*, *42*(9), 3125–3137.

Iwaniec, D. M., Cook, E. M., Davidson, M. J., Berbés-Blázquez, M., & Grimm, N. B. (2020). Integrating existing climate adaptation planning into future visions: A strategic scenario for the central Arizona–Phoenix region. *Landscape and Urban Planning*, *200*, 103820

Jalali, Z., et al. (2022). What we know and do not know about New Zealand's urban microclimate: A critical review. *Energy and Buildings,* 112430

Jay, O., et al. (2021). Reducing the health effects of hot weather and heat extremes: From personal cooling strategies to green cities. *The Lancet*, *398*(10301), 709–724.

Keith, L., Iroz-Elardo, N., Austof, E., Sami, I., & Arora, M. (2021). *The Journal of Climate Change and Health*.

Keith, L., & Meerow, S. (2022). Planning for urban heat resilience. *PAS Report, 600*.

Keith, L., Meerow, S., & Wagner, T. (2019). Planning for extreme heat: A review. *Journal of Extreme Events*, *6*(03n04), 2050003

Khurana, D., Koli, A., Khatter, K., & Singh, S. (2023). Natural language processing: State of the art, current trends and challenges. *Multimedia Tools and Applications*, *82*(3), 3713–3744.

Kotharkar, R., & Surawar, M. (2016). Land use, land cover, and population density impact on the formation of canopy urban heat islands through traverse survey in the Nagpur urban area, India. *Journal of Urban Planning and Development*, *142*(1), 04015003

Kraas, F. (2007). Megacities and global change: Key priorities. *The Geographical Journal*, *173* (1), 79–82.

Krayenhoff, S., Broadbent, A. M., Erell, E., Zhao, L., Georgescu, M., Voogt, J. A., et al. (2018). *Urban Cooling from Heat Mitigation Strategies: Systematic Review of the Numerical Modeling Literature*. In *10th International Conference on Urban Climate/14th Symposium on the Urban Environment*. AMS.

Larson, M., Getz, D., & Pastras, P. (2015). The legitimacy of festivals and their stakeholders: Concepts and propositions. *Event Management*, *19*(2), 159–174.

Mahlkow, N., Lakes, T., Donner, J., Köppel, J., & Schreurs, M. (2016). Developing storylines for urban climate governance by using Constellation Analysis—Insights from a case study in Berlin, Germany. *Urban Climate, 17*, 266–283.

Marschütz, B., et al. (2020). Local narratives of change as an entry point for building urban climate resilience. *Climate Risk Management, 28*, 100223

McGregor, G. R. (2015). *Heatwaves and health: guidance on warning-system development.* World Meteorological Organization.

Meerow, S., & Newell, J. P. (2019). Urban resilience for whom, what, when, where, and why? *Urban Geography, 40*(3), 309–329.

Mees, H. L., Driessen, P. P., & Runhaar, H. A. (2015). "Cool" governance of a "hot" climate issue: Public and private responsibilities for the protection of vulnerable citizens against extreme heat. *Regional Environmental Change, 15*, 1065–1079.

Melbourne C. o. (2023a). *Environmental upgrade finance − City of Melbourne.* <https://www.melbourne.vic.gov.au/business/sustainable-business/1200-buildings/funding-incentives/Pages/environmental-upgrade-finance.aspx> Accessed 12.03.23.

Melbourne C. o (2023b). *Sustainable Melbourne fund. City of Melbourne.* <https://www.melbourne.vic.gov.au/business/sustainable-business/Pages/sustainable-melbourne-fund.aspx> Accessed 12.03.23.

Melbourne City Greens. (2023). Transport. Melbourne City Greens. https://melbournecitygreens.com/transport/ Accessed 08.03.23.

Mishra, V., Ganguly, A. R., Nijssen, B., & Lettenmaier, D. P. (2015). Changes in observed climate extremes in global urban areas. *Environmental Research Letters, 10*(2), 024005

Moser, S. C., & Ekstrom, J. A. (2010). A framework to diagnose barriers to climate change adaptation. *Proceedings of the national academy of sciences, 107*(51), 22026–22031.

Oke, T. R., Mills, G., Christen, A., & Voogt, J. A. (2017). *Urban climates.* Cambridge University Press.

Pelling, M., & Garschagen, M. (2019). Put equity first in climate adaptation. *Nature, 569*(7756), 327–329.

Peters, B. G., & Pierre, J. (2000). Citizens versus the new public manager: The problem of mutual empowerment. *Administration & Society, 32*(1), 9–28.

Sailor, D.J., & Brooks, A. (2009). Quantifying anthropogenic moisture emissions and their potential impact on the urban climate. In: *Eighth Symp. on the Urban Environment.*

Sailor, D. J., & Fan, H. (2004). The importance of including anthropogenic heating in mesoscale modeling of the urban heat island. *Bulletin of the American Meteorological Society*, 397–403.

Sailor, D. J., & Lu, L. (2004). A top−down methodology for developing diurnal and seasonal anthropogenic heating profiles for urban areas. *Atmospheric Environment, 38*(17), 2737–2748.

Sailor, D. J. (2011). A review of methods for estimating anthropogenic heat and moisture emissions in the urban environment. *International Journal of Climatology, 31*(2), 189–199.

Santamouris, M., Cartalis, C., Synnefa, A., & Kolokotsa, D. (2015). On the impact of urban heat island and global warming on the power demand and electricity consumption of buildings—A review. *Energy and Buildings, 98*, 119–124.

State Government of Victoria. (2023). Residential efficiency scorecard. State Government of Victoria. <https://www.energy.vic.gov.au/for-households/save-energy-and-money/residential-efficiency-scorecard> Accessed 12.03.23.

Stewart, I. D. (2019). Why should urban heat island researchers study history? *Urban Climate, 30*, 100484

Stocker, T. F., et al. (2013). *Technical summary. Climate change 2013: The physical science basis. Contribution of working group I to the fifth assessment report of the intergovernmental panel on climate change* (pp. 33–115). Cambridge University Press.

Taylor, J., et al. (2015). Mapping the effects of urban heat island, housing, and age on excess heat-related mortality in London. *Urban Climate, 14,* 517–528.

Torrance, K., & Shun, J. (1976). Time-varying energy consumption as a factor in urban climate. *Atmospheric Environment (1967), 10*(4), 329–337.

Stren, R., & Cameron, R. (2005). Guest editor's preface, metropolitan governance reform: An introduction. *Public Administration and Development, 25*(4), 275–284. Available from http://hdl.handle.net/11427/21653.

Tourism & Transport Forum. (2011). *Tax incentives for sustainable transport* [Online]. Available from <https://www.ttf.org.au/wp-content/uploads/2016/06/TTF-Tax-Incentives-For-Sustainable-Public-Transport-2011.pdf>.

Victoria State Government. (2023a). About PTV. Victoria State Government. <https://www.ptv.vic.gov.au/footer/about-ptv/> Accessed 08.03.23.

Victoria State Government. (2023b). Suburban rail loop. Victoria State Government. <https://bigbuild.vic.gov.au/projects/suburban-rail-loop> Accessed 08.03.23.

Victorian Auditor-General's Office. (2023). Integrated transport planning. Victorian Auditor-General's Office 2023. <https://www.audit.vic.gov.au/report/integrated-transport-planning?section = > Accessed 12.03.23.

Wouters, H., et al. (2017). Heat stress increase under climate change twice as large in cities as in rural areas: A study for a densely populated midlatitude maritime region. *Geophysical Research Letters, 44*(17), 8997–9007.

Yin, Z., & Zhu, S. (2020). Consistencies and inconsistencies in urban governance and development. *Cities (London, England), 106,* 102930

Index